Wayton E Cassat
1955

AN INTRODUCTION TO
PROBABILITY THEORY
AND ITS
APPLICATIONS

WILEY PUBLICATIONS IN STATISTICS

Walter A. Shewhart, Editor

Mathematical Statistics

DWYER—Linear Computations.

FISHER—Contributions to Mathematical Statistics.

WALD—Statistical Decision Functions.

FELLER—An Introduction to Probability Theory and Its Applications, Volume One.

WALD—Sequential Analysis.

HOEL—Introduction to Mathematical Statistics.

Applied Statistics

YOUDEN—Statistical Methods for Chemists.

KEMPTHORNE—Design and Analysis of Experiment (in press).

HALD—Statistics (in press).

MUDGETT—Index Numbers.

TIPPETT—Technological Applications of Statistics.

DEMING—Some Theory of Sampling.

COCHRAN and COX—Experimental Designs.

RICE—Control Charts.

DODGE and ROMIG—Sampling Inspection Tables.

Related Books of Interest to Statisticians

HAUSER and LEONARD—Government Statistics for Business Use.

AN INTRODUCTION TO
PROBABILITY THEORY
AND ITS
APPLICATIONS

BY

WILLIAM FELLER

Eugene Higgins Professor of Mathematics
Princeton University

VOLUME ONE

NEW YORK
JOHN WILEY & SONS, INC.
CHAPMAN & HALL, LTD.
LONDON

COPYRIGHT, 1950
BY
WILLIAM FELLER

All Rights Reserved
This book or any part thereof must not be reproduced in any form without the written permission of the publisher.

Reproduction in whole or in part permitted for any purpose of the United States Government.

COPYRIGHT, CANADA, 1950, INTERNATIONAL COPYRIGHT, 1950
WILLIAM FELLER, PROPRIETOR

All Foreign Rights Reserved
Reproduction in whole or in part forbidden.

D

PRINTED IN THE UNITED STATES OF AMERICA

To
O. E. NEUGEBAUER

PREFACE

It was the author's original intention to write a book on analytical methods in probability theory in which the latter was to be treated as a topic in pure mathematics. Such a treatment would have been more uniform and hence more satisfactory from an aesthetic point of view; it would also have been more appealing to pure mathematicians. However, the generous support by the Office of Naval Research of work in probability theory at Cornell University led the author to a more ambitious and less thankful undertaking of satisfying heterogeneous needs.

It is the purpose of this book to treat probability theory as a self-contained mathematical subject rigorously, avoiding non-mathematical concepts. At the same time, the book tries to describe the empirical background and to develop a feeling for the great variety of practical applications. This purpose is served by many special problems, numerical estimates, and examples which interrupt the main flow of the text. They are clearly set apart in print and are treated in a more picturesque language and with less formality. A number of special topics have been included in order to exhibit the power of general methods and to increase the usefulness of the book to specialists in various fields. To facilitate reading, detours from the main path are indicated by stars. The knowledge of starred sections is not assumed in the remainder.

A serious attempt has been made to unify methods. The specialist will find many simplifications of existing proofs and also new results. In particular, the theory of recurrent events has been developed for the purpose of this book. It leads to a new treatment of Markov chains which permits simplification even in the finite case.

The examples are accompanied by about 340 problems mostly with complete solutions. Some of them are simple exercises, but most of them serve as additional illustrative material to the text or contain various complements. One purpose of the examples and problems is to develop the reader's intuition and art of probabilistic formulation. Several previously treated examples show that apparently difficult problems may become almost trite once they are formulated in a natural way and put into the proper context.

There is a tendency in teaching to reduce probability problems to pure analysis as soon as possible and to forget the specific characteristics of probability theory itself. Such treatments are based on a poorly defined notion of random variables usually introduced at the outset. This book goes to the other extreme and dwells on the notion of sample space, without which random variables remain an artifice.

In order to present the true background unhampered by measurability questions and other purely analytic difficulties this volume is restricted to *discrete sample spaces*. This restriction is severe, but should be welcome to non-mathematical users. It permits the inclusion of special topics which are not easily accessible in the literature. At the same time, this arrangement makes it possible to begin in an elementary way and yet to include a fairly exhaustive treatment of such advanced topics as random walks and Markov chains. The general theory of random variables and their distributions, limit theorems, diffusion theory, etc., is deferred to a succeeding volume.

This book would not have been written without the support of the Office of Naval Research. One consequence of this support was a fairly regular personal contact with J. L. Doob, whose constant criticism and encouragement were invaluable. To him go my foremost thanks. The next thanks for help are due to John Riordan, who followed the manuscript through two versions. Numerous corrections and improvements were suggested by my wife who read both the manuscript and proof.

The author is also indebted to K. L. Chung, M. Donsker, and S. Goldberg, who read the manuscript and corrected various mistakes; the solutions to the majority of the problems were prepared by S. Goldberg. Finally, thanks are due to Kathryn Hollenbach for patient and expert typing help; to E. Elyash, W. Hoffman, and J. R. Kinney for help in proofreading.

WILLIAM FELLER

Cornell University
January 1950

CONTENTS

INTRODUCTION: THE NATURE OF PROBABILITY THEORY . 1
  1. The Background 1
  2. Procedure 3
  3. "Statistical" Probability 4
  4. Historical Note 6

CHAPTER 1 THE SAMPLE SPACE 8
  1. The Empirical Background 8
  2. Illustrative Examples 10
  3. The Sample Space. Events 12
  4. Relations among Events 13
  5. Discrete Sample Spaces 16
  6. Probabilities in Discrete Sample Spaces 17
  7. Problems for Solution 21

CHAPTER 2 ELEMENTS OF COMBINATORIAL ANALYSIS. STIRLING'S FORMULA 23
  1. Preliminaries 23
  2. Samples 24
  3. Examples 26
  4. Partitions 30
  5. The Hypergeometric Distribution 33
  6. Binomial Coefficients 40
  7. Stirling's Formula 41
  8. Problems for Solution: Combinatorial 44
  9. Problems for Solution: Binomial Coefficients and Stirling's Formula 47

CHAPTER 3 THE SIMPLEST OCCUPANCY AND ORDERING PROBLEMS 51
  1. Combinatorial Lemmas 51
  2. Bose-Einstein and Fermi-Dirac Statistics 53
  3. The Classical Occupancy Problem 54
  4. Runs 56
  5. Problems for Solution 58

CHAPTER 4 COMBINATION OF EVENTS 60
  1. Union of Events 60
  2. Examples 62
  3. The Realization of m among N Events 64
  4. Application to Matching and Guessing 66
  5. Application to the Classical Occupancy Problem 69
  6. Miscellany 74
  7. Problems for Solution 75

CONTENTS

CHAPTER 5 CONDITIONAL PROBABILITY. STATISTICAL INDEPENDENCE ... 78

1. Conditional Probability ... 78
2. Compound Experiments ... 81
3. Statistical Independence ... 85
4. Repeated Trials ... 88
4a. A Guide to Abstract Language ... 91
5. Applications to Genetics ... 92
6. Sex-linked Characters ... 96
7. Selection ... 99
8. Problems for Solution ... 100

CHAPTER 6 THE BINOMIAL AND THE POISSON DISTRIBUTIONS ... 104

1. Bernoulli Trials ... 104
2. The Binomial Distribution ... 105
3. The Central Term ... 109
4. The Poisson Approximation ... 110
5. The Poisson Distribution ... 115
6. Observations Fitting the Poisson Distribution ... 119
7. The Multinomial Distribution ... 124
8. Problems for Solution ... 125

CHAPTER 7 THE NORMAL APPROXIMATION TO THE BINOMIAL DISTRIBUTION ... 129

1. The Normal Distribution ... 129
2. The DeMoivre-Laplace Limit Theorem ... 133
3. The Law of Large Numbers ... 141
4. Relation to the Poisson Approximation ... 143
5. Large Deviations ... 144
6. Problems for Solution ... 145

CHAPTER 8 UNLIMITED SEQUENCES OF BERNOULLI TRIALS 149

1. Infinite Sequences of Trials ... 149
2. Systems of Gambling ... 151
3. The Borel-Cantelli Lemmas ... 154
4. The Strong Law of Large Numbers ... 155
5. The Law of the Iterated Logarithm ... 157
6. Interpretation in Number Theory Language ... 161
7. Problems for Solution ... 163

CHAPTER 9 RANDOM VARIABLES; EXPECTATION ... 164

1. Random Variables ... 164
2. Expectations ... 171
3. Examples and Applications ... 173
4. The Variance ... 177
5. Covariance; Variance of a Sum ... 179
6. Chebyshev's Inequality ... 183
7. Kolmogorov's Inequality ... 184
8. The Correlation Coefficient ... 186
9. Problems for Solution ... 187

CONTENTS

CHAPTER 10 LAWS OF LARGE NUMBERS 191

1. Identically Distributed Variables 191
2. Proof of the Law of Large Numbers 195
3. The Theory of "Fair" Games 196
4. The Petersburg Game 199
5. Variable Distributions 201
6. Applications to Combinatorial Analysis 204
7. The Strong Law of Large Numbers 207
8. Problems for Solution 209

CHAPTER 11 INTEGRAL VALUED VARIABLES. GENERATING FUNCTIONS 212

1. Generalities 212
2. Convolutions 214
3. The Geometric and the Pascal Distributions 217
4. Relation to Holding or Waiting Times 218
5. Compound Distributions 221
6. Chain Reactions 223
7. Partial Fraction Expansions 227
8. The Continuity Theorem 232
9. Problems for Solution 235

CHAPTER 12 RECURRENT EVENTS: THEORY 238

1. Definition 238
2. Recurrence Times 241
3. Fundamental Theorems 243
4. Application of the Central Limit Theorem 248
5. Fluctuations in the Coin-tossing Game; the Arc Sine Law 249
6. Proof of the Theorems of Section 5 255
7. Proof of Theorem 3 of Section 3 259
8. Problems for Solution 261

CHAPTER 13 RECURRENT EVENTS: APPLICATIONS TO RUNS AND RENEWAL THEORY 264

1. Success Runs 264
2. More General Patterns 268
3. Numerical Estimates 269
4. The Renewal Equation 272
5. Examples 275
6. Problems for Solution 277

CHAPTER 14 RANDOM WALK AND RUIN PROBLEMS 279

1. General Orientation 279
2. The Gambler's Ruin 282
3. Expected Duration of the Game 286
4. Generating Functions for the Duration of the Game and First Passage Times 288
5. Explicit Expressions 290
6. Passage to the Limit; Diffusion Processes 293
7. Random Walks in the Plane and Space 297

CONTENTS

8. The Generalized One-dimensional Random Walk (Sequential Sampling) 300
9. Problems for Solution 304

CHAPTER 15 MARKOV CHAINS 307

1. Definition 307
2. Illustrative Examples 310
3. Higher Transition Probabilities 317
4. Irreducible Chains 318
5. Classification of States 320
6. Ergodic Properties of Aperiodic Chains; Stationary Distributions . . 324
7. Periodic Chains 329
8. Transient States; Absorption Probabilities 332
9. Application to Card Shuffling 335
10. The General Markov Process 337
11. Miscellany 341
12. Problems for Solution 344

CHAPTER 16 ALGEBRAIC TREATMENT OF FINITE MARKOV CHAINS 347

1. General Theory 347
2. Examples 351
3. Random Walk with Reflecting Barriers 355
4. Transient States; Absorption Probabilities 358
5. Application to Recurrence Times 362

CHAPTER 17 THE SIMPLEST TIME-DEPENDENT STOCHASTIC PROCESSES 363

1. General Orientation 363
2. The Poisson Process 364
3. The Pure Birth Process 367
4. Divergent Birth Processes 369
5. The Birth and Death Process 371
6. Exponential Holding Times 375
7. Waiting Line and Servicing Problems 377
8. The Backward (Retrospective) Equations 384
9. Generalization; The Kolmogorov Equations 386
10. Degenerate Processes 391
11. Problems for Solution 394

ANSWERS TO PROBLEMS 397

INDEX 411

INTRODUCTION

THE NATURE OF PROBABILITY THEORY

1. The Background

Probability is a mathematical discipline with aims akin to those of, for example, geometry or analytical mechanics. In each discipline, we must be careful to distinguish three aspects of the theory: (a) the formal logical content, (b) the intuitive background, (c) applications.

(a) **Formal Logical Contents.** A salient feature of mathematics is that it is concerned solely with relations among undefined things. This point is well illustrated by the game of chess. It is impossible to "define" chess otherwise than by stating a set of rules. The conventional shape of the pieces may be described to some extent, but it will not always be obvious to the inexperienced player which piece is intended for "king." The important thing is to know how the pieces move and act, meaning a set of rules. As a matter of fact, chess can be played without pieces and without chessboard. It is played in writing, with the sixty-four fields represented by as many symbols (A, 1), $\cdots$, (H, 8) in much the same way as analytical geometry describes geometrical points by their coordinates.

As it does not make sense in chess to ask what the "true nature" of a pawn or king is, so geometry does not care what a point and a straight line "really are." They remain undefined notions, and the axioms of geometry specify the relations among them: two points determine a line, etc. These are the rules of the game. The mathematician studies several non-Euclidean geometries in the same way as a chess player may play different variants of the chess game. The various geometries can be studied independently of their relations to reality, and, similarly, it is possible in mechanics to study how bodies would move if Newton's law of attraction were replaced by another one.

(b) **The Intuitive Background.** An essential difference between chess and geometry is that the rules of chess are arbitrary, whereas the axioms of geometry refer in an obvious manner to an intuitive background. In fact, geometrical intuition is so strong that it is likely to run ahead of logical reasoning. The extent to which logic, intuition, and physical experience are interdependent is a difficult problem of philosophy into which we need not enter. It is certain that intuition can be trained and developed. In chess, the bewildered beginner moves cautiously, recalling the individual rules, whereas the experienced player absorbs a complicated situation at a glance and is often unable to account rationally for his intuition. Similarly, it is possible to develop an intuitive feeling for relations, say, in a four-dimensional space. Further, the collective intuition of mankind appears to make rapid progress. Newton's notion of a field of force and of action at a distance, and Maxwell's concept of electromagnetic waves travelling through space, were at first described as "unthinkable" and contrary to intuition. With the popularization of mechanics and radio an education in the history of ideas is now required to understand why those theories originally seemed strange and unacceptable. When the theory of probability was new, it had to struggle against prejudiced intuitions and types of reasoning which are no longer cultivated, so that it is hard for us to understand the initial difficulties. Nowadays small boys are betting and shooting dice, newspapers report on samples of public opinion, and the magic of statistics embraces all phases of life to the extent that young girls anxiously watch the statistics of their chances to get married. Thus everyone has acquired an intuitive feeling for the meaning of statements like "the chances for this event are three in five." This intuition suffices as a background for the first few formal rules of probability. It will be trained and developed as the theory progresses and acquaintance is made with a variety of more sophisticated applications.

(c) **Applications.** In applications of geometry and mechanics theoretical concepts are identified with certain physical objects, but the method is flexible and varies from occasion to occasion so that no general rules can be given. The concept of a rigid body is useful and essential to mechanics, and yet no physical objects meet the specifications. Only experience teaches us which bodies can, with a satisfactory approximation, be treated as rigid. Rubber is usually given as a typical example of a non-rigid body, but in discussing the motion of automobiles most textbooks treat the wheels, including rubber tires, as rigid bodies. This is an example of how theoretical models are chosen and varied according to convenience or needs. Depending on our purposes, we feel free to disregard atomic theories and treat the sun as a tremendous ball of continuous matter or, on another occasion, as a single mass point. We must always remember that mathematics deals with abstract models and that different models can describe the same empirical situation with various degrees of approximation and simplicity. The manner in which mathematical theories are applied does not depend on preconceived ideas and is not a matter of logic: it is a purposeful technique which depends on, and changes with, experience. Of course, every phase of human activities is of interest to the philosopher, and a philosophical analysis of applications of mathematics is a legitimate study. However, such an analysis is not within the realm of mathematics, physics, or statistics. A philosophical discussion of the foundations of probability must be divorced from the mathematical theory and its applications to the same extent as a discussion of our intuitive space concept is now divorced from geometry (though this has not always been so).

2. Procedure

The history of probability (and of mathematics in general) shows a stimulating interplay of theory and applications: progress in theory opens new fields of applications, and each new application creates new theoretical problems and influences the direction of research. Today applications of the theory of probability extend over many fields of different natures, and the number of applications is steadily increasing. Only a general mathematical theory is flexible enough to provide proper tools for such a variety of problems, and we must withstand the temptation to keep our notions, pictures, and terms too close to one particular field of experience. We require a rigorous mathematical theory proceeding along the lines which are generally accepted in geometry and mechanics.

We shall start from the simplest experiences such as tossing a coin or throwing dice, where all statements have an obvious intuitive meaning. This intuition will be translated into an abstract model which will then be generalized to take care of more complicated situations. Illustrative examples should explain the empirical background of the theories and develop the reader's intuition, but the theory itself will have a mathematical character. We shall no more attempt to explain the "true meaning" of probability than the modern physicist dwells on the real meaning of mass and energy or the geometer explains the nature of a point. Instead, we shall prove exact theorems and show how they are applied in actual practice.

Originally, the theory of probability was developed to describe a very limited set of experiences connected with games of chance. The main objective of the theory in this connection was the calculation of certain probabilities. We shall follow the historical path and start from games of chance. We do so not for their importance or general interest, but because games of chance provide the most intuitive background and because, for the benefit of non-mathematical readers, we wish to postpone the use of special analytical tools as long as possible. Accordingly, in the first few chapters we shall calculate a few typical probabilities, but the general theory is independent of particular numerical values. Its object is to discover general laws and relations and to construct abstract models which can to some extent describe physical facts. Probabilities play for us the same role as masses in mechanics: it is possible to discuss the motion of the planetary system without knowing the individual masses and without contemplating methods for their actual measurement; it is also possible to study the hypothetical motion of a non-existent planetary system. We are usually interested in general laws and rarely in specific numerical computations. Useful probability models often refer to non-existing worlds. Thus the systems used in automatic telephone exchanges are based on simple probability considerations, and billions of dollars have been invested on this basis. It is clear that the underlying theory must compare various potential systems of exchanges, the majority of which will never exist, since the theory proves them to be inferior. Similarly, in insurance, probability theory is used to avoid certain undesirable situations which, as a consequence, can never be observed. When actual observations and numerical estimates are desired, it is usually necessary to use refined methods which form a chapter of mathematical statistics and lead beyond probability theory in the proper sense of the word.

3. "Statistical" Probability

The success of the modern mathematical theory of probability is bought at a price: the theory is limited to one particular aspect of the matter. The intuitive notion of probability is connected with the general inductive reasoning and with judgments such as "Paul is probably a happy man," "probably this book will be a failure," "Fermat's conjecture is probably false." Probability judgments of this sort are of interest to the philosopher and the logician, and they are also a legitimate object of a mathematical theory.[1] It must be understood, however, that we are concerned not with modes of inductive reasoning but with something that might be called *physical* or *statistical probability*. In a rough way we may characterize this concept by saying that our probabilities do not refer to judgments but to possible outcomes of a *conceptual experiment*. Before we speak of probabilities, we must agree on an idealized model of a particular conceptual experi-

[1] A modern axiomatic system was given by B. O. Koopman, The Axioms and Algebra of Intuitive Probability, *Annals of Mathematics* (2), vol. 41 (1940), pp. 269–292, and The Bases of Probability, *Bulletin of the American Mathematical Society*, vol. 46 (1940), pp. 763–774.

ment such as tossing a coin, sampling kangaroos on the moon, observing a particle under diffusion, counting the number of telephone calls. At the outset we must agree on the possible outcomes of this experiment (our *sample space*) and our probabilities will be associated with these and nothing else. This is in analogy with the procedure in mechanics where fictitious models involving two, three, or seventeen mass points are introduced, and these points are devoid of individual properties. Similarly, if we agree to analyze the coin-tossing game, then we speak of sequences like "head, head, tail, head, tail, ..." and nothing else. There is no place in our system for speculations concerning the probability that the sun will rise tomorrow. Before speaking of such a probability we would have to agree on an (idealized) model of an experiment, and this would presumably run along the lines "out of infinitely many worlds one is selected at random ...." Little imagination is required to construct such a model, but it appears both uninteresting and meaningless.

The astronomer speaks of measuring the temperature at the center of the sun or of travel to Sirius. These operations seem impossible, and yet it is not senseless to contemplate them. By the same token, we shall not worry whether or not our conceptual experiments can be performed; we shall analyze abstract models. In the back of our minds we keep an intuitive interpretation of probability which gains operational meaning in certain applications. We *imagine* the experiment performed a great many times. An event with probability 0.6 should be expected, in the long run, to occur sixty times out of a hundred. This description is deliberately vague but supplies a picturesque intuitive background sufficient for the more elementary applications. It can be rendered more precise only by a more elaborate theory. It does by no means describe typical applications. In fact, experiments to which probability theory is usefully applied are not necessarily repeated "a great many times under identical conditions." The agricultural experimenter has only few replicas, and frequently he refuses to consider their conditions as identical (because of trends in fertility, etc.). The quality-control engineer applies probability theory to one particular machine and draws inferences (or guesses) concerning a set of circumstances which will never repeat itself. The telephone engineer applies probability theory in order to compare various trunking systems of which ultimately only one comes into existence.

The truth is that, like all mathematics, the theory of probability builds theoretical models which are applied in many and variable ways. The technique of applications can be understood only after the theory. The intuition develops with the theory. Our statistical description of probability suffices as an intuitive background for the beginning. It is vague, but, to use a simile of T. E. Lawrence, it is "like the bow of the Mauretania: the bow has so much weight behind it that it does not need to be sharp like a razor."

4. Historical Note

Measure theory is a rather new branch of mathematics, but without its concepts and tools it would be impossible to treat probability theory along strictly mathematical lines. Until quite recently it was therefore necessary to admit non-mathematical elements, and most textbooks on probability are reminiscent of older books on mechanics in that they contain much philosophy and many attempts to define actual things rather than relations.

The advent of modern statistics put new requirements on probability theory, and certain classical concepts proved inadequate. Great progress was achieved when R. A. Fisher and R. von Mises developed the notion of statistical probability with a definite operational meaning. To von Mises is due the notion of a sample space [2] representing all conceivable outcomes of an experiment. This notion filled a gap in the measure theoretical approach which was gradually emerging in the twenties under the influence of many authors (among them prominently Rademacher and Steinhaus). A complete axiomatic treatment of the foundations of probability theory was given by A. Kolmogorov,[3] who emphasized the influence of von Mises' ideas. Conceptually we shall follow this line, although in the first volume we deal only with discrete probabilities where all relations are so simple that the term "measure theory" appears too solemn.

An unfortunate publicity was given to discussions of the so-called foundations of probability, and thus the erroneous impression was created that essential disagreement can exist among mathematicians. Actually these discussions concern only minor points which are of interest to but few specialists. The formulation of certain properties of randomness ranks high among von Mises' contributions to probability theory. To von Mises these properties served as axioms from which he undertook to develop the whole theory of probability. This enterprise was illuminating and served its historical purpose, but the theory encountered unexpected difficulties which completely destroyed

---

[2] Cf. his book, *Wahrscheinlichkeitsrechnung*, Leipzig and Wien, 1931, with references to his original papers dating back to about 1921. The German word is Merkmalraum (or label space).

[3] A. Kolmogoroff, *Grundbegriffe der Wahrscheinlichkeitsrechnung*, fasc. 3 of vol. 2. of *Ergebnisse der Mathematik*, Berlin, 1933.

its original simplicity. On the other hand, in the measure theoretical approach von Mises' randomness properties can be proved with surprising ease. In this respect there exists no disagreement concerning mathematical facts, and the argument revolves about the question of whether certain statements should occupy the place of axioms or of theorems.

More serious is the contention that the modern approach is too general and embraces subjects which should be kept outside. It is true that the same objection is raised against other mathematical fields, but in probability theory the criticism was aimed specifically against the study of infinitely prolonged games. The so-called St. Petersburg paradox of the classical theory is supposed to show that a rational theory must exclude such cases and that only special sample spaces should be permitted. Now it turns out that certain physical problems connected with absorption probabilities and recurrence times are exactly of the St. Petersburg type, and the suggested limitation of the theory of probability would seriously reduce its usefulness. Actually, the measure theoretical approach leads to no paradoxes or difficulties, and its greatest advantage is that it substitutes theorems for vague discussions of paradoxes. It is easy to condemn and to decry theories as impractical. The foundations of practical things of today were so decried only yesterday, and the theories which will be practical tomorrow are branded as valueless abstract games by the practical men of today.

CHAPTER 1

THE SAMPLE SPACE

1. The Empirical Background

The mathematical theory of probability gains practical value and an intuitive meaning in connection with real or conceptual experiments or phenomena such as tossing a coin once, tossing a coin 100 times, throwing a die, throwing three dice, arranging a deck of cards, matching two decks of cards, playing craps, playing roulette, observing the length of life of a radioactive atom or of a person, selecting a random sample of people and observing the number of left-handers in it, the sex of a newborn baby, crossing two species of plants and observing the phenotypes of the offspring, the number of busy trunklines in a telephone exchange, the number of calls on a telephone, random noise in an electrical communication system, routine quality control of a production process, frequency of accidents, the number of double stars in a region of the skies, the position of a particle under diffusion. All these descriptions are rather vague, and, in order to render the theory meaningful, we have to agree on what we mean by *possible results of the experiment or observation in question*.

A coin does not necessarily fall heads or tails; it can roll away or stand on its edge. Nevertheless, we shall agree to regard "head" and "tail" as the only possible events following the tossing of a coin. This convention simplifies the theory without affecting its applicability. Similar idealizations are frequently unavoidable. It is impossible to measure the length of life of an atom or of a person without some error, but for theoretical purposes it is expedient to imagine that these quantities are exact numbers. The question then arises as to which numbers can actually represent the life span of a person. Is there a maximal age beyond which life is impossible, or is any age conceivable? We hesitate to admit that man can grow 1000 years old, and yet current actuarial practice admits no such limit to life. According to formulas on which modern mortality tables are based, the proportion of men surviving 1000 years is of the order of magnitude of one in $10^{10^{36}}$—a number with 10^{27} billions of digits. This statement does not make sense from a biological or sociological point of view, but considered exclusively from a statistical standpoint it certainly does not contradict any experience. There are fewer than 10^{10} people born in a century. To test the contention statistically, more than $10^{10^{35}}$ centuries would be required, which is considerably more than $10^{10^{34}}$ lifetimes of the earth. Obviously, such extremely small probabilities are compatible with our notion of impossibility. One might think that their use is utterly absurd. Actually, it does no harm and is convenient in simplifying many formulas. Moreover, if we were seriously to discard the possibility of living 1000 years, we would be confronted with even worse difficulties, since we would have to accept the existence of a maximum age. However, the assumption that it should be possible to live x years and impossible to live x years and 2 seconds is as unappealing as the idea of unlimited life.

Any theory necessarily involves idealization, and usually the latter is so natural that it goes without saying. Our first idealization concerns the possible outcomes of an "experiment" or "observation." Only these possible outcomes are objects of the mathematical theory. If we want to construct an abstract model of the experiment, we must at the outset reach a decision as to what constitutes a possible outcome of the (idealized) experiment.

For uniform terminology, the results of experiments or observations will be called *events*. Thus we shall speak of the event that of 5 coins tossed more than three fell heads. Similarly, the "experiment" of distributing the cards in bridge [1] may result in the "event" that North has two aces. The composition of a sample ("two left-handers in a sample of 85") and the result of a measurement ("temperature 120°," "seven trunk lines busy") will each be called an event. Now a single observation may refer simultaneously to several events. Thus, if a throw with 2 dice resulted in the event "3 and 3," then the *same* trial resulted *also* in the following events: "two odd faces," "sum 6," "no ace," etc. These events are not mutually exclusive, so that several of them can occur simultaneously. They are *compound events* in the sense that they can be further decomposed into *simple events*. "Sum 6" means "(1, 5), or (2, 4), or (3, 3), or (4, 2), or (5, 1)," while "two

---

[1] *Definition of bridge and poker.* A deck of bridge cards consists of 52 cards arranged in four suits of 13 each. There are thirteen face values (2, 3, $\cdots$, 10, jack, queen, king, ace) in each suit. The four suits are called spades, clubs, hearts, diamonds. The last two are red, the first two black. Cards of the same face value are called of the same kind. For our purposes, playing bridge means distributing the cards to four players, to be called North, South, East, and West (or N, S, E, W, for short) so that each receives 13 cards. Playing poker, by definition, means selecting 5 cards out of the pack. We shall also study the composition of a hand of r bridge cards without reference to any particular game.

odd faces" is an abbreviation for "(1, 1), or (1, 3), or (1, 5), or (3, 1), or ...." In poker, every individual hand of 5 cards constitutes a simple (indecomposable) event; there are 2,598,960 such simple events or hands. The event "a hand contains two aces and no more" is a compound event and could be described by an enumeration of the 103,776 hands containing exactly two aces. Every particular numerical value x for a temperature represents a simple event, while the statement "the temperature is in the fifties" describes the compound event $50 \leq x < 60$. Every compound event can be decomposed into simple events, that is to say, a compound event is an *aggregate of certain simple events*.

If we want to speak about "experiments" or "observations" in a theoretical way and without ambiguity, we must first agree on the simple (not further decomposable) events representing the thinkable outcomes; they define the idealized experiment. It is usual to refer to these simple events as *sample points*, or *points* for short. By definition, *every indecomposable result of the (idealized) experiment is represented by one, and only one, sample point*. The aggregate of all sample points will be called the *sample space*. All events connected with a given (idealized) experiment can be described in terms of sample points.

2. Illustrative Examples

(a) *Sampling*. Suppose that a sample of 100 people is taken in order to estimate how many people smoke. The only property of the sample which is of interest in this connection is the number x of smokers; this may be any integer between 0 and 100. In this case we may agree that our sample space consists of the 101 "points" $x = 0$, 1, $\cdots$, 100. The result of every particular sample or observation is completely described by stating the corresponding point x. An example of a compound event is the result that "the majority of the people sampled are smokers." This means that the experiment resulted in one of the fifty events $x = 51, 52, \cdots, 100$, but it is not stated in which. Similarly, every property of the sample can be described in enumerating the corresponding cases or sample points. For uniform terminology we speak of events rather than properties of the sample. Mathematically, an event is simply the aggregate of the corresponding sample points.

(b) *Sampling (Continued)*. Suppose now that the 100 people in our sample are classified not only as smokers or non-smokers but also as males or females. The sample may now be characterized by a quadruple (M_s, F_s, M_n, F_n) of integers giving in order the number of male and female smokers, male and female non-smokers. We can take for sample space the aggregate of all quadruples of integers lying between 0 and 100 and adding up to 100. Stating that in a sample "relatively more males than females smoke" means that in our sample the ratio M_s/M_n is greater than F_s/F_n. The point (73, 2, 8, 17) has this property, but (0, 1, 50, 49) has not. Our event can be described in principle by enumerating all quadruples with the desired property.

(c) *Arrangements of Distinguishable Objects.* Consider four objects a, b, c, d. We conceive of their order as the result of an experiment. There are 24 possible arrangements or sample points, namely, *abcd, abdc, acbd, acdb,* $\cdots$, *dcab, dcba*. The event A, "a at the first place," can be described by enumerating the six cases in which it occurs: A is the aggregate of the six sample points *abcd, abdc, acbd, acdb, adbc, adcb*. We shall say that A consists of these six points. Consider now the three analogous events B, C, D defined by the property that b, c, d occupy the second, third, and fourth place, respectively. The event B consists of the six sample points *abcd, abdc, cbad, cbda, dbac, dbca*. There are two points common to A and B, namely, *abcd* and *abdc*; if one of these two arrangements was observed, we would say that both A and B had occurred. The point *abcd* is common to all four events A, B, C, D. Finally, consider the event E that "two and no more letters occupy their alphabetical place." This event consists of the six points *abdc, adcb, acbd, dbca, cbad, bacd*.

(d) *Arrangements of Indistinguishable Objects.* In the above example, the objects may be two red and two black balls, and we may consider balls of the same color as indistinguishable. Denoting the two colors by r and b, we are concerned with all possible arrangements of the symbols r, r, b, b. Our sample space now has only six points, namely, *rrbb, rbrb, rbbr, brrb, brbr, bbrr*. As examples of compound events, let us consider two events R and B defined by the condition that the symbols r and b, respectively, are not separated. The event R contains the sample points *rrbb, brrb, bbrr*, that is to say, R is realized if, and only if, one of these three arrangements occurs. Similarly, the event B contains *rrbb, rbbr, bbrr*. Both R and B occur simultaneously in the arrangements *rrbb* and *bbrr*; the event "both R and B occur" contains these two points and no more. Finally, the event "either R or B or both" consists of all points except *rbrb* and *brbr* where neither R nor B occurs.

Note that the question of whether or not actual balls of the same color are distinguishable is not an object of our theory: so far as the mathematics is concerned, this is a matter of definition or agreement. In practice, the same experiment may be described on either assumption, depending on the purpose of the theory. For example, in certain games the face values of playing cards may be irrelevant; in such cases we shall agree not to distinguish cards of the same suit. Similarly, people are actually distinguishable, but if we are interested only in their grouping according to sex, we shall naturally consider people of the same sex as indistinguishable.

(e) *Coin Tossing.* If a coin is tossed three times, the sample space will consist of eight points which may conveniently be represented by $HHH, HHT, HTH, THH, HTT, THT, TTH, TTT$. The event A, "two or more heads," can be described as the aggregate of the first four points. The event B "just one tail," means either HHT, or HTH, or THH; we say that B contains these three points.

(f) *Ages of a Couple.* An insurance company is interested in the age distribution of couples. Let x stand for the age of the husband, y for the age of the wife. Each observation results in a number-pair (x, y). For the sample space corresponding to a single observation we take the first quadrant of the x, y-plane so that each point $x > 0$, $y > 0$ is a sample point. The event A, "husband is older than 40," is represented by all points to the right of the line $x = 40$; the event B, "husband is older than wife," is represented by the angular region between the x-axis and the bisector $y = x$, that is to say, by the aggregate of points with $x > y$; the event C, "wife is older than 40," is represented by the portion of the first quadrant above the line $y = 40$. For a geometric representation of the joint age distributions of two couples we would require a four-dimensional space.

(g) *Phase Space.* In statistical mechanics, each possible "state" of a system is called a "point in phase space." This is only a difference in terminology. The phase space is simply our sample space; its points are our sample points.

3. The Sample Space. Events

It should be clear from the preceding that we shall never speak of probabilities except in relation to a given sample space (or physically: in relation to a certain conceptual experiment). *We start with the notion of sample space and its points; from now on they will be considered given. They are the primitive and undefined notions of the theory* precisely as the notions of "points" and "straight line" remain undefined in an axiomatic treatment of Euclidean geometry. The nature of the sample points does not enter our theory. The sample space provides a model of an ideal experiment in the sense that, by definition, *every thinkable outcome of the experiment is completely described by one, and only one, sample point.* It is meaningful to talk about an event A only when it is clear for *every* outcome of the experiment whether the event A has or has not occurred. The collection of all those sample points which represent outcomes where A has occurred completely describes the event. Conversely, any given aggregate A containing one or more sample points can be spoken of as an event; it does, or does not, occur according as the outcome of the experiment is, or is not, represented by a point of the aggregate. We therefore define the word *event* to mean the same as an aggregate of sample points. We shall say that an *event A consists of* (*or contains*) *certain points*, namely, those representing outcomes of the ideal experiment in which A occurs.

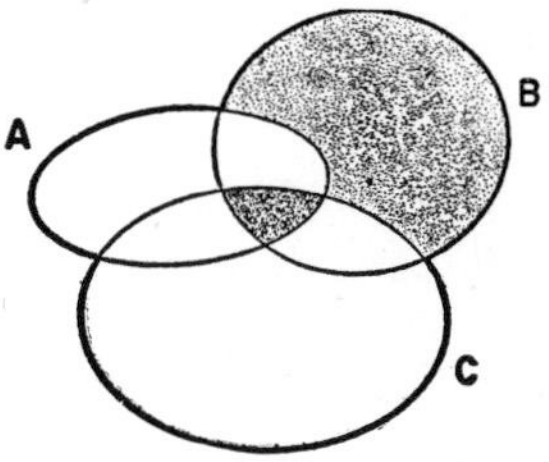

The terms "event" and "sample point" have an intuitive appeal, but these notions are equivalent to point sets and points in all parts of mathematics.

FIGURE 1. *Unions and Intersections of Events.* The domain within heavy boundaries is the union $A \cup B \cup C$. The triangular (*heavily shaded*) domain is the intersection ABC. The moon-shaped (*lightly shaded*) domain is the intersection of B with the complement of $A \cup C$.

4. Relations among Events

We shall now suppose that an arbitrary, but fixed, sample space $\mathfrak{S}$ is given. To every event A there corresponds another event defined by the condition "A does not occur." It contains all points not contained in A.

Definition 1. The event consisting of all points not contained in the event A will be called the complementary event (*or negation*) *of A and will be denoted by $\overline{A}$.*

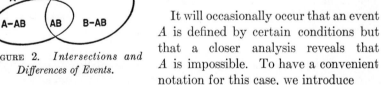

FIGURE 2. *Intersections and Differences of Events.*

It will occasionally occur that an event A is defined by certain conditions but that a closer analysis reveals that A is impossible. To have a convenient notation for this case, we introduce

Definition 2. We shall use the notation

(4.1) $$A = 0$$

to express that the event A contains no sample points (*is impossible*). The zero in (4.1) must be interpreted in a symbolic sense and not as the number.

With any two events A and B we can associate two new events defined by the conditions *"both A and B occur"* and *"either A or B or both occur."* These events will be denoted by AB and $A \cup B$, respectively. The event AB contains all sample points which are common to A and B. If A and B exclude each other, then there are no points common to A and B and the event AB is impossible; analytically, this situation is described by the equation

$$(4.2) \qquad AB = 0$$

which should be read "*A* and *B* are *mutually exclusive.*" The event $A\bar{B}$ means that both A and $\bar{B}$ occur or, in other words, that A but not B occurs. Similarly, $\bar{A}\,\bar{B}$ means that neither A nor B occurs. The event $A \cup B$ means that at least one of the events A and B occurs; it contains all sample points except those which belong neither to A nor to B.

Examples. (*a*) In the example (2.*c*), the event AB is defined by the conditions that a occupies the first and b the second place; thus, AB contains the two points *abcd* and *abdc*. The event $A \cup B$ contains the ten points *abcd, abdc, acbd, acdb, adbc, adcb, cbad, cbda, dbac, dbca*. The event $A\bar{B}$ consists of *acbd, acdb, adbc, adcb*.

(*b*) In the example (2.*d*), the event $R \cup B$ contains all points except *rbrb* and *brbr*, while RB consists of *rrbb* and *bbrr*. The event $R\bar{B}$ contains only the point *brrb*.

(*c*) In the example (2.*f*), the event AB means that the husband is older than 40 *and* older than his wife, while $A\bar{B}$ means that he is older than 40 but *not* older than his wife. Then AB is represented by the infinite trapezoidal region between the x-axis and the lines $x = 40$ and $y = x$. The event $A\bar{B}$ is represented by the angular domain between the lines $x = 40$ and $y = x$, the latter boundary included. The event AC means that both husband and wife are older than 40. The event $A \cup C$ means that at least one of them is older than 40, while $A \cup B$ means that the husband is either older than 40 or, if not that, at least older than his wife (in official language "husband's age exceeds 40 years or wife's age, whichever is smaller").

In the theory of probability we can describe the event AB in words as the simultaneous occurrence of A and B. In standard mathematical terminology AB is called the (logical) intersection of A and B. Similarly, $A \cup B$ is the union of A and B. Our notion carries over to the case of events $A, B, C, D, \ldots$. We can still define two new events $ABCD \ldots$ and $A \cup B \cup C \cup D \cup \ldots$ consisting, respectively, in the simultaneous realization of all events $A, B, C, D, \ldots$ and in the realization of at least one among the events $A, B, C, D, \ldots$.

Definition 3. *To every collection $A, B, C, \ldots$ of events we define two new events as follows.* The aggregate of the sample points which belong to all the given sets will be denoted by $ABC \ldots$ and called the *intersection* (or *simultaneous realization*) of $A, B, C, \ldots$. The aggregate of sample points which belong to at least one of the given sets will be denoted by $A \cup B \cup C \ldots$ and called the *union* (or *realization of at least one*) of the given events. The events $A, B, C, \ldots$ are *mutually exclusive* if no two have a point in common, that is, if $AB = 0, AC = 0, \cdots, BC = 0, \ldots$.

Example. (*d*) *Bridge* (cf. footnote on p. 9). Let A, B, C, D be the events, respectively, that North, South, East, West have at least one ace. It is clear that at least one player has an ace, so that at least one of the four events must occur. Hence $A \cup B \cup C \cup D = \mathfrak{S}$ is the whole sample space. The event $ABCD$ occurs if, and only if, each player has an ace. The event "West has all four aces" means that none of the three events A, B, C has occurred; this is the same as the simultaneous occurrence of $\bar{A}$ and $\bar{B}$ and $\bar{C}$ or the event $\bar{A}\,\bar{B}\,\bar{C}$.

We still require a symbol to express the statement that A cannot occur without B occurring or that the occurrence of A implies the occurrence of B. This means that every point of A is contained in B. One should think of intuitive analogies like the aggregate of all mothers, which forms a part of the aggregate of all women: all mothers are women but not all women are mothers.

Definition 4. *If every point of A is contained in B, we shall write $A \subset B$ and say that A implies B. Alternatively, we shall also write $B \supset A$ and say that B is implied by A. In that case, we shall also write $B - A$ instead of $B\bar{A}$ to denote the event that B but not A occurs.*

The event $B - A$ contains all those points which are in B but not in A. With this notation we can write $\bar{A} = \mathfrak{S} - A$ and $A - A = 0$.

Examples. (*e*) If A and B are mutually exclusive, then the occurrence of A implies the non-occurrence of B and vice versa. Thus $AB = 0$ means the same as $A \subset \bar{B}$ and as $B \subset \bar{A}$.

(*f*) The event $A - AB$ means the occurrence of A but not of both A and B. Thus $A - AB = A\bar{B}$.

(*g*) In the example (2.*c*), if a, b, and c are in their natural places, so is d. Hence $ABC \subset D$. The event $D - ABC$ means that d occupies the fourth place, but not all three remaining letters are in their natural places. This event consists of the five arrangements *acbd, cbad, bacd, bcad, cabd*.

(h) In the example (2.f) we have $BC \subset A$; in words "if husband is older than wife (B) and wife is older than 40 (C), then husband is older than 40 (A)." How can the event $A - BC$ be described in words?

5. Discrete Sample Spaces

The simplest sample spaces are those which contain only a finite number, n, of points. If n is fairly small (as in the case of tossing a few coins), it is easy to visualize the space. The space of distributions of cards in bridge is more complicated. However, one may imagine each sample point represented on a chip and may then consider the collection of these chips as representing the sample space. An event A (like "North has two aces") is represented by a certain set of chips; $\bar{A}$, by the remaining ones. It takes only one step from here to imagine a bowl with infinitely many chips or a sample space with an infinite sequence of points $E_1, E_2, E_3, \ldots$.

Examples. (a) Let us agree to toss a coin as often as necessary to turn up one head. The points of the sample space are then $E_1 = H$, $E_2 = TH$, $E_3 = TTH$, $E_4 = TTTH$, etc. We may or may not consider as thinkable the possibility that H never appears. If we do, this possibility should be represented by a point E_0.

(b) *Craps.* This game requires the player to throw 2 dice until a decision has been reached under the following condition. He wins if the first throw results in the sums 7 or 11 or, alternatively, if the first sum is 4, 5, 6, 8, 9, or 10, and the same sum reappears before the 7 has appeared. This implies that the player loses if the first throw results in the sum 2, 3, or 12 or, alternatively, if the first sum is 4, 5, 6, 8, 9, or 10, and the sum 7 appears before the first sum reappears. There are clearly infinitely many possible results, since it is conceivable that the game will not end in n trials, however large n may be. Nevertheless, it is easy to enumerate all sample points according to the number of throws required. *One* throw decides the game if the sum is 2, 3, 7, 11, or 12. *Two* throws decide if they result in (4, 4), (4, 7), (5, 5), (5, 7), $\cdots$, (10, 10), (10, 7), etc. Thus a systematic enumeration of all sample points is tedious, but possible. The rules of the game would be senseless if we were not convinced that the game will be decided sooner or later so that we need not worry about the possibility of an unending game.

Definition. A sample space is called discrete if it contains only finitely many points or infinitely many points which can be arranged into a simple sequence $E_1, E_2, \ldots$.

Not every sample space is discrete. It is a known theorem (due to G. Cantor) that the sample space consisting of all positive numbers is not discrete. We are here confronted with a distinction familiar in mechanics, where one usually first considers discrete mass points, with each individual point carrying a finite mass. This concept contrasts with continuous mass distribution, where each individual point has zero mass. In the first case, the mass of a system is obtained simply by adding the masses of the individual points; in the second case, masses are computed by integration over mass densities. Quite similarly, the probabilities of events in discrete sample spaces are obtained by mere additions, whereas in other spaces integrations are necessary. Except for the technical tools required, there is no essential difference between the two cases. In order to present actual probability considerations unhampered by technical difficulties, we shall first take up only discrete sample spaces. It will be seen that even this special case leads to many interesting and important results.

In this volume we shall consider only discrete sample spaces.

6. Probabilities in Discrete Sample Spaces

The probabilities of the various events are numbers of the same nature as distances in geometry or masses in mechanics. The theory assumes that they are given but need assume nothing as to their actual numerical value or how they are measured in practice. Some of the most important applications are of a qualitative nature and independent of numerical values; the general conclusions of the theory are applied in many ways exactly as the theorems of geometry serve as a basis for physical theories and engineering applications. In the relatively few cases where numerical values for probabilities are required the methods of procedure vary as widely as do the methods of determining distances. There is little in common in the practices of the carpenter, the practical surveyor, the pilot, and the astronomer, when they measure distances. In our context, we may consider the diffusion constant, which is a notion of the theory of probability. To find its numerical value physical considerations relating it to other theories are required; a direct measurement is impossible. By contrast, mortality tables are constructed from rather crude observations. In most actual applications the determination of probabilities, or the comparison of theory and observation, requires rather sophisticated statistical methods which in turn are based on a refined probability theory. In other words, the intuitive meaning of probability is clear, but only as the theory proceeds shall we be able to see how it is applied. All possible "definitions" of probability fall far short of the actual practice.

When tossing a "good" coin we do not hesitate to associate probability 1/2 with either head or tail. This amounts to saying that when a coin is tossed n times all 2^n possible results have the same probability. From a theoretical standpoint, this is a *convention*. Frequently, it has been contended that this convention is logically unavoidable and the only possible one. Yet there have been philosophers and statisticians defying the convention and starting from contradictory assumptions (uniformity or non-uniformity in nature). It has also been claimed that the probabilities 1/2 are due to experience. As a matter of fact, whenever refined statistical methods have been used to check on actual coin tossing, the result has been invariably that head and tail are *not* equally likely. And yet we stick to our model of an "ideal" coin, even though no good coins exist. We preserve the model not merely for its logical simplicity, but essentially for its usefulness and applicability. In many applications it is sufficiently accurate to describe reality. More important is the empirical fact that departures from our scheme are always coupled with phenomena such as an eccentric position of the center of gravity. In this way our idealized model can be extremely useful even if it never applies exactly. For example, in modern statistical quality control based on Shewhart's methods, idealized probability models are used to discover "assignable causes" for flagrant departures from these models and thus to remove impending machine troubles and process irregularities at an early stage.

Similar remarks apply to other cases. The number of possible distributions of cards in bridge is almost 10^{30}. Usually, we agree to consider them as equally probable. For a check of this convention more than 10^{30} experiments would be required—a billion of billion of years if every living person played one game every second, day and night. However, consequences of the assumption can be verified experimentally, for example, by observing the frequency of multiple aces in the hands at bridge. It turns out that for crude purposes the idealized model describes experience sufficiently well, provided that the card shuffling is done better than is usual. It is more important that the idealized scheme, when it does not apply, permits the discovery of "assignable causes" for the discrepancies, for example, the reconstruction of the mode of shuffling. These are examples of limited importance, but they indicate the usefulness of assumed models. More interesting cases will appear only as the theory proceeds.

Fundamental Convention. *Given a discrete sample space* $\mathfrak{S}$ *with the sample points* $E_1, E_2, \ldots,$ *we shall assume that with each point* E_j *there* is associated a number, called the *probability of E_j* and denoted by $Pr\{E_j\}$. It is to be non-negative and such that

(6.1) $$Pr\{E_1\} + Pr\{E_2\} + \ldots = 1.$$

Note that we do not exclude the possibility that a point has probability zero. This convention may appear artificial but is necessary to avoid complications. In discrete sample spaces probability zero is in practice interpreted as an impossibility, and any sample point known to have probability zero can, with impunity, be eliminated from the sample space. However, frequently the numerical values of the probabilities are not known in advance, and involved considerations are required to decide whether or not a certain sample point has positive probability.

Definition. The probability $Pr\{A\}$ of any event A is the sum of the probabilities of all sample points in it.

The fundamental equation (6.1) states that the probability of the entire sample space $\mathfrak{S}$ is unity, or $Pr\{\mathfrak{S}\} = 1$. It follows that for any event A

(6.2) $$0 \leq Pr\{A\} \leq 1.$$

Consider now two arbitrary events A_1 and A_2. To compute the probability $Pr\{A_1 \cup A_2\}$ that either A_1 or A_2 or both occur, we have to add the probabilities of all sample points contained either in A_1 or in A_2, but each point is to be counted only once. We have, therefore,

(6.3) $$Pr\{A_1 \cup A_2\} \leq Pr\{A_1\} + Pr\{A_2\}.$$

Now, if E is any point contained both in A_1 and in A_2, then $Pr\{E\}$ occurs twice in the right-hand member but only once in the left-hand member. Therefore, the right side exceeds the left side by the amount $Pr\{A_1A_2\}$, and we have the simple but important

Theorem. For any two events A_1 and A_2 the probability that either A_1 or A_2 or both occur is given by

(6.4) $$Pr\{A_1 \cup A_2\} = Pr\{A_1\} + Pr\{A_2\} - Pr\{A_1A_2\}.$$

If $A_1A_2 = 0$, that is, if A_1 and A_2 are mutually exclusive, then (6.4) reduces to

(6.5) $$Pr\{A_1 \cup A_2\} = Pr\{A_1\} + Pr\{A_2\}.$$

Example. A coin is tossed twice. For sample space we take the four points HH, HT, TH, TT, and associate with each probability $1/4$. Let A_1 and A_2 be, respectively, the events "head at first and second trial." Then A_1 consists of HH and HT, and A_2 of TH and HH. Furthermore $A = A_1 \cup A_2$ contains the three points HH, HT, and TH, whereas $A_1 A_2$ consists of the single point HH. Thus $Pr\{A_1 \cup A_2\}$
$$= \frac{1}{2} + \frac{1}{2} - \frac{1}{4} = \frac{3}{4}.$$

The probability $Pr\{A_1 \cup A_2 \cup \cdots \cup A_n\}$ of the realization of at least one among n events can be computed by a formula analogous to (6.4); this will be taken up in Chapter 4, section 1. Here we note only that the inequality (6.3) obviously holds in general. Thus *for arbitrary events* $A_1, A_2, \ldots$ *the inequality*

(6.6) $\qquad Pr\{A_1 \cup A_2 \cup \cdots\} \leq Pr\{A_1\} + Pr\{A_2\} + \ldots$

holds. In the special case where the events $A_1, A_2, \ldots$ *are mutually exclusive, we have*

(6.7) $\qquad Pr\{A_1 \cup A_2 \cup \cdots\} = Pr\{A_1\} + Pr\{A_2\} + \ldots.$

Occasionally (6.6) is referred to as Boole's inequality.

We shall first investigate the simple special case where the sample space has a finite number, N, of points each having probability $1/N$. In this case, the probability of any event A equals the number of points in A divided by N. In the older literature, the points of the sample space were called "cases" and the points of A, "favorable" cases (favorable for A). *If* all points have the same probability, then the probability of an event A is the ratio of the number of favorable cases to the total number of cases. Unfortunately, this statement has been much abused to provide a "definition" of probability. It is often contended that in *every* finite sample space probabilities of all points are equal. This is not so. For a single throw of an untrue coin, the sample space still contains only the two points, head and tail, but they may have arbitrary probabilities p and q, with $p + q = 1$. A newborn baby is a boy or girl, but in applications we have to admit that the two possibilities are not equally likely. In many applications in physics and technics, we have simple alternatives with different probabilities. The usefulness of sample spaces in which all sample points have the same probability is restricted almost entirely to the study of games of chance and to combinatorial analysis.

7. Problems for Solution

1. Among the digits 1, 2, 3, 4, 5 first one is chosen, and then a second selection is made among the remaining four digits. Assume that all twenty possible results have the same probability. Find the probability that an odd digit will be selected (a) the first time, (b) the second time, (c) both times.

2. A coin is tossed until for the first time the same result appears twice in succession. To every possible outcome requiring n tosses attribute probability $1/2^n$. Describe the sample space. Find the probability of the following events: (a) the experiment ends before the sixth toss, (b) an *even* number of tosses is required.

3. Two dice are thrown. Let A be the event that the sum of the faces is odd, B the event of at least one ace. Describe the events AB, $A \cup B$, $A\bar{B}$. Find their probabilities, assuming that all 36 sample points have equal probabilities.

4. In the example (2.f), discuss the meaning of the following events: (a) ABC, (b) $A - AB$, (c) $A\bar{B}C$.

5. In the example (2.f), verify that $A\bar{C} \subset B$.

6. *Bridge* (cf. footnote on p. 9). For $k = 1, 2, 3, 4$, let N_k be the event that North has at least k aces. Let S_k, E_k, W_k be the analogous events for South, East, West. What can be said about the number x of aces in West's possession in the events (a) $\bar{W}_1$, (b) $N_2 S_2$, (c) $\bar{N}_1 \bar{S}_1 \bar{E}_1$; (d) $W_2 - W_3$, (e) $N_1 S_1 E_1 W_1$, (f) $N_3 W_1$, (g) $(N_2 \cup S_2) E_2$.

7. In the preceding problem verify that (a) $S_3 \subset S_2$, (b) $S_3 W_2 = 0$, (c) $N_2 S_1 E_1 W_1 = 0$, (d) $N_2 S_2 \subset \bar{W}_1$, (e) $(N_2 \cup S_2) W_3 = 0$, (f) $W_4 = \bar{N}_1 \bar{S}_1 \bar{E}_1$.

8. Verify the following relations [2]:

(a) $\overline{A \cup B} = \bar{A}\bar{B}$.
(b) $(A \cup B) - B = A - AB = A\bar{B}$.
(c) $AA = A \cup A = A$.
(d) $(A - AB) \cup B = A \cup B$.
(e) $(A \cup B) - AB = A\bar{B} \cup \bar{A}B$.
(f) $\overline{A \cup B} = \overline{AB}$.
(g) $(A \cup B)C = AC \cup BC$.

9. Find simple expressions for (a) $(A \cup B)(A \cup \bar{B})$, (b) $(A \cup B)(\bar{A} \cup B)(A \cup \bar{B})$, (c) $(A \cup B)(B \cup C)$.

10. State which of the following relations are correct and which incorrect:

(a) $(A \cup B) - C = A \cup (B - C)$.
(b) $ABC = AB(C \cup B)$.
(c) $A \cup B \cup C = A \cup (B - AB) \cup (C - AC)$.
(d) $A \cup B = (A - AB) \cup B$.
(e) $AB \cup BC \cup CA \supset ABC$.
(f) $(AB \cup BC \cup CA) \subset (A \cup B \cup C)$.
(g) $(A \cup B) - A = B$.
(h) $\overline{AB}C \subset A \cup B$.
(i) $\overline{A \cup B \cup C} = \bar{A}\bar{B}\bar{C}$.
(j) $\overline{(A \cup B)}C = \bar{A}C \cup \bar{B}C$.
(k) $\overline{(A \cup B)}C = \bar{A}\bar{B}C$.
(l) $\overline{(A \cup B)}C = C - C(A \cup B)$.

[2] Note that $\overline{A \cup B}$ denotes the complement of $A \cup B$ which is not the same as $\bar{A} \cup \bar{B}$. Similarly, $\overline{AB}$ is not the same as $\bar{A}\bar{B}$.

11. Let A, B, C be three arbitrary events. Find expressions for the events that of A, B, C:

(a) Only A occurs.
(b) Both A and B, but not C, occur.
(c) All three events occur.
(d) At least one occurs.
(e) At least two occur.
(f) One and no more occurs.
(g) Two and no more occur.
(h) None occurs.
(i) Not more than two occur.

12. The union $A \cup B$ of two events can be expressed as the union of two mutually exclusive events, thus: $A \cup B = A \cup (B - AB)$. Express in a similar way the union of three events A, B, C.

CHAPTER 2

ELEMENTS OF COMBINATORIAL ANALYSIS
STIRLING'S FORMULA

The following two chapters on combinatorial analysis are a necessary interruption of the main course of the book. We want to derive a few important formulas in a way appropriate for our purposes. The advanced reader may pass on directly to Chapter 4, where the thread of Chapter 1 is taken up again.

In the study of simple games of chance, sampling procedures, occupancy and order problems, etc., we are usually dealing with finite sample spaces in which the same probability is attributed to all points. To compute the probability of an event A we have then only to divide the number of sample points in A ("favorable cases") by the total number of sample points ("possible cases"). This is facilitated by a systematic use of a few rules which we now proceed to review. Simplicity and economy of thought can be achieved by adhering to a few standard tools, and we shall follow this procedure instead of describing the shortest computational method in each special case.[1]

1. Preliminaries

Pairs. *With m elements $a_1, \cdots, a_m$ and n elements $b_1, \cdots, b_n$ it is possible to form mn pairs (a_j, b_k) containing one element from each group.*

Proof. Arrange the pairs in a rectangular array in the form of a multiplication table with m rows and n columns so that (a_j, b_k) stands at the intersection of the jth row and kth column. Then each pair appears once and only once, and the assertion becomes obvious.

Examples. (*a*) *Bridge Cards* (cf. footnote to Chapter 1, section 1). As sets of elements take the four suits and the thirteen face values,

---

[1] The interested reader will find many topics of elementary combinatorial analysis treated in the classical textbook, *Choice and Chance*, by W. A. Whitworth, fifth edition, London, 1901, reprinted by G. E. Stechert, New York, 1942. The companion volume by the same author, *DCC Exercises*, reprinted New York, 1945, contains **700** problems with complete solutions.

respectively. Each card is defined by its suit and its face value, and there exist $4 \cdot 13 = 52$ such combinations, or cards.

(b) *"Seven-way Lamps."* Some floor lamps so advertised contain 3 ordinary bulbs and also an indirect lighting fixture which can be operated on three levels but need not be used at all. Each of these four possibilities can be combined with 0, 1, 2, or 3 bulbs. Hence there are $4 \cdot 4 = 16$ possible combinations of which one, namely (0, 0), means that no bulb is on. There remain fifteen (not seven) ways of operating the lamps.

Multiplets. *Given n_1 elements $a_1, \cdots, a_{n_1}$, and n_2 elements $b_1, \cdots, b_{n_2}$, etc., up to n_r elements $x_1, \cdots, x_{n_r}$; it is possible to form $n_1 \cdot n_2 \cdots n_r$ combinations $(a_{j_1}, b_{j_2}, \cdots, x_{j_r})$ containing one element of each kind.*

Proof. If $r = 2$, the assertion reduces to the first rule. If $r = 3$, take the pair (a_i, b_j) as element of a new kind. There are $n_1 n_2$ such pairs and n_3 elements c_k. Each triple (a_i, b_j, c_k) is itself a pair consisting of (a_i, b_j) and an element c_k; the number of triplets is therefore $n_1 n_2 n_3$. Proceeding by induction, the assertion follows for every r.

Examples. (c) *Multiple Classifications.* Suppose that people are classified according to sex, marital status, and profession. The various categories play the role of elements. If there are seventeen professions, then we have $2 \cdot 2 \cdot 17 = 68$ classes in all.

(d) In an agricultural experiment three different treatments are to be tested (for example, the application of a fertilizer, a spray, and temperature). If these treatments can be applied on r_1, r_2, and r_3 levels or concentrations, respectively, then there are $r_1 r_2 r_3$ combinations, or ways of treatment.

2. Samples

Consider the set or *"population"* of n elements $a_1, a_2, \cdots, a_n$. Any ordered arrangement of r symbols is called a *sample of size r* drawn from our population. It is understood that the sample is ordered, and we write it in the form $(a_{j_1}, a_{j_2}, \cdots, a_{j_r})$. For an intuitive picture we can imagine that the elements are selected one by one. Two procedures are then possible. First, *sampling with replacement;* here each selection is made from the entire population, so that the same element can be drawn more than once. The samples are then arrangements in which repetitions are permitted. Second, *sampling without replacement;* here an element once chosen is removed from the population, so that the sample becomes an arrangement without repetitions. Obviously, in this case, the sample size r cannot exceed the population size n.

In sampling with replacement each of the r elements can be chosen in n ways: the number of possible samples is therefore n^r, as can be seen from the last theorem with $n_1 = n_2 = \cdots = n$. In sampling without replacement we have n possible choices for the first element, but only $n - 1$ for the second, $n - 2$ for the third, etc. Using the same rule, we see that without replacement the number of samples is $n(n-1) \cdots (n-r+1)$. Such products appear so often that it is convenient to introduce the notation [2]

(2.1) $$(n)_r = n(n-1) \cdots (n-r+1).$$

Clearly $(n)_r = 0$ whenever $r > n$. We have thus the following

Theorem. The number of different possible samples of size r from a population of n elements is n^r if the sampling is with replacement, and $(n)_r$ if it is without replacement.

Mr. and Mrs. Smith form a sample of size two drawn from the human population; at the same time, they form a sample of size one drawn from the population of all couples. This example shows that the sample size is defined only in relation to a given population. Tossing a coin r times is one way of obtaining a sample of size r drawn from the population of the two letters, H and T. The same arrangement of r letters H and T is a single sample point in the space corresponding to the experiment of tossing a coin r times.

Drawing r elements from a population of size n is an experiment whose possible outcomes are samples of size r. Their number is n^r or $(n)_r$, depending on whether or not replacement is used. In either case, our conceptual experiment is described by a sample space in which each individual point represents a sample of size r.

So far we have not spoken of probabilities associated with our samples. Usually we shall assign equal probabilities to all of them and then speak of random samples. The word "random" is not well defined, but when applied to samples or selections it has a unique meaning. Whenever we speak of *random samples of fixed size r, the adjective random is to imply that all possible samples have the same probability*, namely, n^{-r} in sampling with replacement and $1/(n)_r$ in sampling without replacement, n denoting the size of the population from which the sample is drawn. If n is large and r relatively small, the ratio $(n)_r/n^r$ is near unity. This suggests that for large populations and relatively small samples the difference between the two ways of sampling is negligible [cf. Chapter 6, example (2.g)].

We have introduced a practical terminology but have made no

---

[2] The notation $(n)_r$ is not standard, but it will be used consistently in this book.

statements as to the applicability of our model of random sampling to reality. Tossing coins, throwing dice, and similar activities may be interpreted as experiments in practical random sampling with replacements, and our probabilities are numerically close to frequencies observed in long-run experiments even though perfectly balanced coins or dice do not exist. Random sampling without replacement is typified by successive drawings of cards from a shuffled deck (provided that shuffling is done much better than is usual). In sampling human populations the statistician encounters considerable and often unpredictable difficulties, and bitter experience has shown that it is difficult to obtain even a crude image of randomness.

3. Examples

(a) *Random Sampling Numbers.* To compare our model of coin tossing with actual experiments, we require the records of a long series of trials. We could then compare various probabilities with the observed frequencies of corresponding events. Records of the desired kind are not readily available, but Kendall and Smith [3] have published 100,000 random digits, that is, a record of 100,000 trials representing random sampling in the population of the digits 0, 1, $\cdots$, 9. At any point of these tables the next r digits should represent a random sample as closely as one can expect observations to correspond to a theoretical model.

As an example, consider the event that the digit is 7. The probability is 1/10, and we expect that among r digits the 7 will occur approximately $r/10$ times. The observed frequencies of the occurrence of 7 in the first 100 batches of 100 digits each (first 100 columns in the tables) are given in Table 1. As would be expected, the frequencies fluctuate around 0.1. If we take larger samples, the fluctuation of the frequencies is smaller. This is borne out by the last column of Table 1 which records the average frequencies of the digit 7 for ten batches of 1000 digits each. It goes without saying that a more advanced theory is required to judge to what extent empirical data like those in Table 1 agree with our abstract model. We shall return to this material in Chapter 3, and again in Chapter 7, example (3.*a*), when discussing the theory of independent trials and observations. There we shall find that in about four out of ten counts of 10,000 ideal random digits the frequency of the digit 7 must be expected to fall outside the interval 0.10 ± 0.0032. This means that an agreement as

---

[3] M. G. Kendall and Babington Smith, *Tables of Random Sampling Numbers*, Tracts for Computers No. 24, Cambridge, 1940. Older and smaller tables under the same title by L. H. C. Tippett are in the same series, No. 16.

TABLE 1

Number of Occurrences of the Digit 7 among the First 100 Groups of 100 Digits Each in the Kendall-Smith Tables

Thousand	Hundred										Average Frequency
	1	2	3	4	5	6	7	8	9	10	
1	12	8	5	14	12	9	10	7	10	8	0.095
2	15	5	8	6	12	8	5	15	3	11	.088
3	7	8	12	13	7	10	10	8	9	11	.095
4	14	12	8	16	16	6	9	11	10	10	.112
5	8	10	10	12	20	5	7	6	11	6	.095
6	7	15	8	8	9	8	8	7	12	17	.099
7	8	4	5	10	7	7	11	14	7	9	.082
8	7	5	12	16	10	8	8	11	6	6	.089
9	11	7	17	8	9	14	13	14	8	10	.111
10	12	12	10	9	10	9	11	10	11	8	.102
1–10,000											.0968

good or better than that in Table 1 can be expected only in about six out of ten cases.

As a second example, we take *pairs* of random digits and the event that both digits of the pair are 7. The probability that a pair is (7, 7) is 1/100. Table 2 gives the observed frequencies for 5000 pairs (10,000 digits).

(b) *Random Sampling Numbers (Continued).* As an application of the theorem of section 2, we shall calculate the *probability p that 5 consecutive random digits are all different.* There are 10^5 possible arrangements of which $(10)_5 = 10 \cdot 9 \cdot 8 \cdot 7 \cdot 6$ are without repetition. Hence

(3.1) $$p = \frac{(10)_5}{10^5} = 0.3024.$$

One expects intuitively that in large mathematical tables having many decimal places the last five digits will have many properties of randomness. (In ordinary logarithmic and other smaller tables the tabular difference is nearly constant and the last digit therefore varies

TABLE 2

Number of Occurrences of the Combination (7, 7) among the First 5000 Pairs (10,000 Digits) in the Kendall-Smith Tables

Group of 500 Pairs	1–100	101–200	201–300	301–400	401–500	Total	Average Frequency from Beginning
1	0	1	0	1	0	2	0.004
2	1	1	0	0	2	4	.006
3	0	3	1	1	1	6	.008
4	0	2	2	2	1	7	.0095
5	1	1	2	0	1	5	.0096
6	2	1	1	0	1	5	.0097
7	0	0	0	1	0	1	.0086
8	0	3	3	1	0	7	.0093
9	0	3	1	1	3	8	.0100
10	0	1	1	0	1	3	.0096
11	2	0	2	0	1	5	.0096
12	3	2	0	0	0	5	.0097
13	1	2	0	1	1	5	.0097
14	0	1	2	1	1	5	.0097
15	5	0	1	0	0	6	.0099
16	1	0	2	1	3	7	.0101
17	1	0	0	0	0	1	.0096
18	2	1	0	0	0	3	.0094
19	2	1	0	0	2	5	.0095
20	0	0	2	0	3	5	.0095

regularly.) As an experiment, sixteen-place tables [4] were selected and entries were counted where the last five digits are all different. In the first twelve batches of 100 entries each, the frequencies of this event are as follows:

$$0.30, \quad 0.27, \quad 0.30, \quad 0.34, \quad 0.26, \quad 0.32,$$
$$.37, \quad .36, \quad .26, \quad .31, \quad .36, \quad .32.$$

These numbers fluctuate around the value 0.3024, and small-sample theory shows that the magnitude of the fluctuations is well within the

[4] *Tables of Probability Functions*, vol. I, National Bureau of Standards, 1941.

expected limits. The average frequency is 0.3142, which is rather close to the theoretical probability, 0.3024 [cf. Chapter 7, example (3.b)].

Consider next the number $e = 2.71828\ldots$. The first 800 decimals [5] form 160 groups of five digits each, which we arrange in sixteen batches of ten each. In these sixteen batches the numbers of groups in which all five digits are different are as follows:

$$3, \ 1, \ 3, \ 4, \ 4, \ 1, \ 4, \ 4,$$
$$4, \ 2, \ 3, \ 1, \ 5, \ 4, \ 6, \ 3.$$

The frequencies oscillate, as they should, around the value 0.3024, and small-sample theory confirms that the magnitude of the fluctuations is not larger than should be expected. The overall frequency of our event in the 160 groups is $52/160 = 0.325$, which is reasonably close to $p = 0.3024$.

(c) *Birthdays*. The birthdays of r people form a sample of size r from the population of all days in the year. The years are not of equal length, and we know that the birth rates are not quite constant throughout the year. However, in a first approximation, we may take a random selection of people as equivalent to a random selection of birthdays. Furthermore, we shall, for simplicity, ignore the existence of leap years and shall consider random samples of r birthdays in a year of 365 days.

Let us calculate the *probability*, p, *that all r birthdays are different*. The number of arrangements is 365^r. If the r birthdays are different, they form a sample without replacement, and there exist $(365)_r$ such samples. Hence

$$(3.2) \qquad p = \frac{(365)_r}{365^r}.$$

This formula is not very suggestive. For *numerical calculations* it is preferable to write it in the form

$$(3.3) \qquad p = \left(1 - \frac{1}{365}\right)\left(1 - \frac{2}{365}\right)\left(1 - \frac{3}{365}\right)\cdots\left(1 - \frac{r-1}{365}\right).$$

If r is small, we can neglect all cross products and have in a crude approximation

$$(3.4) \qquad p \approx 1 - \frac{1 + 2 + \cdots + (r-1)}{365} = 1 - \frac{r(r-1)}{730}.$$

For $r = 10$ the correct value is $p = 0.883\ldots$, while (3.4) gives the approximation 0.877.

For larger r we obtain a much better approximation by passing to logarithms. The Taylor expansion [cf. (6.9)] shows that $\log(1 - x) \approx -x$, provided x is small. Thus from (3.3)

$$(3.5) \qquad \log p \approx -\frac{1 + 2 + \cdots + (r-1)}{365} = -\frac{r(r-1)}{730};$$

[5] *Intermédiaire des recherches mathématiques*, vol. 2, 1946, p. 112.

the error term can be calculated from the remainder of the Taylor expansion, and it turns out to be approximately $r^3/6 \cdot 365^2$. For $r = 30$ the approximation (3.4) gives a negative value; formula (3.5) gives 0.3037 instead of the correct value $p = 0.294$. For $r \le 40$ the error in p is less than $1/12$.

How large should r be to have $p \approx 1/2$? We have to solve the equation $\log p = -\log 2$. The above consideration shows that, instead, we may solve $r(r-1)/730 = 0.7$. The root is $r = 22.6$. With $r = 22$ *people the probability that no two have the same birthday just exceeds* $1/2$, *but for* 23 *people this probability is already smaller than* $1/2$. [Cf. example (4.c) of Chapter 6.]

(d) *Elevator.* An elevator starts with $r = 7$ passengers and stops at $n = 10$ floors. What is the probability p that no two passengers leave at the same floor? To render the question precise, we assume that all arrangements of discharging the passengers have the same probability (which is presumably only a crude approximation). There exist 10^7 arrangements, of which $(10)_7$ are arrangements without repetition. Therefore $p = 10^{-7}(10)_7 = (10 \cdot 9 \cdot 8 \cdot 7 \cdot 6 \cdot 5 \cdot 4)10^{-7} = 0.06048$. When the event was once observed, the occurrence was deemed remarkable and odds of 1000 to 1 were offered against a repetition.

4. Partitions

If the sample size equals the population size n, then a sample without repetitions is the same as an ordering of the n elements of the population. By the theorem of section 2, the number of such orderings is $(n)_n = n(n-1) \cdots 3 \cdot 2 \cdot 1$. Instead of $(n)_n$ we use the usual notation

$$(4.1) \qquad n! = 1 \cdot 2 \cdot 3 \cdots (n-1) \cdot n.$$

We found

Theorem 1. The number of different orderings of n elements is $n!$.

Let $r < n$ and consider the samples without repetition of size r from a population of n elements. There are $(n)_r$ such samples. The elements in each can be ordered in $r!$ ways, which means that there exist $r!$ samples having the same elements. If the order is disregarded, these $r!$ samples become indistinguishable, and the number of distinguishable arrangements becomes $(n)_r/r!$. This number is known as the *binomial coefficient*

$$(4.2) \qquad \binom{n}{r} = \frac{(n)_r}{r!} = \frac{n(n-1) \cdots (n-r+1)}{1 \cdot 2 \cdots (r-1) \cdot r}.$$

Theorem 2. Out of n elements a group of size $r \le n$ can be selected in $\binom{n}{r}$ different ways (two groups are different if one contains an element not contained in the other).

An alternative way of writing (4.2) is

(4.3) $$\binom{n}{r} = \frac{n!}{r!(n-r)!}.$$

This formula shows that

(4.4) $$\binom{n}{r} = \binom{n}{n-r}.$$

Now $\binom{n}{n} = 1$ while $\binom{n}{0}$ is meaningless. Formula (4.4) suggests defining

(4.5) $$\binom{n}{0} = 1.$$

This amounts to the same thing as defining $0! = 1$.

Examples. (a) *Bridge* (cf. footnote on p. 9, Chapter 1). Let a hand of 13 cards be selected at random from a full deck. The order within a hand is irrelevant. The number of different hands is, therefore, $\binom{52}{13} = 635{,}013{,}559{,}600$. Let us now calculate the *probability p that a hand contains all thirteen face values*, assuming, of course, that all hands have the same probability. For each face value we have the free choice of a suit. The hand may be considered as a sample with repetitions of size 13 of the population of four suits. By the theorem of section 2 there exist 4^{13} such samples, and hence

$$p = 4^{13} \div \binom{52}{13} = 0.0001057 \ldots .$$

(b) *Poker* (cf. footnote on p. 9, Chapter 1). The number of different hands at poker is $\binom{52}{5} = 2{,}598{,}960$. To find the number of hands containing five different face values we note that these face values can be chosen in $\binom{13}{5}$ different ways. As in the preceding example we see that the corresponding suits can be selected in 4^5 ways. Hence, if the hand is chosen at random, *the probability that the 5 cards have five different face values is*

$$p = 4^5 \binom{13}{5} \div \binom{52}{5} \approx 0.5071.$$

(c) Each of the 48 states has two senators. We consider the events that in a committee of 48 senators chosen at random: (1) a given state is represented, (2) all states are represented.

In the first case it is better to calculate the probability q of the complementary event, namely, that the given state is not represented. There are 96 senators, and 94 not from the given state. Hence,

$$q = \binom{94}{48} \div \binom{96}{48} = \frac{48 \cdot 47}{96 \cdot 95},$$

and the probability that the given state is represented is $1 - q = 0.75263\ldots$. Now the theorem of section 2 shows that a committee including one senator from each state can be chosen in 2^{48} different ways. The probability that *all* states are included in the committee is, therefore, $p = 2^{48} \div \binom{96}{48}$. Using Stirling's formula (cf. section 7), it can be shown that $p \approx (3\pi)^{1/2} 2^{-46} \approx 4 \cdot 10^{-14}$.

Theorem 3. Let $r_1, r_2, \cdots, r_k$ be non-negative integers such that

(4.6) $$r_1 + r_2 + \cdots + r_k = n.$$

The number of ways in which n objects can be divided into k groups of which the first contains r_1 objects, the second r_2 objects, etc., is

(4.7) $$\frac{n!}{r_1! r_2! \cdots r_k!}.$$

(Here, the order of the groups is essential, but no attention is paid to the order within the groups. The numbers (4.7) are called *multinomial coefficients*.)

Proof. Consider first the case $k = 2$. To partition n objects into two groups is the same as to select r_1 objects to go into the first group; the $r_2 = n - r_1$ remaining objects form the second group. For $k = 2$ the number of partitionings is therefore $\binom{n}{r_1} = n!/(r_1! r_2!)$. To perform the partitioning in the general case we start by selecting the first group of size r_1; of the remaining $(n - r_1)$ objects we select a group of size r_2, etc. After forming the $(k - 1)$st group there remain $n - r_1 - r_2 - \cdots - r_{k-1} = r_k$ objects, which form the last group. We conclude that the number of partitions is:

(4.8) $$\binom{n}{r_1}\binom{n - r_1}{r_2}\binom{n - r_1 - r_2}{r_3} \cdots \binom{n - r_1 \cdots - r_{k-2}}{r_{k-1}}$$

which reduces to (4.7) if all binomial coefficients are expressed in accordance with (4.3).

Examples. *(d) Bridge.* The $n = 52$ cards are distributed among players with $r_1 = r_2 = r_3 = r_4 = 13$. Hence *the number of different situations at a bridge table is:*

$$(4.9) \qquad \frac{52!}{(13!)^4} = (5.3645\ldots)10^{28}.$$

Let us now calculate the probability that each player has an ace. The four aces can be ordered in $4! = 24$ ways, and each order represents one possibility of giving one ace to each player. The remaining 48 cards can be distributed in $(48!)(12!)^{-4}$ ways. Hence the required probability is

$$24 \cdot \frac{48!(13)^4}{52!} = 0.105\ldots.$$

(e) Dice. A throw of 12 dice can result in 6^{12} different outcomes to all of which we attribute equal probabilities. The event that each face appears twice can occur in as many ways as 12 dice can be arranged in six groups of two each. Hence the probability of the event is $12!/(2^6 \cdot 6^{12}) = 0.003438\ldots.$

5. The Hypergeometric Distribution

Many combinatorial problems can be reduced to the following form. In a population of n elements n_1 are red and $n_2 = n - n_1$ are black. A group of r elements is chosen at random (without replacement and without regard to order). We seek the probability q_k that the group so chosen will contain exactly k red elements. Here k can be any integer between zero and n_1 or r, whichever is smaller.

To find q_k, we note that the chosen group contains k red and $r - k$ black elements. The red ones can be chosen in $\binom{n_1}{k}$ different ways and the black ones in $\binom{n - n_1}{r - k}$ ways. Since any choice of k red elements may be combined with any choice of black ones, we find

$$(5.1) \qquad q_k = \frac{\binom{n_1}{k}\binom{n - n_1}{r - k}}{\binom{n}{r}}.$$

The system of probabilities so defined is called the *hypergeometric distribution*.[6] Using formula (4.3), it is possible to rewrite (5.1) in the form

$$(5.2) \qquad q_k = \frac{\binom{r}{k}\binom{n-r}{n_1-k}}{\binom{n}{n_1}}.$$

Note. The probabilities q_k are defined only for k not exceeding r or n_1. However, from the definition (4.2) it follows that $\binom{a}{b} = 0$ whenever $b > a$. Therefore, formulas (5.1) and (5.2) give $q_k = 0$ if either $k > n_1$ or $k > r$. Accordingly, the definitions (5.1) and (5.2) may be used for all $k \geq 0$, provided that the relation $q_k = 0$ is interpreted as impossibility.

Examples. (a) *Quality Inspection.* In industrial quality control, lots of size n are subjected to sampling inspection. The defective items in the lot play the role of "red" elements. Their number n_1 is, of course, unknown. A sample of size r is taken, and the number k of defective items in it is determined. Formula (5.1) then permits us to draw inferences as to the likely magnitude of n_1; this is a typical problem of statistical estimation which is beyond the scope of the present book.

(b) In example (4.c), the population consists of $n = 96$ senators of whom $n_1 = 2$ represent the given state (are "red"). A group of $r = 48$ senators is chosen at random. It may include $k = 0, 1,$ or 2 senators from the given state. From (5.2) we find [remembering that $\binom{n}{0} = 1$ by (4.5)]

$$q_0 = \frac{48 \cdot 47}{96 \cdot 95} = 0.24737 \ldots,$$

$$q_1 = \frac{48}{95} = 0.50527 \cdots,$$

$$q_2 = \frac{48 \cdot 47}{96 \cdot 95} = 0.24737 \ldots.$$

The value q_0 was obtained in a different way in example (4.c).

---

[6] The name is explained by the fact that the generating function (cf. Chapter 11) of $\{q_k\}$ can be expressed in terms of hypergeometric functions.

(c) *Distribution of Aces among r Bridge Cards.* Let $p_0(r)$, $p_1(r)$, $\cdots$, $p_4(r)$ denote the probabilities that among r bridge cards drawn at random there are 0, $\cdots$, 4 aces, respectively. For $r = 5$ we obtain the probability that a poker hand (cf. footnote on p. 9) contains 0, 1, $\cdots$, 4 aces. We get $p_k(r)$ from (5.2) with $n = 52$, $n_1 = 4$. A simple calculation shows that

(5.3)
$$p_0(r) = \frac{(52-r)_4}{(52)_4},$$

$$p_1(r) = \frac{4r(52-r)_3}{(52)_4},$$

$$p_2(r) = \frac{6r(r-1)(52-r)_2}{(52)_4},$$

$$p_3(r) = \frac{4r(r-1)(r-2)(52-r)}{(52)_4},$$

$$p_4(r) = \frac{(r)_4}{(52)_4}.$$

Table 3 gives the values $p_k(r)$ for all possible combinations of k and r. In using it, note that the probability of having 0, 1, $\cdots$, 4 aces among r cards is the same as the probability of having 4, 3, $\cdots$, 0 aces among the remaining $52 - r$ cards. In other words, we have

(5.4) $\quad p_0(r) = p_4(52-r), \quad p_1(r) = p_3(52-r), \quad p_2(r) = p_2(52-r)$

as can be verified from (5.3).

(d) *A Waiting Time Problem.* We shall for a moment deviate from our path in order to discuss an alternative interpretation of the probabilities of Table 3. Suppose that card after card is drawn from a deck of bridge cards and that all 52! possible orders of picking have equal probabilities. Then $p_k(r)$ is the probability that among the first r cards there will be exactly k aces. In particular, $p_0(r)$ is the probability that no ace turns up in the first r drawings or that more than r are required for the first ace to turn up. In a similar way $p_0(r) + p_1(r)$ is the probability that in r drawings 0 or 1 ace turns up, which means that it takes more than r drawings before the second ace turns up. Also, $p_0(r) + p_1(r) + p_2(r)$ becomes the probability that the third ace turns up sometime after the rth drawing.

For each $k \leq 4$ there exists a number r_k of drawings such that the probabilities that the kth ace will turn up before or after the r_kth drawing will both be closest to $1/2$. This number r_k is called the

TABLE 3
Probabilities (5.3)

r	$p_0(r)$	$p_1(r)$	$p_2(r)$	$p_3(r)$	$p_4(r)$	
1	0.92308	0.07692	0.00000	0.00000	0.00000	51
2	.85068	.14480	.00452	.00000	.00000	50
3	.78262	.20416	.01303	.00018	.00000	49
4	.71874	.25555	.02500	.00071	.00000	48
5	.65884	.29947	.03993	.00174	.00002	47
6	.60277	.33643	.05735	.00340	.00006	46
7	.55036	.36690	.07679	.00582	.00013	45
8	.50144	.39136	.09784	.00910	.00026	44
9	.45585	.41027	.12008	.01334	.00047	43
10	.41344	.42405	.14312	.01862	.00078	42
11	.37407	.43313	.16659	.02499	.00122	41
12	.33757	.43794	.19016	.03250	.00183	40
13	.30382	.43885	.21349	.04120	.00264	39
14	.27266	.43625	.23630	.05109	.00370	38
15	.24396	.43051	.25831	.06218	.00504	37
16	.21758	.42198	.27925	.07447	.00672	36
17	.19341	.41099	.29890	.08791	.00879	35
18	.17130	.39786	.31705	.10248	.01130	34
19	.15115	.38291	.33350	.11812	.01432	33
20	.13283	.36642	.34810	.13475	.01790	32
21	.11622	.34867	.36070	.15229	.02211	31
22	.10123	.32993	.37117	.17065	.02702	30
23	.08773	.31043	.37942	.18971	.03271	29
24	.07563	.29042	.38537	.20933	.03925	28
25	.06483	.27011	.38896	.22938	.04673	27
26	.05522	.24970	.39016	.24970	.05522	26
	$p_4(r)$	$p_3(r)$	$p_2(r)$	$p_1(r)$	$p_0(r)$	r

$p_k(r)$ is the probability that among r bridge cards selected at random there will be exactly k aces.

median for the kth ace. From Table 3 we find that the four medians are 8, 20, 32, 44, respectively.

Define now a new game by the following rule: "Cards are drawn as long as necessary for the first ace to turn up." Note that this rule defines a new sample space. We get a picture of the sample points if in each of the 52! different orderings of the deck of cards we remove all cards which come after the first ace. Naturally, several orderings will lead to the same new sample point, and we have here the first example of a sample space in which we attribute different probabilities to the various points.

(e) *A Sampling Problem.* Suppose that 1000 fish caught in a lake are marked by red spots and released. After a while a new catch of 1000 fish is made, and it is found that 100 among them have red spots. What conclusions can be drawn concerning the number of fish in the lake? This is a typical problem of *statistical estimation*. It would lead us too far to describe the various methods that a modern statistician might use, but we shall show how the hypergeometric distribution gives us a clue to the solution of the problem. We assume naturally that the two catches may be considered as random samples from the population of all fish in the lake. (In practice this assumption excludes situations where the two catches are made at one locality and within a short time.) We also suppose that the number of fish in the lake does not change between the two catches.

We generalize the problem by admitting arbitrary sample sizes. Let

n = the (unknown) number of fish in the lake.
n_1 = the number of fish in the first catch. They play the role of red balls.
r = the number of fish in the second catch.
k = the number of red fish in the second catch.
$q_k(n)$ = the probability that the second catch contains exactly k red fish.

In this formulation it is rather obvious that $q_k(n)$ is given by (5.1). In practice n_1, r, and k can be observed, but n is unknown. Note, incidentally, that n is a fixed number which in no way depends on chance. It is, therefore, meaningless to ask for the probability that n is greater than, say, 6000. We know that $n_1 + r - k$ different fish were caught, and therefore $n \geq n_1 + r - k$. This is all that can be said with certainty. In our example we had $n_1 = r = 1000$ and $k = 100$, and it is conceivable that the lake contains only 1900 fish. However, starting from this hypothesis, we are led to the conclusion that an event of a fantastically small probability has occurred. In fact, the probability that two samples of size 1000 will exhaust an entire population of 1900 fish is, by (5.1),

$$\frac{\binom{1000}{100}\binom{900}{900}}{\binom{1900}{1000}} = \frac{(1000!)^2}{100!\,1900!}.$$

Stirling's formula (cf. section 7) shows this probability to be of the order of magnitude 10^{-430}, and we therefore reject our hypothesis as unreasonable. A similar reasoning would induce us to reject the hypothesis that n is very large, say, a million. This consideration leads us to seek the particular value of n for which $q_k(n)$ attains its largest value, since for that n our observation would have the greatest probability. For any particular set of observations n_1, r, k the value of n for which $q_k(n)$ is largest is denoted by $\hat{n}$ and is called the *maximum likelihood estimate* of n. This notion was introduced by R. A. Fisher. To find $\hat{n}$ consider the ratio

$$(5.5) \qquad \frac{q_k(n)}{q_k(n-1)} = \frac{(n-n_1)(n-r)}{(n-n_1-r+k)n}.$$

A simple calculation shows that this ratio is greater than or smaller than unity, according as $nk < n_1 r$ or $nk > n_1 r$. This means that with increasing n the sequence $q_k(n)$ first increases and then decreases; it reaches its maximum when n is the largest integer short of $n_1 r/k$, so that

$$(5.6) \qquad \hat{n} \approx \frac{n_1 r}{k}.$$

In our particular example the maximum likelihood estimate of the number of fish is $\hat{n} = 10{,}000$.

The true number n may be larger or smaller, and we may ask for limits within which we may reasonably expect n to lie. For this purpose let us test the hypothesis that n is smaller than 8500. We substitute in (5.1) $n = 8500$, $n_1 = r = 1000$, and calculate the probability that the second sample contains 100 or fewer red fish. This probability is $x = q_0 + q_1 + \cdots + q_{100}$. A direct evaluation is cumbersome, but using the normal approximation of Chapter 7, we find easily that $x \approx 0.04$. Similarly, if $n = 12{,}000$, the probability that the second sample contains 100 or more red fish is about 0.03. These figures would justify a bet that the true number n of fish lies somewhere between 8500 and 12,000. There exist other ways of formulating these conclusions and other methods of estimation, but we do not propose to discuss the details.

From the definition of the probabilities q_k it follows that $q_0 + q_1 + q_2 + \cdots = 1$. Formula (5.2) therefore implies that for any positive integers n, n_1, and r

$$(5.7) \quad \binom{r}{0}\binom{n-r}{n_1} + \binom{r}{1}\binom{n-r}{n_1-1}$$

$$+ \binom{r}{2}\binom{n-r}{n_1-2} + \cdots + \binom{r}{n_1}\binom{n-r}{0} = \binom{n}{n_1}.$$

This identity is frequently useful. We have proved it only for positive integers n and r, but it holds true without this restriction for arbitrary positive or negative numbers n and r (it is meaningless if n_1 is not a positive integer). (An indication of two proofs is given in section 9, problems 5 and 6; for application cf. problems 7–13.)

The hypergeometric distribution can easily be generalized to the case where the original population of size n contains several classes of elements. For example, let the population contain three classes of sizes n_1, n_2, and $n - n_1 - n_2$, respectively. If a sample of size r is taken, the probability that it contains k_1 elements of the first, k_2 elements of the second, and $r - k_1 - k_2$ elements of the last class is, by analogy with (5.1),

$$(5.8) \quad \frac{\binom{n_1}{k_1}\binom{n_2}{k_2}\binom{n-n_1-n_2}{r-k_1-k_2}}{\binom{n}{r}}.$$

It is, of course, necessary that $k_1 \leq n_1$, $k_2 \leq n_2$, and $r - k_1 - k_2 \leq n - n_1 - n_2$. [For further properties of the hypergeometric distribution cf. section 8, problem 42; Chapter 6, example (4.g); Chapter 7, problem 10; and Chapter 9, example (5.e) and problem 11.]

Example. (f) *Bridge.* The population of 52 cards consists of four classes, each of thirteen elements. The probability that a hand of 13 cards consists of five spades, four hearts, three diamonds, and one club is

$$\frac{\binom{13}{5}\binom{13}{4}\binom{13}{3}\binom{13}{1}}{\binom{52}{13}}.$$

The probability that a hand contains 5 cards of some suit, 4 of another, 3 of a third, and 1 of the last suit is 4! times as large, or 0.1293 ....

6. Binomial Coefficients

The numbers $(n)_r$ and $\binom{n}{r}$, defined in (2.1) and (4.2), have been defined only if n and r are positive integers, but it is convenient to extend their definition. Since r denotes the number of factors, it must be an integer. However, the number

(6.1) $$(x)_r = x(x-1) \cdots (x-r+1)$$

is well defined for all real x provided only that r is a positive integer. For $r = 0$ we shall put $(x)_0 = 1$. Then

(6.2) $$\binom{x}{r} = \frac{(x)_r}{r!} = \frac{x(x-1) \cdots (x-r+1)}{r!}$$

defines the binomial coefficients for all values of x and all positive integers r. For $r = 0$ we put, as in (4.5), $\binom{x}{0} = 1$ *and* $0! = 1$. *For negative integers r we put*

(6.3) $$\binom{x}{r} = 0 \qquad (r < 0).$$

We shall never use the symbol $\binom{x}{r}$ *if r is not an integer.*

With this definition we have, for example,

$$\binom{-1}{3} = -1, \quad \binom{7}{11} = 0, \quad \binom{-2}{4} = 5,$$

$$\binom{3/2}{4} = \frac{3}{2} \cdot \frac{1}{2} \cdot \frac{-1}{2} \cdot \frac{-3}{2} \cdot \frac{1}{4!} = \frac{3}{128},$$

$$\binom{-1/3}{3} = \frac{-1}{3} \cdot \frac{-4}{3} \cdot \frac{-7}{3} \cdot \frac{1}{6} = -\frac{14}{81}.$$

Three important properties will be used in the sequel. First, *for any positive integer n*

(6.4) $$\binom{n}{r} = 0 \quad \text{if either} \quad r > n \quad \text{or} \quad r < 0.$$

Second, *for any number x and any integer r*

(6.5) $$\binom{x}{r-1} + \binom{x}{r} = \binom{x+1}{r}.$$

These relations are easily verified from the definition. The third relation will be assumed known from calculus textbooks: *for any number a and all values $-1 < t < 1$ we have Newton's binomial formula*

(6.6) $$(1 + t)^a = 1 + \binom{a}{1} t + \binom{a}{2} t^2 + \binom{a}{3} t^3 + \ldots.$$

If a is a positive integer, all terms to the right containing powers higher than t^a vanish automatically; then the formula is correct for all t. If a is not a positive integer, the right side represents an *infinite* series.

As an example consider the case $a = -1$. We find easily

(6.7) $$\binom{-1}{r} = (-1)^r$$

whence (6.6) reduces to the *geometric series*

(6.8) $$\frac{1}{1+t} = 1 - t + t^2 - t^3 + t^4 - + \ldots,$$

an expansion which is valid for $-1 < t < 1$. Integrating (6.8), we obtain another formula which will be useful in the sequel, namely, the *Taylor expansion of the natural logarithm*

(6.9) $$\log(1 + t) = t - \tfrac{1}{2} t^2 + \tfrac{1}{3} t^3 - \tfrac{1}{4} t^4 + \ldots.$$

This expansion is again valid for $-1 < t < 1$. It has been used in (3.5).

Many other relations can be deduced from (6.6). For example, for any positive integer n we find, letting $t = 1$,

(6.10) $$\binom{n}{0} + \binom{n}{1} + \binom{n}{2} + \cdots + \binom{n}{n} = 2^n$$

and letting $t = -1$,

(6.11) $$\binom{n}{0} - \binom{n}{1} + \binom{n}{2} - + \cdots \pm \binom{n}{n} = 0.$$

These two identities will be used in the sequel. (For further identities see the problems of section 9.)

7. Stirling's Formula

A direct numerical evaluation of expressions involving $n!$ is usually inexpedient, and it is therefore important to find simple approximations to $n!$ It is clear that n^n increases faster than $n!$, and we shall obtain information concerning $n!$ from a study of the ratio $a_n = n!/n^n$. Obviously

(7.1) $$\frac{a_{n+1}}{a_n} = \frac{1}{\left(1 + \dfrac{1}{n}\right)^n}.$$

Now the basis of the natural logarithms $e = 2.71828\ldots$ is defined as the limit of the denominator in (7.1), so that $a_{n+1}/a_n \to 1/e$. Thus for large n the sequence a_n behaves essentially like a geometric sequence with ratio e^{-1}. This suggests replacing n^n by $(n/e)^n$ and studying the sequence

(7.2) $$b_n = n!\left(\frac{e}{n}\right)^n = a_n e^n.$$

To investigate the growth of b_n, we first pass to logarithms, then use (7.1) and the Taylor expansion (6.9) to find

(7.3) $$\log \frac{b_{n+1}}{b_n} = 1 + \log \frac{a_{n+1}}{a_n} = 1 - n \log\left(1 + \frac{1}{n}\right)$$
$$= \frac{1}{2n} - \frac{1}{3n^2} + \cdots.$$

The right-hand member is positive, so that b_{n+1}/b_n is greater than 1. Hence the sequence $\{b_n\}$ increases, which means that $n!$ grows faster than $(n/e)^n$. On the other hand, if we replace the exponent by $n + 1/2$ and put

(7.4) $$\beta_n = n!\left(\frac{e}{n}\right)^{n+\frac{1}{2}}$$

then (7.3) is to be replaced by

(7.5) $$\log \frac{\beta_{n+1}}{\beta_n} = 1 - \left(n + \frac{1}{2}\right)\log\left(1 + \frac{1}{n}\right)$$
$$= -\frac{1}{12n^2} + \frac{1}{12n^3} - \cdots.$$

The right side is negative for all n, and hence $\beta_{n+1} < \beta_n$. We see, therefore, that the sequence β_n decreases so that a limit $\beta = \lim \beta_n$ exists. From (7.4) we have then

(7.6) $$n!\left(\frac{e}{n}\right)^{n+\frac{1}{2}} \to \beta.$$

2.7] STIRLING'S FORMULA

For a satisfactory approximation to $n!$ we now require only the numerical value of β. In principle β could be calculated to any desired number of decimals by computing the left side in (7.6) for some n sufficiently large. Fortunately we can avoid this uninspiring procedure, since the theory developed in Chapter 7 will show that $\beta = (2\pi e)^{1/2}$. Up to then the exact value of β will play no role, and we are free to postpone the proof that $\beta = (2\pi e)^{1/2}$ (cf. Chapter 7, section 2). We can now rewrite (7.6) in the final form known as

Stirling's Formula

(7.7) $$n! \sim (2\pi)^{1/2} n^{n+1/2} e^{-n}.$$

Here the sign $\sim$ is used to indicate that the ratio of the two sides tends to 1.

It is true that the difference of the two sides in (7.7) increases over all bounds, but it is the *percentage error* which really matters. The percentage error decreases steadily (since β_n decreases), and Table 4 shows that Stirling's approximation is remarkably accurate even for small n. For $n = 5$ the error is 2 per cent, and for $n \geq 9$ it has dropped below 1 per cent.

TABLE 4

STIRLING'S APPROXIMATIONS TO $n!$

n	$n!$	Approximation from (7.7)	Per Cent Error	Approximation from (7.9)	Per Cent Error
1	1	0.922	8	1.0023	0.2
2	2	1.919	4	2.0007	.04
5	120	118.019	2	120.01	.01
10	$(3.62880)10^6$	$(3.5986\ldots)10^6$	0.8	$(3.62881\ldots)10^6$	....
100	$(9.3326\ldots)10^{157}$	$(9.3249\ldots)10^{157}$	0.08	$(9.3326\ldots)10^{157}$	....

We saw that β_n decreases so that $\beta_n > \beta$. This implies that Stirling's approximation (7.7) somewhat *underestimates* $n!$. It is easy to improve on (7.7) and, in particular, to find alternative approximations which will overestimate $n!$. For that purpose it suffices to modify the definition (7.4) of β_n so that the right side of (7.5) will be positive. Now, if we put $\gamma_n = \beta_n e^{-1/12n}$, then $\gamma_n \to \beta$ and

(7.8) $$\log \frac{\gamma_{n+1}}{\gamma_n} = \log \frac{\beta_{n+1}}{\beta_n} + \frac{1}{12n(n+1)}.$$

A comparison with (7.5) shows that the right side in (7.8) starts with terms of the magnitude n^{-4}, and we can, therefore, expect that

(7.9) $$n! \sim (2\pi)^{1/2} n^{n+1/2} e^{-n+1/(12n)}.$$

will give an even better approximation to $n!$ than (7.7). The last two columns of Table 4 confirm this assumption. Even for $n = 2$ formula (7.9) yields an error of less than 4 hundredths of a per cent. An additional feature of (7.9) is that it always *overestimates* $n!$. To prove this we have only to show that γ_n increases or that (7.8) is positive. Note that (7.8) is symmetric in n and $n + 1$, so that it is best to express the formula in terms of the mean $\mu = n + \tfrac{1}{2}$. Now $n = \mu - \tfrac{1}{2}$, and $n + 1 = \mu + \tfrac{1}{2}$, and, using (7.5), we can rewrite (7.8) in the form

$$(7.10) \quad \log \frac{\gamma_{n+1}}{\gamma_n} = 1 - \mu \left\{ \log\left(1 + \frac{1}{2\mu}\right) - \log\left(1 - \frac{1}{2\mu}\right) \right\} + \frac{1}{12\mu^2} \cdot \frac{1}{1 - 1/(4\mu^2)}.$$

Formulas (6.8) and (6.9) provide convenient expansions for the last fraction and the two logarithms. We get

$$(7.11) \quad \log \frac{\gamma_{n+1}}{\gamma_n} = 1 - \mu \left\{ \frac{1}{\mu} + \frac{1}{12\mu^3} + \frac{1}{80\mu^5} + \frac{1}{448\mu^7} + \cdots \right\}$$

$$+ \frac{1}{12\mu^2} \left\{ 1 + \frac{1}{4\mu^2} + \frac{1}{16\mu^4} + \cdots \right\}$$

$$= \frac{1}{120\mu^4} + \frac{1}{336\mu^6} + \cdots.$$

The coefficient of μ^{-2k} is $2^{-2k} \left(\dfrac{1}{3} - \dfrac{1}{2k+1} \right) > 0$, and hence all terms are positive.

This accomplishes the proof of the *theorem: the right side in* (7.9) *overestimates* $n!$, *while the right side in* (7.7) *underestimates* $n!$.

Formula (7.9) is called *Stirling's second-term approximation to* $n!$. In the same way we see that a third approximation would add a term $e^{-1/360\mu^3}$, etc.

8. Problems for Solution: Combinatorial

Note: Assume in each case that all arrangements have the same probability.

1. How many different sets of initials can be formed if every person has (a) exactly two, (b) at most two, (c) at most three, given names?

2. In how many ways can two rooks of different colors be put on a chessboard so that they can take each other?

3. Letters in the Morse alphabet are formed by a succession of dashes and dots with repetitions permitted. How many letters is it possible to form with ten symbols or less?

4. Each domino piece is marked by two numbers. The pieces are symmetrical so that the number-pair is not ordered. How many different pieces can be made using the numbers $1, 2, \cdots, n$?

5. The numbers $1, 2, \cdots, n$ are arranged in random order. Find the probability that the digits (a) 1 and 2, (b) 1, 2, and 3, appear as neighbors in the order named.

6. Find the probability that among three random digits there occur 2, 1, or 0 repetitions.

7. Do problem 6 for the case of four random digits.

8. Find the probabilities p_r that in a sample of r random digits no two are equal. Estimate the numerical value of p_{10}, using Stirling's formula.

9. What is the probability that among k random digits (a) 0 does not appear; (b) 1 does not appear; (c) neither 0 nor 1 appears; (d) at least one of the two digits

0 and 1 does not appear? Let A and B represent the events in (a) and (b). Express the other events in terms of A and B.

10. What is the probability that among k random digits 0 appears (a) exactly three times; (b) three times or less?

11. Suppose that each of n sticks is broken into one long and one short part. The $2n$ parts are arranged into n pairs from which new sticks are formed. Find the probability (a) that the parts will be joined in the original order, (b) that all long parts are paired with short parts.[7]

12. *Testing a statistical hypothesis.* A Cornell professor got a ticket twelve times for illegal overnight parking. All twelve tickets were given either Tuesdays or Thursdays. Find the probability of this event. (Was his renting a garage only for Tuesdays and Thursdays justified?)

13. *Continuation.* Of twelve police tickets none was given on Sunday. Is this evidence that no tickets are given on Sundays?

14. From a population of 100 people a sample of size n is taken. Find the probability that none of r given people will be included in the sample, assuming the sampling to be (a) without, (b) with replacement. Compare the numerical values for $n = r = 3$ and $n = r = 10$.

15. A box contains 90 good and 10 defective screws. If 10 screws are used, what is the probability that none is defective?

16. From the population of five symbols a, b, c, d, e, a sample of size 25 is taken. Find the probability that the sample will contain five symbols of each kind. Check the result in tables of random numbers,[8] identifying the digits 0 and 1 with a, the digits 2 and 3 with b, etc.

17. If n men, among whom are A and B, stand in a row, what is the probability that there will be exactly r men between A and B? If they stand in a ring instead of in a row, show that the probability is independent of r and hence $1/(n-1)$.

18. What is the probability that 2 throws with 3 dice each will show the same configuration if, (a) the dice are distinguishable, (b) they are not.

19. Show that it is more probable to get at least one ace with 4 dice than at least one double ace in 24 throws of 2 dice. (The answer is known as de Méré's paradox. Chevalier de Méré, a gambler, thought that the two probabilities ought to be equal and blamed mathematics for his losses.)

20. How many dice have to be thrown to render the probability of no ace less than 1/3?

21. What is the probability that the birthdays of twelve people will fall in 12 different calendar months? (Assume equal probabilities for the 12 months.)

22. What is the probability that the birthdays of six people will fall in exactly 2 calendar months?

23. Given 30 people, find the probability that among the 12 months there are 6 containing two, and 6 containing three, birthdays.

---

[7] When cells are exposed to harmful radiation, some chromosomes break and play the role of our "sticks." The "long" side is the one containing the so-called centromere. If two "long" or two "short" parts unite, the cell dies. Cf. D. G. Catcheside, The Effect of X-ray Dosage upon the Frequency of Induced Structural Changes in the Chromosomes of *Drosophila Melanogaster*, *Journal of Genetics*, vol. 36 (1938), pp. 307-320.

[8] They are occasionally extraordinarily obliging: cf. J. A. Greenwood and E. E. Stuart, Review of Dr. Feller's Critique, *Journal for Parapsychology*, vol. 4 (1940) pp. 298-319, in particular p. 306.

24. A closet contains n pairs of shoes. If $2r$ shoes are chosen at random (with $2r < n$), what is the probability that there will be no complete pair among them?

25. In the preceding problem find the probabilities that among the $2r$ shoes there will be (a) exactly one complete pair, (b) exactly two complete pairs.

26. A group of $2N$ boys and $2N$ girls is divided into two equal groups. Find the probability p that each group will be equally divided into boys and girls. Estimate p, using Stirling's formula.

27. Find the probability that a hand of 13 bridge cards contains exactly k red cards, $k = 0, 1, \cdots, 13$.

28. Find the probability that a hand of 13 bridge cards contains exactly k spades.

29. In bridge, prove that the probability p of West's receiving exactly k aces is the same as the probability that an arbitrary hand of 13 cards contains exactly k aces. (This is intuitively clear. Note, however, that the two probabilities refer to two different experiments, since in the second case 13 cards are chosen at random and in the first case all 52 are distributed.)

30. The probability that in a bridge game East receives m and South n spades is the same as the probability that of two hands of 13 cards each, drawn at random from a deck of bridge cards, the first contains m and the second n spades.

31. What is the probability that the bridge hands of North and South together contain exactly k aces, where $k = 0, 1, 2, 3, 4$?

32. Let a, b, c, d be four non-negative integers such that $a + b + c + d = 13$. Find the probability $p(a, b, c, d)$ that in a bridge game the players North, East, South, West have a, b, c, d spades, respectively.

33. Using the result of problem 32, find the probability that some player receives a, another b, a third c, and the last d, spades if (a) $a = 5, b = 4, c = 3, d = 1$; (b) $a = b = c = 4, d = 1$; (c) $a = b = 4, c = 3, d = 2$.
(Note that the three cases are essentially different.)

34. Let a, b, c, d be integers with $a + b + c + d = 13$. Find the probability $q(a, b, c, d)$ that a hand at bridge will consist of a spades, b hearts, c diamonds, and d clubs.

35. *Distribution of aces at bridge.* For every possible combination of integers a, b, c, d with $a + b + c + d = 4$, find the probability that one player will have a aces, another b aces, a third c aces, and the last d aces.

36. Find the probability that each of two hands contains exactly k aces if the two hands are composed of r bridge cards each, and are drawn (a) from the same deck, (b) from two decks.

37. Show that when $r = 13$ the probability in part (a) of problem 36 is the probability that two preassigned bridge players receive exactly k aces each.

38. Find the probability for a *poker* hand to be a [9]
   (a) Royal flush (ten, jack, queen, king, ace in a single suit).
   (b) Straight flush (any five in sequence in a single suit).
   (c) Four of a kind (four cards of equal face values).
   (d) Full house (one pair and one triple of cards with equal face values).
   (e) Flush (five cards in a single suit).

[9] For the definition cf. footnote on p. 9. Note that our definition is a simplification of the usual one: for example, what we call flush can at the same time be a straight flush. In practice such a hand would be classified as straight flush. The difference of the probabilities (e) minus (b) is the probability of a flush in the gambler's sense of the word.

(f) Straight (five cards in sequence regardless of the suit).
(g) Three of a kind (three equal face values plus two extra cards).
(h) Two pairs (two pairs of equal face values plus one other card).
(i) One pair (one pair of equal face values plus three different cards).

39. In the example (3.d) the elevator starts with seven passengers and stops at ten floors. The various arrangements of discharge may be denoted by symbols like (3, 2, 2), to be interpreted as the event that three passengers leave together at a certain floor, two other passengers at another floor, and the last two at still another floor. Find the probabilities of the fifteen possible arrangements ranging from (7) to (1, 1, 1, 1, 1, 1, 1).

40. *Bridge-bingo.* A deck of 52 bridge cards is dealt in the usual way to four players. The cards of the deck are then called in random order, and whichever player has the card removes it from his hand. The player who is first without cards wins. Let $p_1(r)$, $p_2(r)$, $p_3(r)$ be the probabilities that, after the rth card is called, one, two, or three preassigned players are without cards. Calculate these probabilities. Give numerical values.

41. A population of n elements includes np red ones and nq black ones ($p + q = 1$). A random sample of size r is taken with replacement. Show that the probability of its including exactly k red elements is

(*) $$\binom{r}{k} p^k q^{r-k}.$$

42. *A limit theorem for the hypergeometric distribution.* If n is large and $n_1/n = p$, then the probability q_k given by (5.1) and (5.2) is close to the expression (*) of the preceding problem. More precisely,

$$\binom{r}{k}\left(p - \frac{k}{n}\right)^k \left(q - \frac{r-k}{n}\right)^{r-k} < q_k < \binom{r}{k} p^k q^{r-k} \left(1 - \frac{r}{n}\right)^{-r}.$$

A comparison of this and the preceding problem shows: *for large populations there is practically no difference between sampling with or without replacement* [cf. Chapter 6, example (2g)].

9. Problems for Solution: Binomial Coefficients and Stirling's Formula

Note: All following identities are of interest and frequently used.

1. For any $a > 0$

(9.1) $$\binom{-a}{k} = (-1)^k \binom{a+k-1}{k}.$$

Prove this directly and also by differentiation of the geometric series $\Sigma x^k = (1-x)^{-1}$.

2. Prove that

(9.2) $$\binom{2n}{n} 2^{-2n} = (-1)^n \binom{-\frac{1}{2}}{n}.$$

3. Prove that

(9.3) $$\binom{n}{0}\binom{n}{k} - \binom{n}{1}\binom{n-1}{k-1} + \binom{n}{2}\binom{n-2}{k-2} \cdots \pm \binom{n}{k}\binom{n-k}{0} = 0.$$

4. For integral $n \geq 2$

$$\binom{n}{1} + 2\binom{n}{2} + 3\binom{n}{3} + \cdots = n2^{n-1},$$

(9.4) $$\binom{n}{1} - 2\binom{n}{2} + 3\binom{n}{3} - + \cdots = 0,$$

$$2\cdot 1\binom{n}{2} + 3\cdot 2\binom{n}{3} + 4\cdot 3\binom{n}{4} + \cdots = n(n-1)2^{n-2}.$$

(*Hint:* Use the binomial formula.)

5. In section 5 we remarked that the terms of the hypergeometric distribution should add to unity. This amounts to saying that for any positive integers a, b, n,

(9.5) $$\binom{a}{0}\binom{b}{n} + \binom{a}{1}\binom{b}{n-1} + \binom{a}{2}\binom{b}{n-2} + \cdots + \binom{a}{n}\binom{b}{0} = \binom{a+b}{n}.$$

Prove this by induction. [*Hint:* Prove first that (9.5) holds for $a = 1$ and all b.]

6. *Continuation.* By a comparison of the coefficients of t^n on both sides of

(9.6) $$(1+t)^a(1+t)^b = (1+t)^{a+b}$$

prove more generally that (9.5) is true for arbitrary numbers a, b (and integral n).

7. Using (9.5), prove that

(9.7) $$\sum_k \binom{a}{k+r}\binom{b}{k} = \binom{a+b}{a-r}$$

for all integral a, b, and r.

8. Using (9.5), prove that

(9.8) $$\binom{n}{0}^2 + \binom{n}{1}^2 + \binom{n}{2}^2 + \cdots + \binom{n}{n}^2 = \binom{2n}{n}.$$

9. Using (9.8), prove that

(9.9) $$\sum_{\nu=0}^{n} \frac{(2n)!}{(\nu!)^2(n-\nu)!^2} = \binom{2n}{n}^2.$$

10. Prove the identity for integers $0 < a < b$

(9.10) $$\sum_{k=1}^{a}(-1)^{a-k}\binom{a}{k}\binom{b+k}{b+1} = \binom{b}{a-1}.$$

[*Hint*: Use first (9.1), then (9.5), then again (9.1). Alternatively, compare coefficients of t^{a-1} in $(1-t)^a(1-t)^{-b-2} = (1-t)^{a-b-2}$.]

11. Using (9.5), prove that

(9.11) $$\sum_{j=0}^{k}\binom{a+k-j-1}{k-j}\binom{b+j-1}{j} = \binom{a+b+k-1}{k}.$$

[*Hint*: Apply (9.1) back and forth.]

12. Prove that for $0 < k < n$

(9.12) $$\binom{n}{k} - \binom{n}{k-1} + \binom{n}{k-2} - \cdots \pm \binom{n}{1} \mp 1 = \binom{n-1}{k}.$$

[*Hint*: Compare the coefficients of t^k on either side of $(1+t)^n(1+t)^{-1} = (1+t)^{n-1}$. Cf. problem 6.]

13. From (9.12) deduce

(9.13) $$\binom{n}{r} - \binom{n}{r+1} + \binom{n}{r+2} - \cdots \mp 1 = \binom{n-1}{r-1}.$$

14. Prove by induction on a that for integers $a \geq r > 0$

(9.14) $$\sum_{k=0}^{a-1} \binom{a-k}{r} = \binom{a+1}{r+1}.$$

15. Prove by induction

(9.15) $$\binom{n}{1}\frac{1}{1} - \binom{n}{2}\frac{1}{2} + \binom{n}{3}\frac{1}{3} - + \cdots + (-1)^{n-1}\binom{n}{n}\frac{1}{n}$$
$$= 1 + \frac{1}{2} + \frac{1}{3} + \cdots + \frac{1}{n}.$$

[*Hint*: Remember (6.11).]

16. Show that for any positive integer m

(9.16) $$(x+y+z)^m = \sum \frac{m!}{a!b!c!} x^a y^b z^c$$

where the summation extends over all non-negative integers a, b, c, such that $a+b+c = m$.

17. Using Stirling's formula (7.7), prove that

(9.17) $$\binom{2n}{n} \sim (\pi n)^{-\frac{1}{2}} 2^{2n}.$$

18. Prove that for any positive integers a and b

(9.18) $$\frac{(a+1)(a+2)\cdots(a+n)}{(b+1)(b+2)\cdots(b+n)} \sim \frac{b!}{a!} n^{a-b}.$$

19. The gamma function is defined by

(9.19) $$\Gamma(x) = \int_0^\infty z^{x-1} e^{-z} \, dz$$

where $x > 0$. Show that $\Gamma(x) \sim (2\pi)^{\frac{1}{2}} e^{-x} x^{x-\frac{1}{2}}$. [Note that if $x = n$ is an integer, $\Gamma(n) = (n-1)!$.]

20. Let a and r be arbitrary positive numbers and n a positive integer. Show that

(9.20) $$a(a+r)(a+2r) \cdots (a+nr) \sim C r^{n+1} n^{n+(a/r)+\frac{1}{2}} e^{-n}.$$

[The constant C is equal to $(2\pi)^{\frac{1}{2}}/\Gamma(a/r)$.]

21. Using the results of the preceding problem, show that

(9.21) $$\frac{a(a+r)(a+2r) \cdots (a+nr)}{b(b+r)(b+2r) \cdots (b+nr)} \sim \frac{\Gamma(b/r)}{\Gamma(a/r)} n^{(a-b)/r}.$$

22. Prove the following *alternative form of Stirling's formula:*

(9.22) $$n! \sim (2\pi)^{1/2}(n + 1/2)^{n+1/2} e^{-(n+1/2)}.$$

23. *Continuation.* Using the method of the text, show that

$$(2\pi)^{1/2}(n + 1/2)^{n+1/2} e^{-(n+1/2) - 1/24(n+1/2)} < n! < (2\pi)^{1/2}(n + 1/2)^{n+1/2} e^{-(n+1/2)}.$$

24. Extending Stirling's formula, prove that

(9.24) $$n! \sim (2\pi)^{1/2} n^{n+1/2} \exp\left\{-n + \frac{1}{12n} - \frac{1}{360n^3} + \cdots\right\}.$$

25. Prove similarly

(9.25) $$n! \sim (2\pi)^{1/2}\left(n+\frac{1}{2}\right)^{n+1/2}\exp\left\{-\left(n+\frac{1}{2}\right) - \frac{1}{24(n+1/2)} + \frac{7}{2880(n+1/2)^3} + \cdots\right\}.$$

* CHAPTER 3

THE SIMPLEST OCCUPANCY AND ORDERING PROBLEMS

This chapter represents a digression from the main path of the book, and a knowledge of its results is not required in the sequel. We shall treat a few special problems, partly for their intrinsic interest, and partly for their importance in applications. Among these, the Bose and Fermi statistics provide an instructive illustration of the physicist's use of probability models.

1. Combinatorial Lemmas

We want to study random distributions of r objects in n cells or compartments, with no restrictions imposed on the number of objects in any particular cell. Each object can be placed in n different ways, and therefore (cf. Chapter 2, section 2) the number of different distributions is n^r. However, if the objects are indistinguishable, then an exchange of two objects is not observable, and the number of *distinguishable* distributions is smaller than n^r.

Examples. (a) With two objects and two cells we have four arrangements which may conveniently be represented by $(AB\,|\,-)$, $(A\,|\,B)$, $(B\,|\,A)$, and $(-\,|\,AB)$. However, with two A's we have only the three distinguishable arrangements $(AA\,|\,-)$, $(A\,|\,A)$, and $(-\,|\,AA)$, which are preferably represented by $(2\,|\,0)$, $(1\,|\,1)$, and $(0\,|\,2)$. For three indistinguishable things and two cells we have the four distributions $(3\,|\,0)$, $(2\,|\,1)$, $(1\,|\,2)$, $(0\,|\,3)$. The first and the last correspond to unique arrangements of three distinct things, while $(2\,|\,1)$, and similarly $(1\,|\,2)$, may stand for any of the three arrangements $(AB\,|\,C)$, $(AC\,|\,B)$, $(BC\,|\,A)$.

(b) The six faces of a die may be interpreted as cells; a throw with r dice then puts each die into one of the six cells. There are 6^r different arrangements, and they could be made distinguishable, for example, by using dice of different colors or size. Usually, however, we are unable (or unwilling) to identify the individual dice. With 6 dice the event "no two faces alike" can occur in $6! = 720$ different ways, but they are considered indistinguishable.

---

* Starred chapters treat special topics and may be omitted at first reading.

(c) Situations of a similar nature are common. For example, the days of the year correspond to cells, and people's birthdays to things placed into the cells. The birthdays of r people can be distributed in 365^r ways [cf. Chapter 2, example (3.c)]. For purposes of birthday statistics it does not matter whether Peter or Paul has a birthday, and we agree to treat birthdays as indistinguishable. Again, in the elevator example [(3.d) of Chapter 2] the floors are cells into which passengers are placed, and it is natural not to distinguish between passengers. When a boy collects coupons found in cereal packages, the different kinds of coupons correspond to cells. The collector sees only the number of coupons of each kind, which means that he does not distinguish between all n^r arrangements.

To find the number of distinguishable arrangements of r indistinguishable objects in n cells, we shall represent the cells by the spaces between $n+1$ bars and the objects by A's. Thus $|\ AAA\ |\ A\ |\ |\ |\ |\ AAAA\ ||$ is used to indicate that there are seven cells in all and that they contain 3, 1, 0, 0, 0, 4, 0 things, respectively. The arrangement always starts and ends with a bar, but we may put the remaining $n-1$ bars and r letters in an arbitrary order. The arrangement is fixed by selecting the r places in which the letters stand. Hence there are as many distinguishable arrangements as there are ways of selecting r places out of $r+n-1$, namely, $\binom{n+r-1}{r} = \binom{n+r-1}{n-1}$. We have thus proved

Lemma 1. *There are n^r different ways of putting r objects into n cells. With indistinguishable things the number of distinguishable arrangements is* $\binom{n+r-1}{r} = \binom{n+r-1}{n-1}$.

Examples. (d) A throw with r dice can result in $\binom{r+5}{5}$ distinguishable arrangements. For a throw with r coins there are $r+1$ distinguishable results; in fact, the number of heads is $0, 1, \cdots, r$.

(e) *Partial derivatives.* Consider an analytic function of n variables $f(x_1, \cdots, x_n)$. Formally we can calculate its partial derivatives of order r in n^r ways. However, the order of differentiations plays no role, and it matters only how often each variable appears. We have one cell corresponding to each variable, and our lemma shows that there exist $\binom{n+r-1}{r}$ different partial derivatives of rth order. A function of three variables has 15 derivatives of fourth and 21 derivatives of fifth order.

Lemma 2. Let $r \geq n$. *The number of ways in which r indistinguishable things can be put into n cells, none of which is to be empty, is* $\binom{r-1}{n-1}$.

Proof. In the arrangement of letters and bars described above it is now required that at most one bar appears between any two letters. There are $r - 1$ spaces between the letters, and we can choose any $n - 1$ among them as places for bars.

2. Bose-Einstein and Fermi-Dirac Statistics

In the examples of the preceding section it was natural to assume that all n^r arrangements have equal probabilities. We shall now consider cases where facts and experience have compelled physicists to abandon this hypothesis and to assign probabilities in different ways.

Consider a mechanical system of r indistinguishable particles. In statistical mechanics it is usual to subdivide the phase space into a large number, n, of small regions or cells so that each particle is assigned to one cell. In this way the state of the entire system is described in terms of a random distribution of the r particles in n cells. Offhand it would seem that (at least with an appropriate definition of the n cells) all n^r arrangements should have equal probabilities. If this is true, the physicist speaks of *Maxwell-Boltzmann statistics* (the term "statistics" is here used in a sense peculiar to physics). Numerous attempts have been made to prove that physical particles behave in accordance with Maxwell-Boltzmann statistics, but modern theory has shown beyond doubt that this statistics *does not apply to any known particles;* in no case are all n^r arrangements approximately equally probable. Two different probability models have been introduced, and each describes satisfactorily the behavior of one type of particle. The justification of either model depends on its success. Neither claims universality, and it is possible that some day a third model may be introduced for certain kinds of particles.

Remember that we are here concerned only with *indistinguishable* particles. We have r particles and n cells. *By Bose-Einstein statistics we mean that only distinguishable arrangements are considered and that each is assigned probability* $1 \div \binom{n+r-1}{r}$. It is shown in statistical mechanics that this assumption holds true for photons, nuclei, and atoms containing an even number of elementary particles.[1] To describe other particles a third possible assignment of probabilities must be introduced. By *Fermi-Dirac statistics* we understand these hypotheses: (1) *it is impossible for two or more particles to be in the same cell, and* (2) *all distinguishable arrangements satisfying the first condition have equal probabilities.* The first hypothesis requires that $r \leq n$. An arrangement is then completely described by stating which of the n cells contain a particle, and since there are r particles the corresponding cells can be chosen in $\binom{n}{r}$ ways. Hence, with *Fermi-Dirac statistics there are in all* $\binom{n}{r}$ *possible arrangements, each having probability* $\binom{n}{r}^{-1}$. This model applies to electrons, neutrons, and protons. We have here an instructive example of the impossibility of selecting or justifying probability models by *a priori* arguments. In fact, no pure reasoning could tell that photons and protons would not obey the same probability laws. (Essential differences between Maxwell-Boltzmann and Bose-Einstein statistics are discussed in problems 6–10. Cf., in particular, problems 8 and 9.)

Example. Let $n = 5$, $r = 3$. The arrangement $(A \mid - \mid A \mid A \mid -)$ has probability 6/125, 1/35, or 1/10, according to whether Maxwell-Boltzman, Bose-Einstein, or Fermi-Dirac statistics is used.

3. The Classical Occupancy Problem

We return to the random distribution of r objects in n cells and assume again that *each of the n^r possible arrangements has probability n^{-r}.* The probability that a specified cell contains exactly k objects is then

$$(3.1) \qquad p_k = \binom{r}{k} \frac{(n-1)^{r-k}}{n^r}$$

(if $k > r$ the binomial coefficient vanishes so that $p_k = 0$, as is proper). To prove (3.1), it suffices to note that the k objects can be chosen in $\binom{r}{k}$ ways, and the remaining $r - k$ objects can be placed into the remaining $n - 1$ cells in $(n-1)^{r-k}$ ways.

Example. A sequence of 100 random digits represents a distribution of 100 things into ten cells. Accordingly, the probability p_k that the digit 7 appears among 100 random digits exactly k times is given

[1] Cf. H. Margenau and G. M. Murphy, *The Mathematics of Physics and Chemistry*, New York, 1943, Chapter 12.

by (3.1) with $r = 100$, $n = 10$. Table 1 of Chapter 2 gives the actual counts of the occurrences of the digit 7 in 100 sequences of 100 digits each. Interpreting probabilities as long-run frequencies, we should expect the number of sequences in which the 7 occurs exactly k times to be approximately $100p_k$. Table 1 compares the theory with actual counts. Presumably doubts will arise in the reader's mind as to whether the counts confirm the theory. The theory of the chi-square test provides objective means of judging the closeness of observed frequencies to theoretical probabilities. It turns out that under ideal circumstances roughly in two out of three cases chance fluctuations would produce deviations larger than those exhibited in Table 1.

TABLE 1

OCCURRENCE OF 7 AMONG 100 RANDOM DIGITS

k	$100p_k$	N_k	k	$100p_k$	N_k
0	0.003	0	11	11.988	9
1	0.030	0	12	9.879	10
2	0.162	0	13	7.430	2
3	0.589	1	14	5.130	5
4	1.587	1	15	3.268	3
5	3.389	6	16	1.929	3
6	5.958	6	17	1.059	2
7	8.890	11	18	0.543	0
8	11.482	18	19	0.260	0
9	13.042	8	20	0.117	1
10	13.186	14			

p_k is the probability of exactly k occurrences, and N_k the observed number of occurrences in the 100 batches of 100 digits each recorded in Table 1 of Chapter 2.

Formula (3.1) is a special case of the so-called *binomial distribution* which will be taken up in Chapter 6, where we shall examine various properties of (3.1) and see how it can be evaluated when r and n are large. Note that (3.1) does not solve all problems of occupancy. For example, the theory thus far developed does not permit the computation of the probability that k cells will be empty. Tools appropriate for such problems will be developed in section 3 of the next chapter.

For Bose-Einstein statistics the probability q_k of a k-fold occupancy of any specified cell is given by a formula analogous to (3.1). It is worth noting here that the two sequences p_k and q_k exhibit completely different characters (cf. problems 7–9).

4. Runs

In any ordered sequence of elements of two kinds, each maximal subsequence of elements of like kind is called a *run*. For example, the sequence *AAABAABBBA* opens with an *A*-run of length 3; it is followed by runs of length 1, 2, 3, 1, respectively. The *A*- and *B*-runs alternate so that the total number of runs is always one plus the number of *unlike neighbors* in the given sequence.

Examples. (*a*) An observation [2] yielded the following arrangement of empty and occupied seats along a lunch counter: *EOEEOEEEOEEE-OEOE*. Note that no two occupied seats are adjacent. Can this be due to chance? With five occupied and eleven empty seats it is impossible to get more than eleven runs, and this number was actually observed. It will be shown later that if all arrangements were equally probable the probability of eleven runs would be 0.0578.... This small probability to some extent confirms the hunch that the separations observed were intentional. This suspicion cannot be proved by statistical methods, but further evidence could be collected from continued observation. If the lunch counter were frequented by families, there would be a tendency for occupants to cluster together, and this would lead to relatively small numbers of runs. Similarly, counting runs of boys and girls in a classroom might disclose the mixing to be better or worse than random. Improbable arrangements give clues to *assignable causes; an excess of runs points to intentional mixing, a paucity of runs to intentional clustering*. It is true that these conclusions are never foolproof. Even with perfect randomness improbable situations occur and may mislead us into a search for assignable causes. However, this will be a rarity, and with an appropriate criterion we shall in actual practice be misled once in 100 times but find assignable causes 99 out of 100 times.

Examples for statistical applications of the theory of runs occur in industrial quality control as introduced by Shewhart. As washers are produced, they will vary in thickness. Long runs of thick washers may suggest imperfections in the production process and lead to the removal of the causes; thus oncoming trouble may be forestalled and greater homogeneity of product achieved.

In biological field experiments one counts successions of healthy and diseased plants, and long runs are suggestive of contagion. The

[2] F. S. Swed and C. Eisenhart, Tables for Testing Randomness of Grouping in a Sequence of Alternatives, *Annals of Mathematical Statistics*, vol. 14 (1943), pp. 66–87.

meteorologist watches successions of dry and wet months [3] to discover clues to a tendency of the weather to persist.

(b) In physics, the theory of runs is used in the study of cooperative phenomena. In Ising's theory of one-dimensional lattices the energy depends on the number of unlike neighbors, that is, the number of runs. In a more refined theory the lengths of the runs also play a role. Here, as in agricultural experiments, it is desirable to generalize the theory of runs to multidimensional arrangements. At present only one-dimensional arrangements have been investigated in detail.

Many questions relative to runs [4] can be asked, but we shall prove here only the following

Theorem. Suppose that all arrangements of r_1 elements of one kind and r_2 elements of a second kind have equal probabilities, and let P_k be the probability that an arrangement contains exactly k runs. Then, for an even number $k = 2\nu$

$$(4.1) \qquad P_{2\nu} = 2 \binom{r_1 - 1}{\nu - 1} \binom{r_2 - 1}{\nu - 1} \div \binom{r_1 + r_2}{r_1},$$

while for $k = 2\nu + 1$

$$(4.2) \quad P_{2\nu+1} = \left\{ \binom{r_1 - 1}{\nu} \binom{r_2 - 1}{\nu - 1} + \binom{r_1 - 1}{\nu - 1} \binom{r_2 - 1}{\nu} \right\} \div \binom{r_1 + r_2}{r_1}$$

(cf. problems 14–16).

Proof. The $r = r_1 + r_2$ elements can be arranged in $r!$ ways. However, any permutation among the r_1 elements of the first kind, or among the r_2 elements of the second kind, will leave the outer appearance unchanged. Hence there exist

$$(4.3) \qquad \frac{r!}{r_1! r_2!} = \binom{r_1 + r_2}{r_1} = \binom{r_1 + r_2}{r_2}$$

distinguishable orderings, and each represents $r_1! r_2!$ different arrangements of the $r_1 + r_2$ elements. It follows that all distinguishable orderings have equal probabilities, and the denominator in either (4.1)

---

[3] W. G. Cochran, An Extension of Gold's Method of Examining the Apparent Persistence of One Type of Weather, *Quarterly Journal of the Royal Meteorological Society*, vol. 64, No. 277 (1938), pp. 631–634.

[4] For further results and literature see S. S. Wilks, *Mathematical Statistics*, Princeton, 1943, Chapter 10.

or (4.2) is the number of distinguishable orderings. We have to prove that the numerators represent the numbers of distinguishable arrangements with 2ν or $2\nu + 1$ runs, respectively. Consider first the case $k = 2\nu$. Then we have ν runs of the first kind and ν runs of the second kind. Each run represents a cell. By lemma 2 of section 1 the r_1 elements of the first kind can be distributed in $\binom{r_1 - 1}{\nu - 1}$ ways into ν cells none of which is empty. Similarly, the elements of the second kind can be distributed in $\binom{r_2 - 1}{\nu - 1}$ distinguishable ways. Finally, the runs alternate but a run of either kind may be first. This accounts for the numerator in (4.1). If the number of runs is odd, say, $2\nu + 1$, then there are $\nu + 1$ runs of one kind and ν runs of the other, and the same method of counting shows the numerator in (4.2) to be the number of distinguishable orderings with $2\nu + 1$ runs.

In the paper by Swed and Eisenhart quoted above, the probabilities P_k are tabulated for all r_1 and r_2 up to 20.

Examples. If $r_1 = r_2 = 2$, an arrangement may consist of 2, 3, or 4 runs and these possibilities have equal probabilities.

For $r_1 = 2$, $r_2 = 3$ the number of runs equals 2, 3, 4, or 5 with probabilities 0.2, 0.3, 0.4, 0.1, respectively.

For $r_1 = r_2 = 3$, the probabilities of 2, 3, 4, 5, 6 runs are 0.1, 0.2, 0.4, 0.2, 0.1.

In the lunch-counter example (4.a) we have $r_1 = 5$ and $r_2 = 11$. The probability of 11 runs $\binom{10}{5} \div \binom{16}{5} = \dfrac{3}{52}$.

5. Problems for Solution

1. If r_1 indistinguishable things of one kind and r_2 indistinguishable things of a second kind are placed into n cells, find the number of distinguishable arrangements.

2. If r_1 dice and r_2 coins are thrown, how many results can be distinguished?

3. In how many different distinguishable ways can r_1 white, r_2 black, and r_3 red balls be arranged?

In problems 4–6 let n be the number of cells and r the number of objects; assume that the objects are distinguishable, and that all n^r distributions are equally probable.

4. When $r = n$, find the probabilities that (a) no cell, (b) only one cell remains empty.

5. Find the probability that the first cell contains k_1 things, the second k_2 things, etc., where $k_1 + k_2 + \cdots + k_n = r$.

6. The most probable number of things in any given cell is the integer ν_0 with $(r - n + 1)/n < \nu_0 \leq (r + 1)/n$. More precisely the probabilities (3.1) satisfy the relation $p_0 < p_1 < p_2 < \cdots < p_{\nu-1} \leq p_\nu > p_{\nu+1} > \cdots > p_r$ (cf. problem 8).

In problems 7–10 *r and n have the same meaning as above, but we assume that the objects are indistinguishable and that all distinguishable arrangements have equal probabilities (Bose-Einstein statistics).*

7. The probability that a given cell contains exactly k things is

(5.1) $$q_k = \binom{n+r-k-2}{r-k} \div \binom{n+r-1}{r}.$$

8. Show that when $n > 2$ zero is the most probable number of things in any cell, or more precisely, $q_0 > q_1 > q_2 > \ldots$ (cf. problem 6).

9. *Limit theorem.* Let $n \to \infty$ and $r \to \infty$, so that the average number of particles per cell, r/n, tends to λ. Then

(5.2) $$q_k \to \frac{\lambda^k}{(1+\lambda)^{k+1}}.$$

[The right side is known as the *geometric distribution*. The corresponding limiting form for (3.1) is the *Poisson distribution*; cf. Chapter 6.]

10. The probability that exactly m cells remain empty is

(5.3) $$\rho_m = \binom{n}{m}\binom{r-1}{n-m-1} \div \binom{n+r-1}{r}.$$

(Cf. problem 4. A similar general formula for the case where all n^r arrangements are equally probable will be given in Chapter 4, section 5.)

11. From the meanings of the probabilities p_k, q_k, and ρ_ν [cf. (3.1), (5.1), and (5.3)] it follows that $\Sigma p_k = \Sigma q_k = \Sigma \rho_k = 1$. Prove this algebraically from identities which were given in Chapter 2 (either in the text or as problems in section 9).

Further theorems on runs: In the following problems we consider arrangements of r_1 *alphas and r_2 betas and assume that all arrangements are equally probable.*

12. The probability that the arrangement starts with an alpha run of length ν is

$$\frac{(r_1)_\nu r_2}{(r_1+r_2)_{\nu+1}} = \frac{\binom{r_1+r_2-\nu-1}{r_1-\nu}}{\binom{r_1+r_2}{r_1}}.$$

13. The probability that the arrangement starts with a beta is $r_2/(r_1+r_2)$.

14. From the theorem of section 4 deduce that the most probable number of runs is an integer k with $\dfrac{2r_1 r_2}{r_1+r_2} < k \leq \dfrac{2r_1 r_2}{r_1+r_2} + 3$.

15. The probability of having exactly k runs of alphas is

$$\pi_k = \binom{r_1-1}{k-1}\binom{r_2+1}{k} \div \binom{r_1+r_2}{r_1}.$$

(*Hint:* Use the theorem of section 4 and note that the betas are arranged in $k-1$, k, or $k+1$ runs. Cf. Chapter 7, problem 11, and Chapter 9, problem 12.)

16. The probability for the alphas to be arranged in k runs of which k_1 are of length 1, k_2 of length 2, $\cdots$, k_ν of length ν (with $k_1 + \cdots + k_\nu = k$) is

$$\frac{k!}{k_1! k_2! \cdots k_\nu!} \binom{r_2+1}{k} \div \binom{r_1+r_2}{r_1}.$$

CHAPTER 4

COMBINATION OF EVENTS

This chapter is concerned with events A which are defined in terms of certain other events $A_1, A_2, \cdots, A_N$. Thus in a game of bridge the event A, "at least one player has a complete suit," is the union of the four events A_k, "player number k has a complete suit" ($k = 1, 2, 3, 4$). Of the events A_k one, two, or more can occur simultaneously, and, because of this overlap, the probability of A is not the sum of the four probabilities $Pr\{A_k\}$. Given a set of events $A_1, \cdots, A_N$, we shall show how to compute the probabilities that $0, 1, 2, 3, \ldots$ among them occur. The formulas are useful for certain applications and are also of theoretical interest. However, in this book the formulas will not be used explicitly, and the present chapter can therefore be omitted at a first reading. As a compromise it is suggested to study only section 1.

The material of this chapter and a variety of applications are covered in a recent monograph by M. Fréchet[1], to which the reader is referred for further information.

1. Union of Events

If A_1 and A_2 are two events, then $A = A_1 \cup A_2$ denotes the event that either A_1 or A_2 or both occur. By formula (6.4) of Chapter 1 we have

(1.1) $\qquad Pr\{A\} = Pr\{A_1\} + Pr\{A_2\} - Pr\{A_1 A_2\}.$

We want to generalize this formula to the case of N events $A_1, A_2, \cdots, A_N$, that is, we wish to compute the probability of the event that at least one among the A_k occurs. In symbols this event is $A = A_1 \cup A_2 \cup \cdots \cup A_N$. For our purpose it is not sufficient to know the probabilities of the individual events A_k, but we must be given complete information concerning all possible overlaps. This means that for every pair (i, j), every triple (i, j, k), etc., we must know the probability of A_i and A_j, or A_i, A_j, and A_k, etc., occurring

[1] Les probabilités associées à un système d'événements compatibles et dépendants, *Actualités scientifiques et industrielles*, nos. 859 and 942, Paris, 1940 and 1943.

simultaneously. For convenience of notation we shall denote these probabilities by the letter p with appropriate subscripts. Thus

(1.2) $p_i = Pr\{A_i\}, \quad p_{i,j} = Pr\{A_i A_j\}, \quad p_{i,j,k} = Pr\{A_i A_j A_k\}, \quad \ldots$.

The order of the subscripts is irrelevant, but for uniqueness we shall always *write the subscripts in increasing order;* thus, we write $p_{3,7,11}$ and not $p_{7,3,11}$. Two subscripts are never equal. For the sum of all p's with r subscripts we shall write S_r, that is, we define

(1.3) $S_1 = \Sigma p_i, \quad S_2 = \Sigma p_{i,j}, \quad S_3 = \Sigma p_{i,j,k}, \quad \ldots$.

Here $i < j < k < \cdots \leq N$, so that in the sums each combination appears once and only once; hence S_r has $\binom{N}{r}$ terms. The last sum, S_N, reduces to the single term $p_{1,2,3,\cdots,N}$, which is the probability of the simultaneous realization of all N events. For $N = 2$ we have only the two terms S_1 and S_2, and formula (1.1) can be written

(1.4) $Pr\{A\} = S_1 - S_2$.

The generalization to an arbitrary number N of events is given in the following

Theorem. *The probability P_1 of the realization of at least one among the events $A_1, A_2, \cdots, A_N$ is given by*

(1.5) $P_1 = S_1 - S_2 + S_3 - S_4 + - \cdots \pm S_N$.

Proof. We prove (1.5) by the so-called method of inclusion and exclusion. To compute P_1 we should add the probabilities of all sample points which are contained in at least one of the A_i, but each point should be taken only once. To proceed systematically we first take the points which are contained in only one A_i, then those contained in exactly two events A_i, and so forth, and finally the points (if any) contained in all A_i. Now let E be any sample point contained in exactly n among our N events A_i. Without loss of generality we may number the events so that E is contained in $A_1, A_2, \cdots, A_n$ but not contained in $A_{n+1}, A_{n+2}, \cdots, A_N$. Then $Pr\{E\}$ appears as a contribution to those $p_i, p_{ij}, p_{ijk}, \ldots$ whose subscripts range from 1 to n. Hence $Pr\{E\}$ appears n times as a contribution to S_1, and $\binom{n}{2}$ times as a contribution to S_2, etc. In all, when the right-hand side of (1.5) is expressed in terms of the probabilities of sample points we find

$Pr\{E\}$ with the factor

$$(1.6) \qquad n - \binom{n}{2} + \binom{n}{3} - + \cdots \pm \binom{n}{n}.$$

To prove the theorem we have to show that this number equals 1. This follows at once on comparing (1.6) with the binomial expansion of $(1 - 1)^n$ [cf. (6.11) of Chapter 2]. The latter starts with 1, and the terms of (1.6) follow with reversed sign. Hence for every $n \geq 1$ the expression (1.6) equals 1, and this proves the theorem.

2. Examples

(a) In a game of bridge let A_i be the event "player number i has a complete suit." Then $p_i = 4/\binom{52}{13}$; the event that both player i and player j have complete suits can occur in $4 \cdot 3$ ways and has probability $p_{i,j} = 12/\binom{52}{13}\binom{39}{13}$; similarly $p_{i,j,k} = 24/\binom{52}{13}\binom{39}{13}\binom{26}{13}$.

Finally, $p_{1,2,3,4} = p_{1,2,3}$, since whenever three players have a complete suit so does the fourth. The probability that *some* player has a complete suit is therefore by (1.5)

$$(2.1) \qquad P_1 = \frac{16\binom{39}{13}\binom{26}{13} - 72\binom{26}{13} + 72}{\binom{52}{13}\binom{39}{13}\binom{26}{13}}.$$

Using Stirling's formula, we see that $P_1 = \frac{1}{4} 10^{-10}$ approximately. In this particular case P_1 is very nearly the sum of the probabilities of A_i, but this is the exception rather than the rule.

(b) *Matches (Coincidences).* The following problem with many variants and a surprising solution goes back to Montmort (1708). It has been generalized by Laplace and many other authors.

Two similar decks of N different cards each are put into random order and matched against each other. If a card occupies the same place in both decks, we speak of a *match* (*coincidence* or *rencontre*). Matches may occur at any of the N places and at several places simultaneously. This experiment may be described in more amusing forms. For example, the two decks may be represented by a set of N letters and their envelopes, and a capricious secretary may perform the random matching. Alternatively we may imagine the hats in a checkroom mixed and distributed at random to the guests. A match occurs if a person gets his own hat. It is instructive to venture guesses as to how the probability of a match depends on N: how does the probability of a match of hats in a diner with 8 guests compare with the corresponding probability at a gathering of 10,000 people? It seems surprising that the probability is practically independent of N and roughly 2/3. (For less frivolous applications cf. problems 10 and 11.)

The probabilities of having exactly 0, 1, 2, 3, ... matches will be calculated in section 4. Here we shall find only the probability P_1 of at least 1 match. For simplicity of expression let us renumber the cards 1, 2, $\cdots$, N in such a way that one deck appears in its natural order, and assume that each permutation of the second deck has probability $1/N!$. Let A_k be the event that a match occurs at the kth place. This means that card number k is at the kth place while the remaining $N-1$ cards may be in an arbitrary order. Clearly $p_k = (N-1)!/N! = 1/N$. Similarly, for every combination i, j, we have $p_{i,j} = (N-2)!/N! = 1/N(N-1)$, etc. The sum S_r contains $\binom{N}{r}$ terms each of which equals $(N-r)!/N!$. Hence $S_r = 1/r!$, and from (1.5) we find the required probability to be

$$(2.2) \qquad P_1 = 1 - \frac{1}{2!} + \frac{1}{3!} - + \cdots \pm \frac{1}{N!}.$$

Note that $1 - P_1$ represents the first $N+1$ terms in the expansion

$$(2.3) \qquad e^{-1} = 1 - 1 + \frac{1}{2!} - \frac{1}{3!} + \frac{1}{4!} - + \cdots.$$

Therefore we have with a good approximation

$$(2.4) \qquad P_1 \approx 1 - e^{-1} = 0.63212\ldots.$$

The degree of approximation is shown in the following table of correct values of P_1:

$N =$	3	4	5	6	7
$P_1 =$	0.66667	0.62500	0.63333	0.63196	0.63214

(c) *A Sampling Problem.* A pack of cards consists of s identical series, each containing n cards numbered 1, 2, $\cdots$, n. A random sample of $r \geq n$ cards is drawn from the pack without replacement. We require the probability u_r that each number is represented in the sample. As a particular case consider a deck of bridge cards. For $s = 4$, $n = 13$ we get the probability that a hand of r cards contains all thirteen values (cf. footnote on p. 9). For $s = 13$, $n = 4$ we get the probability that all four suits are represented.

To calculate u_r let A_ν be the event that number ν does not occur in the sample. Remembering that r cards out of m can be selected in $(m)_r = m(m-1)\cdots(m-r+1)$ ways, we find

$$(2.5) \qquad p_i = \frac{(ns-s)_r}{(ns)_r}, \quad p_{ij} = \frac{(ns-2s)_r}{(ns)_r}, \quad \ldots$$

Since the number of events is $N = n$, we have $S_1 = np_i$, $S_2 = \binom{n}{2} p_{ij}$ etc. Substituting into (1.5), we find for the probability that some number does not occur

$$(2.6) \qquad 1 - u_r = \sum_{k=1}^{n} (-1)^{k-1} \binom{n}{k} \frac{(ns-ks)_r}{(ns)_r}$$

[cf. problems 12–14 and Chapter 9, example (3.d)].

If in (2.6) we let the number of series $s \to \infty$, then clearly

$$(2.7) \qquad u_r \to 1 - \sum_{k=1}^{n} (-1)^{k-1} \binom{n}{k} \left(1 - \frac{k}{n}\right)^r.$$

It is intuitively clear that in the limit our sampling becomes sampling with replacement from the population of the numbers $1, 2, \cdots, n$. The right side of (2.7) is then *the probability that in a random sample of size r each element appears at least once*. This is also the probability of having no empty boxes if r balls are put into n boxes [cf. equation (5.4) with $m = 0$]. Finally, (2.7) answers the question of the collector of coupons as to how many coupons he will have to acquire before having a complete series of n coupons. (Cf. problems 12–14.)

3. The Realization of *m* among *N* Events

Theorem. *For any integer m with $1 \leq m \leq N$ the probability $P_{[m]}$ that exactly m among the N events $A_1, \cdots, A_N$ occur simultaneously is given by*

$$(3.1) \quad P_{[m]} = S_m - \binom{m+1}{m} S_{m+1} + \binom{m+2}{m} S_{m+2} - + \cdots \pm \binom{N}{m} S_N.$$

Note: According to (1.5), the probability $P_{[0]}$ that none among the A_j occurs is

$$(3.2) \qquad P_{[0]} = 1 - P_1 = 1 - S_1 + S_2 - S_3 \pm \cdots \mp S_N.$$

This shows that (3.1) gives the correct value also for $m = 0$ provided that we put $S_0 = 1$.

Proof. We proceed as in the proof of (1.5). Let E be an arbitrary sample point, and suppose that it is contained in exactly n among the N events A_j. Then $Pr\{E\}$ appears as a contribution to $P_{[m]}$ only if $n = m$. To investigate how $Pr\{E\}$ contributes to the right side of (3.1), note that $Pr\{E\}$ appears in the sums $S_1, S_2, \cdots, S_n$ but not in $S_{n+1}, \cdots, S_N$. It follows that $Pr\{E\}$ does not contribute to the right side in (3.1) if $n < m$. If $n = m$, then $Pr\{E\}$ appears in one and only one term of S_m. To complete the proof of the theorem it remains to show that for $n > m$ the contributions of $Pr\{E\}$ to the terms S_m, $S_{m+1}, \cdots, S_n$ on the right in (3.1) cancel. Now out of the n events containing E we can form $\binom{n}{k}$ k-tuplets; hence $Pr\{E\}$ appears in S_k with the factor $\binom{n}{k}$. For $n > m$ the total contribution of $Pr\{E\}$ to the right side in (3.1) is therefore

$$(3.3) \quad \binom{n}{m} - \binom{m+1}{m}\binom{n}{m+1} + \binom{m+2}{m}\binom{n}{m+2} - + \cdots \mp \binom{n}{m}\binom{n}{n}.$$

However, $\binom{m+\nu}{m}\binom{n}{m+\nu} = \binom{n}{m}\binom{n-m}{\nu}$, and hence (3.3) reduces to

$$(3.4) \quad \binom{n}{m}\left\{\binom{n-m}{0} - \binom{n-m}{1} + - \cdots \mp \binom{n-m}{n-m}\right\}.$$

Within braces we have the binomial expansion of $(1-1)^{n-m}$ so that (3.3) vanishes, as asserted.

Example. *Quadruples in a Bridge Hand.* By a quadruple we shall understand 4 cards of the same face value, so that a bridge hand of 13 cards may contain 0, 1, 2, or 3 quadruples. A hand can be selected in $\binom{52}{13}$ ways, and we attribute probability $p = \binom{52}{13}^{-1}$ to each way. We have here $N = 13$ possible quadruples, or events. Four aces and 9 other cards can be chosen in $\binom{48}{9}$ ways. For reasons of symmetry we get similarly for all $i, j, k \ldots$

$$p_i = \binom{48}{9} p, \quad p_{ij} = \binom{44}{5} p, \quad p_{ijk} = \binom{40}{1} p$$

and hence

$$S_1 = 13 \binom{48}{9} p, \quad S_2 = \binom{13}{2}\binom{44}{5} p, \quad S_3 = \binom{13}{3} 40 p$$

while $S_4 = S_5 = \cdots = S_{13} = 0$, since it is impossible that more than 3 quadruples occur. The probabilities of 0, 1, 2, or 3 quadruples are therefore

$$P_{[0]} = 1 - \left\{ 13\binom{48}{9} - \binom{13}{2}\binom{44}{5} + \binom{13}{3} 40 \right\} p = 0.9657997 \cdots$$

$$P_{[1]} = \left\{ 13\binom{48}{9} - 2\binom{13}{2}\binom{44}{5} + 3\binom{13}{3} 40 \right\} p = 0.0340669 \cdots$$

$$P_{[2]} = \left\{ \binom{13}{2}\binom{44}{5} - 3\binom{13}{3} 40 \right\} p = 0.0001334 \cdots$$

$$P_{[3]} = \binom{13}{3} 40 p \approx 2 \cdot 10^{-9}$$

4. Application to Matching and Guessing

In example (2.b) we considered the matching of two decks of cards and found that $S_k = 1/k!$. Substituting into (3.1), we find the following result.

In a random matching of two similar decks of N distinct cards the probability $P_{[m]}$ of having exactly m matches is given by

$$P_{[0]} = 1 - 1 + \frac{1}{2!} - \frac{1}{3!} + - \cdots \pm \frac{1}{(N-3)!} \mp \frac{1}{(N-2)!}$$

$$\pm \frac{1}{(N-1)!} \mp \frac{1}{N!}$$

(4.1) $\quad P_{[1]} = 1 - 1 + \dfrac{1}{2!} - \dfrac{1}{3!} + - \cdots \pm \dfrac{1}{(N-3)!} \mp \dfrac{1}{(N-2)!}$

$$\pm \frac{1}{(N-1)!}$$

4.4] APPLICATION TO MATCHING AND GUESSING

$$P_{[2]} = \frac{1}{2!}\left\{1 - 1 + \frac{1}{2!} - \frac{1}{3!} + -\cdots \pm \frac{1}{(N-3)!} \mp \frac{1}{(N-2)!}\right\}$$

$$P_{[3]} = \frac{1}{3!}\left\{1 - 1 + \frac{1}{2!} - \frac{1}{3!} + -\cdots \pm \frac{1}{(N-3)!}\right\}$$

. . . .

$$P_{[N-2]} = \frac{1}{(N-2)!}\left\{1 - 1 + \frac{1}{2!}\right\}$$

$$P_{[N-1]} = \frac{1}{(N-1)!}\{1 - 1\} = 0$$

$$P_{[N]} = \frac{1}{N!}$$

The last relation is obvious. The vanishing of $P_{[N-1]}$ expresses the impossibility of having $N - 1$ matches without having all N cards in the same order.

The braces to the right in (4.1) contain the initial terms of the expansion of e^{-1}. *For large N we have therefore approximately*

$$(4.2) \qquad P_{[m]} \approx \frac{1}{m!} e^{-1}.$$

In Table 1 the columns headed $P_{[m]}$ give the exact values of $P_{[m]}$ for $N = 3, 4, 5, 6, 10$. The last column gives the limiting values

$$(4.3) \qquad p_m = \frac{e^{-1}}{m!}.$$

It will be noticed that the approximation of p_m to $P_{[m]}$ is rather good even for moderate values of N.

For the numbers p_m defined by (4.3) we have $\Sigma p_k = e^{-1}(1 + 1 + \frac{1}{2!} + \frac{1}{3!} + \ldots) = e^{-1}e = 1$. Accordingly, the p_k may be interpreted as probabilities. As a matter of fact, we shall later see that the important *Poisson distribution* leads, for a particular value (unity) of a parameter, to an assignment of probabilities in accordance with (4.3).

Formulas (4.1) are useful in testing *guessing abilities*. In wine

TABLE 1

PROBABILITIES OF m CORRECT GUESSES IN CALLING A DECK OF N DISTINCT CARDS

	$N = 3$		$N = 4$		$N = 5$		$N = 6$		$N = 10$		p_m
	$P_{[m]}$	b_m	$P_{[m]}$	b_m	$P_{[m]}$	b_m	$P_{[m]}$	b_m	$P_{[m]}$	b_m	
0	0.333	0.296	0.375	0.316	0.367	0.328	0.368	0.335	0.36788	0.34868	0.367879
1	.500	.444	.333	.422	.375	.410	.367	.402	.36788	.38742	.367879
2	...	.222	.250	.211	.167	.205	.187	.201	.18394	.19371	.183940
3	.167	.037	...	.047	.083	.051	.056	.053	.06131	.05733	.061313
4			.042	.004	...	.006	.021	.008	.01534	.01116	.015328
5					.008	.000	...	.001	.00306	.00149	.003066
6							.001	.000	.00052	.00014	.000511
7									.00007	.00001	.000073
8									.00001	.....	.000009
9									.....	.....	.000001
10									.....	.....	.000000

The $P_{[m]}$ are given by (4.1), the b_m by (4.4). The last column gives the Poisson limits (4.3).

tasting, psychic experiments, etc., the subject is asked to call an unknown order of N things, say, cards. Any actual insight on the part of the subject will appear as a departure from randomness. To judge the amount of insight we must appraise the probability of turns of good luck. Now chance guesses can be made according to several systems among which we mention three extreme possibilities. (1) The subject sticks to one card and keeps calling it. With this system he is sure to have one, and only one, correct guess in each series; chance fluctuations are eliminated. (2) The subject calls each card once so that each series of N guesses corresponds to a rearrangement of the deck. If this system is applied without insight, formulas (4.1) should apply. (3) A third possibility is that N guesses are made absolutely independently of each other. There are N^N possible arrangements. It is true that every person has fixed mental habits and is prone to call certain patterns more frequently than others, but in first approximation we may assume all N^N arrangements to be equally probable. Since m correct and $N - m$ incorrect guesses can be arranged in $\binom{N}{m}(N-1)^{N-m}$ different ways, the probability of exactly m correct guesses is now

$$(4.4) \quad b_m = \binom{N}{m} \frac{(N-1)^{N-m}}{N^N}.$$

(This is a special case of the binomial distribution; cf. Chapter 6.)

Table 1 gives a comparison of the probabilities of success when guesses are made in accordance with system (2) or (3). To judge the merits of the two methods we require the theory of mean values and probable fluctuations. It turns out that the average number of correct chance guesses is one under all systems; the chance fluctuations are somewhat larger under system (2) than (3). A glance at Table 1 will show that in practice the differences will not be excessive.

* 5. Application to the Classical Occupancy Problem

We now return to the problem of Chapter 3, section 3, and consider a random distribution of r things in n cells, assuming that each arrangement has probability n^{-r}. We seek the probability $P_{[m]}$ of finding exactly m cells empty.

Let A_k be the event that cell number k is empty ($k = 1, 2, \cdots, n$). In this event all r balls are placed in the remaining $n-1$ cells, and this can be done in $(n-1)^r$ different ways. Similarly, there are $(n-2)^r$ arrangements, leaving two preassigned cells empty, etc. Accordingly

$$(5.1) \quad p_i = \left(1 - \frac{1}{n}\right)^r, \quad p_{ij} = \left(1 - \frac{2}{n}\right)^r, \quad p_{ijk} = \left(1 - \frac{3}{n}\right)^r, \cdots$$

and hence for every $\nu \leq n$

$$(5.2) \quad S_\nu = \binom{n}{\nu}\left(1 - \frac{\nu}{n}\right)^r.$$

Substituting into (3.1), we find

$$(5.3) \quad P_{[m]} = \sum_{\nu=0}^{n-m} (-1)^\nu \binom{m+\nu}{m}\binom{n}{m+\nu}\left(1 - \frac{m+\nu}{n}\right)^r.$$

This formula simplifies by the use of the identity $\binom{m+\nu}{m}\binom{n}{m+\nu}$

$= \binom{n}{m}\binom{n-m}{\nu}$ to

$$(5.4) \quad P_{[m]} = \binom{n}{m}\sum_{\nu=0}^{n-m}(-1)^\nu \binom{n-m}{\nu}\left(1 - \frac{m+\nu}{n}\right)^r.$$

* Starred sections treat special topics and may be omitted at first reading.

Such is *the probability of finding exactly m cells empty.* [For $m = 0$ we find the right side of (2.7)].

We have already used the model of r random digits to illustrate the random distribution of r things in $n = 10$ cells. Empty cells correspond in this case to missing digits: if m cells are empty, $10 - m$ different digits appear in the given sequence. Table 2 provides a numerical illustration.

TABLE 2
PROBABILITIES $P_{[m]}$ ACCORDING TO (5.4) FOR $n = 10$

m	$r = 10$	$r = 18$
0	0.000 363	0.134 673
1	.016 330	.385 289
2	.136 080	.342 987
3	.355 622	.119 425
4	.345 144	.016 736
5	.128 596	.000 876
6	.017 189	.000 014
7	.000 672	.000 000
8	.000 005	.000 000
9	.000 000	.000 000

$P_{[m]}$ is the probability that exactly m of the digits $0, 1, \cdots, 9$ will *not* appear in a sequence of r random digits.

It is clear that a direct numerical evaluation of (5.4) is limited to the case of relatively small n and r. On the other hand, the occupancy problem is of particular interest when n is large. If 10,000 balls are distributed in 1000 cells, is there any chance of finding an empty cell? In a group of 2000 people, is there any chance of finding a day in the year which is not a birthday? Fortunately, questions of this kind can be answered by means of a remarkably simple approximation to $P_{[m]}$ with an error which tends to zero as $n \to \infty$. This approximation and the argument leading to it are typical of many *limit theorems* in probability.

Our purpose, then, is to discuss the limiting form of the formula for $P_{[m]}$ as $n \to \infty$ and $r \to \infty$. The relation between r and n is, in principle, arbitrary. However, the ratio r/n represents the average number of things per cell. If it is excessively large, then we cannot expect any empty cells; in this case $P_{[0]}$ is near unity and all $P_{[m]}$ with $m \geq 1$ are small. On the other hand, if r/n tends to zero, then practically all cells must be empty, and in this case $P_{[m]} \to 0$ for every fixed m. Therefore only the intermediate case is of real interest.

4.5] APPLICATION TO THE CLASSICAL OCCUPANCY PROBLEM

Formula (5.4) for $P_{[m]}$ was derived from the expression (5.2) for S_ν, and we shall derive an approximation to $P_{[m]}$ from an approximation to S_ν, for ν fixed. Since $1 - x < e^{-x}$ for all positive x, we get directly

$$(5.5) \qquad S_\nu = \binom{n}{\nu}\left(1 - \frac{\nu}{n}\right)^r < \frac{n^\nu}{\nu!} e^{-\nu r/n}.$$

Now put for brevity

$$(5.6) \qquad ne^{-r/n} = \lambda.$$

Then (5.5) takes on the form

$$(5.7) \qquad S_\nu < \frac{\lambda^\nu}{\nu!}.$$

We now show that the right side is not only an upper bound for S_ν, but also an approximation to it. More precisely, we show that

$$(5.8) \qquad \frac{1}{\nu!}\lambda^\nu - S_\nu \to 0$$

for every fixed ν. If r varies as function of n so that $\lambda \to 0$, then (5.8) is implied by (5.7). Now λ is small unless r/n^2 is small, and it suffices therefore to consider the case where

$$(5.9) \qquad \frac{r}{n^2} \to 0.$$

The Taylor expansion for e^{-x-x^2} shows that for sufficiently small positive x we have $1 - x > e^{-x-x^2}$. Therefore, at least for sufficiently large n,

$$(5.10) \qquad S_\nu = \binom{n}{\nu}\left(1 - \frac{\nu}{n}\right)^r > \frac{(n-\nu)^\nu}{\nu!} e^{-\nu r/n - \nu^2 r/n^2}$$

$$= \frac{\lambda^\nu}{\nu!}\left\{\left(1 - \frac{\nu}{n}\right) e^{-\nu r/n^2}\right\}^\nu.$$

In view of (5.9) the term within braces tends to unity, so that the right side is asymptotically equivalent to $\lambda^\nu/\nu!$. The two inequalities (5.7) and (5.10) therefore imply (5.8).

We now introduce the approximation (5.8) into the formula (3.1) for $P_{[m]}$. For the νth term on the right side we have found (at least for large m)

$$(5.11) \qquad (-1)^\nu \binom{m+\nu}{m} S_{m+\nu} \approx (-1)^\nu \frac{\lambda^{m+\nu}}{m!\nu!}.$$

This means that the expansion (3.1) for $P_{[m]}$ approaches termwise the series

$$(5.12) \qquad \frac{\lambda^m}{m!}\left\{1 - \frac{\lambda}{1!} + \frac{\lambda^2}{2!} - \frac{\lambda^3}{3!} + - \ldots\right\}.$$

Furthermore, by (5.7) the terms of (3.1) are smaller in absolute value than those of (5.12). If λ is restricted to a finite interval, then the series within braces in (5.12) converges absolutely and uniformly to $e^{-\lambda}$. It follows that in the limit (5.12) represents $P_{[m]}$ and we have thus proved the

Theorem.[2] *If n and r increase so that $\lambda = ne^{-r/n}$ remains bounded, then for every fixed m*

$$(5.13) \qquad P_{[m]} - e^{-\lambda}\frac{\lambda^m}{m!} \to 0.$$

The approximating expressions

$$(5.14) \qquad p(m;\lambda) = e^{-\lambda}\frac{\lambda^m}{m!}$$

define the so-called *Poisson distribution*, which is of great importance and describes a variety of phenomena. For the particular value $\lambda = 1$ we get once more the distribution (4.3).

In practice one may use $p(m, \lambda)$ as an approximation to $P_{[m]}$ whenever n is great. For moderate values of n an estimate of the error is required, but we shall not enter into it.

Examples. (a) Table 3 gives the approximate probabilities of finding m cells empty when the number of cells is 1000 and the number of things varies from 5000 to 9000. For $r = 5000$ the median value of the number of empty cells is six: seven or more empty cells are about as probable as six or fewer. Even with 9000 things in 1000 cells we have one chance in twelve to find an empty cell.

(b) In birthday statistics $n = 365$, and r is the number of people. For $r = 1900$ we find $\lambda = 2$, approximately. *In a village of 1900 people the probabilities of finding m days of the year which are not birthdays are approximately as follows*:

$$P_{[0]} = 0.135, \quad P_{[1]} = 0.271, \quad P_{[2]} = 0.271, \quad P_{[3]} = 0.180,$$

$$P_{[4]} = 0.090, \quad P_{[5]} = 0.036, \quad P_{[6]} = 0.012, \quad P_{[7]} = 0.003.$$

---

[2] Due (with a different proof) to R. von Mises, Über Aufteilungs- und Besetzungswahrscheinlichkeiten, *Revue de la Faculté des Sciences de l'Université d'Istanbul*, N.S., vol. 4 (1939), pp. 145–163.

TABLE 3

Poisson Approximation (5.14) to the Probabilities (5.4) of Finding Exactly m Empty Cells When r Things Are Randomly Distributed in $n = 1000$ Cells

$p(m; \lambda)$

r	λ	$m=0$	$m=1$	$m=2$	$m=3$	$m=4$	$m=5$	$m=6$	$m=7$	$m=8$	$m=9$	$m=10$	$m=11$
5000	6.74	0.0012	0.0080	0.0269	0.0604	0.1017	0.1371	0.1540	0.1482	0.1249	0.0935	0.0630	0.0386
5500	4.09	.0167	.0685	.1400	.1909	.1951	.1596	.1088	.0636	.0325	.0148	.0060	.0023
6000	2.48	.0838	.2077	.2575	.2128	.1320	.0655	.0271	.0096	.0030	.0008	.0002	
6500	1.50	.2231	.3347	.2510	.1255	.0471	.0141	.0035	.0008	.0001			
7000	0.91	.4027	.3661	.1666	.0506	.0115	.0021	.0003					
7500	0.55	.5777	.3163	.0873	.0162	.0023	.0003						
8000	0.34	.7126	.2406	.0414	.0049	.0004							
8500	0.20	.8187	.1637	.0164	.0011	.0001							
9000	0.12	.8869	.1064	.0064	.0003								

The probability of finding exactly m cells each containing exactly k things can be derived in the same way. As von Mises has shown, this probability can again be approximated by the Poisson expression (5.14), only this time λ must be defined by

$$\lambda = ne^{-r/n}\left(\frac{r}{n}\right)^k/k!. \tag{5.15}$$

* 6. Miscellany

(1) The Realization of at Least m Events. With the notations of section 3 *the probability P_m that m or more of the events $A_1, \cdots, A_N$ occur simultaneously is given by*

$$P_m = P_{[m]} + P_{[m+1]} + \cdots + P_{[N]}. \tag{6.1}$$

To find a formula for P_m in terms of S_k it is simplest to proceed by induction, starting with formula (1.5) and using the recurrence relation $P_{m+1} = P_m - P_{[m]}$. One gets for $m \geq 1$

$$P_m = S_m - \binom{m}{m-1}S_{m+1} + \binom{m+1}{m-1}S_{m+2} \tag{6.2}$$
$$- \binom{m+2}{m-1}S_{m+3} + \cdots \pm \binom{N-1}{m-1}S_N.$$

It is also possible to derive (6.2) directly, using the argument which led to (3.1).

(2) Further Identities. The coefficients S_ν can be expressed in terms of either $P_{[k]}$ or P_k as follows

$$S_\nu = \sum_{k=\nu}^{N} \binom{k}{\nu} P_{[k]} \tag{6.3}$$

and

$$S_\nu = \sum_{k=\nu}^{N} \binom{k-1}{\nu-1} P_k. \tag{6.4}$$

Indication of Proof. For given values of $P_{[m]}$ the equations (3.1) may be taken as linear equations in the unknowns S_ν, and we have to prove that (6.3) represents the unique solution. If (6.3) is introduced into the expression (3.1) for $P_{[m]}$, the coefficient of $P_{[k]}$ ($m \leq k \leq N$) to the right is found to be

* Starred sections treat special topics and may be omitted at first reading.

(6.5) $$\sum_{\nu=m}^{k} (-1)^{\nu-m} \binom{\nu}{m}\binom{k}{\nu} = \binom{k}{m} \sum_{\nu=m}^{k} (-1)^{\nu-m} \binom{k-m}{\nu-m}.$$

If $k = m$ this expression reduces to 1. If $k > m$ the sum is the binomial expansion of $(1-1)^{k-m}$ and therefore vanishes. Hence the substitution (6.3) reduces (3.1) to the identity $P_{[m]} = P_{[m]}$. The uniqueness of the solution of (3.1) follows from the fact that each equation introduces only one new unknown, so that the S_ν can be computed recursively. The truth of (6.4) can be proved in a similar way.

(3) Bonferroni's Inequalities. A string of inequalities both for $P_{[m]}$ and for P_m can be obtained in the following way. *If in either* (3.1) *or* (6.2) *only the terms involving* $S_m, S_{m+1}, \cdots, S_{m+r-1}$ *are retained while the terms involving* $S_{m+r}, S_{m+r+1}, \cdots, S_N$ *are dropped, then the error* (*i.e., true value minus approximation*) *has the sign of the first omitted term* [*namely,* $(-1)^r$] *and is smaller in absolute value.* Thus, for $r = 1$ and $r = 2$:

(6.6) $$S_m - (m+1)S_{m+1} \leq P_{[m]} \leq S_m$$

and

(6.7) $$S_m - mS_{m+1} \leq P_m \leq S_m.$$

Indication of Proof. To prove the statement for (3.1) it must be shown that

(6.8) $$\sum_{\nu=t}^{N} (-1)^{\nu-t} \binom{\nu}{m} S_\nu \geq 0,$$

for every t. Now use (6.3) to write the left side as a linear combination of the $P_{[k]}$. For $t \leq k \leq N$ the coefficient of $P_{[k]}$ equals

$$\sum_{\nu=t}^{k} (-1)^{\nu-t} \binom{\nu}{m}\binom{k}{\nu} = \binom{k}{m} \sum_{\nu=t}^{k} (-1)^{\nu-t} \binom{k-m}{\nu-m}.$$

The last sum equals $\binom{k-m-1}{t-m-1}$ and is therefore positive (Chapter 2, section 9, problem 13).

For further inequalities the reader is referred to Fréchet's monograph cited at the beginning of the chapter.

7. Problems for Solution

Note: Assume in each case that all possible arrangements have the same probability.

1. Ten pairs of shoes are in a closet. Four shoes are selected at random. Find the probability that there will be at least one pair among the four shoes selected.

2. Five dice are thrown. Find the probability that at least three of them show the same face. (Verify by the methods of Chapter 3.)

3. Find the probability that in 5 tossings a coin falls heads at least three times in succession.

4. Solve problem 3 for a head-run of at least length 5 in 10 tossings.

5. Solve problems 3 and 4 for ace runs when a die is used instead of a coin.

6. Two dice are thrown r times. Find the probability p_r that each of the six combinations $(1, 1), \cdots, (6, 6)$ appears at least once.

7. *Sampling with replacement.* A sample of size r is taken from a population of n people. Find the probability u_r that N given people will all be included in the sample. (This question applies to collecting coupons.)

8. *Continuation.* Show that if $n \to \infty$ and $r \to \infty$ so that $r/n \to p$, then $u_r \to (1 - e^{-p})^N$.

9. *Sampling without replacement.* Answer problem 7 for this case and show that 8 holds with $u_r \to p^N$.

10. In the general expansion of a determinant of order N the number of terms containing one or more diagonal elements is $N!P_1$ with P_1 defined by (2.2).

11. The number of ways in which 8 rooks can be placed on a chessboard so that none can take another and that none stands on the white diagonal is $8!(1 - P_1)$, where P_1 is defined by (2.2) with $N = 8$.

12. From (2.6) conclude that

$$\sum_{k=0}^{n} (-1)^k \binom{n}{k} (ns - ks)_r = 0$$

if $r < n$ and

$$\sum_{k=0}^{n} (-1)^k \binom{n}{k} (ns - ks)_n = s^n n!.$$

13. Solve problem 12 by evaluating the rth derivative, at $x = 0$, of

$$\frac{1}{(1 - x)^{ns+r+1}} \{1 - (1 - x)^s\}^n.$$

14. In the sampling problem (2.c) find the probability that it will take exactly r drawings to get a sample containing all numbers. Pass to the limit as $s \to \infty$.

15. *Bridge-bingo.* From the table given in the answer section to problem 40 of section 8 in Chapter 2, compute the probabilities $Q_1(r)$, $Q_2(r)$, $Q_3(r)$ that after r drawings exactly 1, 2, 3 of the 4 players are without cards.

16. A cell contains N chromosomes, between any two of which an interchange of parts may occur. If r interchanges occur [which can happen in $\binom{N}{2}^r$ distinct ways], find the probability that exactly m chromosomes will be involved.[3]

17. Find the probability that exactly k suits will be missing in a poker hand (for definition cf. footnote on p. 9).

18. Find the probability that a hand of 13 bridge cards contains the ace-king pairs of exactly k suits.

19. *Multiple matching.* Two similar decks of N distinct cards each are matched simultaneously against a similar target deck. Find the probability u_m of having

---

[3] For $N = 6$ cf. D. G. Catcheside, D. E. Lea, and J. M. Thoday, Types of Chromosome Structural Change Introduced by the Irradiation of *Tradescantia* Microspores, *Journal of Genetics*, vol. 47 (1945–46), pp. 113–149.

exactly m double matches. Show that $u_0 \to 1$ as $N \to \infty$ (which implies that $u_m \to 0$ for $m \geq 1$).

20. *Multiple matching.* The procedure of the preceding problem is modified as follows. Out of the $2N$ cards N are chosen at random, and only these N are matched against the target deck. Find the probability of no match. Prove that it tends to $1/e$ as $N \to \infty$.

21. *Multiple matching.* Answer problem 20 if r decks are used instead of two.

22. *To section 5.* Prove that for $m \geq 1$ the probability of m or more cells remaining empty is

(7.1) $$P_m = \binom{n}{m} \sum_{\nu=0}^{n-m} (-1)^\nu \binom{n-m}{\nu} \left(1 - \frac{m+\nu}{n}\right)^r \frac{m}{m+\nu}.$$

23. From (7.1) deduce the identity [4]

(7.2) $$\sum_{k=0}^{n} (-1)^k \binom{n}{k} k^r = 0.$$

valid for $n > r$. Verify (7.2) by considering the rth derivative of $(1 - e^x)^n$ at $x = 0$.

24. In the classical occupancy problem, the probability $P_{[m]}(k)$ of finding exactly m cells occupied by exactly k things is

$$P_{[m]}(k) = \frac{(-1)^m n! r!}{m! n^r} \sum_j (-1)^j \frac{(n-j)^{r-jk}}{(j-m)!(n-j)!(r-jk)!(k!)^j},$$

the summation extending over those $j \geq m$ for which $j \leq n$ and $kj \leq r$.

25. Prove the last statement of section 5 for the case $k = 1$.

26. Using (3.1), derive the probability of finding exactly m empty cells in the case of Bose-Einstein statistics.

27. Verify that the formula obtained in 26 checks with Chapter 3, formula (5.3).

[4] In the notations of the calculus of differences (7.2) can be written $(-1)^n \Delta^n 0^r = 0$.

CHAPTER 5

CONDITIONAL PROBABILITY
STATISTICAL INDEPENDENCE

1. Conditional Probability

The following example leads in a natural way to the notion of conditional probability. Suppose a population of N people includes N_A colorblind people and N_H females. Let the events that a person chosen at random is colorblind and a female be A and H, respectively. Then (cf. the definition of random choice, Chapter 2, section 2)

(1.1) $$Pr\{A\} = \frac{N_A}{N}, \quad Pr\{H\} = \frac{N_H}{N}.$$

Instead of the entire population, we may investigate the female subpopulation and require the probability that a female chosen at random is colorblind. This probability is N_{HA}/N_H, where N_{HA} is the number of colorblind females. We have here no new notion, but we need a new notation to designate which particular subpopulation is under investigation. The most widely adopted symbol is $Pr\{A|H\}$; it may be read "the probability of the event A (colorblindness), assuming the event H (that the person chosen is female)." The formula

(1.2) $$Pr\{A \mid H\} = \frac{N_{AH}}{N_H} = \frac{Pr\{AH\}}{Pr\{H\}}$$

suggests the following general definition whose usefulness and plausibility will be illustrated by further examples.

Definition. Let H be an event with positive probability. For an arbitrary event A we shall write

(1.3) $$Pr\{A \mid H\} = \frac{Pr\{AH\}}{Pr\{H\}}.$$

The quantity so defined will be called the conditional probability of A on the hypothesis H (or for given H). When all sample points have equal probabilities, $Pr\{A \mid H\}$ is the ratio N_{AH}/N_H of the number of sample points common to A and H, to the number of points in H.

Conditional probabilities remain undefined when the hypothesis has zero probability. This is of no consequence in the case of discrete sample spaces, but is important in the general theory.

Though the symbol $Pr\{A \mid H\}$ itself is practical, its phrasing in words is so unwieldy that in practice less formal descriptions are used. Thus in our introductory example we referred to the probability of a female being colorblind instead of saying "the conditional probability of a person chosen at random being colorblind on the hypothesis that the person is a female." Often the phrase "on the hypothesis H" is replaced by "if it is known that H occurred." In short, our formulas and symbols are unequivocal, but phrasings in words are often informal and must be properly interpreted.

Whenever convenient for stylistic clarity one speaks of *absolute probabilities* in contradistinction to conditional ones. Strictly speaking, the adjective "absolute" is redundant and will be omitted (as has been done in Chapter 1).

Examples. (*a*) An urn contains r red and b black balls. Two balls are chosen at random without replacements. If the first ball is red, what is the probability that the second ball is red also? The hypothesis H, "first ball red," can occur in $r(r + b - 1)$ ways; the event AH, "both balls red," can occur in $r(r - 1)$ ways. Therefore, $Pr\{A \mid H\} = (r - 1)/(r + b - 1)$. This formula expresses the fact that the second choice refers to an urn with $r - 1$ red and b black balls.

(*b*) *Distribution of Sexes.* Consider families with exactly two children. Letting b and g stand for boy and girl, respectively, and the first letter for the older child, we have four possibilities: *bb, bg, gb, gg*. These are the four sample points, and we associate probability 1/4 with each. Given that a family has a boy (event H), what is the probability that both children are boys (event A)? The event AH means *bb*, and H means *bb*, or *bg*, or *gb*. Therefore, $Pr\{A \mid H\} = 1/3$: in about one-third of the families with the characteristic H we can expect that A also will occur. It is interesting that most people expect the answer to be 1/2. This is the correct answer to a different question, namely: a boy is chosen at random and found to come from a family with two children; what is the probability that the other child is a boy? The difference may be explained empirically. With our original problem we might refer to a card file of families, with the second to a file of males. In the latter, each family with two boys will be represented twice, and this explains the difference between the two results.

(*c*) *Bridge.* If North has no ace (hypothesis H), what is the probability that South has no ace either? Assuming all arrangements equally probable, we find $Pr\{H\} = \binom{48}{13} \div \binom{52}{13} = 0.329 \ldots$. The event AH means "neither North nor South has an ace" and hence

$$Pr\{AH\} = \binom{48}{13}\binom{35}{13} \div \binom{52}{13}\binom{39}{13} = \binom{48}{26} \div \binom{52}{26}.$$

Therefore

(1.4) $\qquad Pr\{A \mid H\} = \binom{35}{13} \div \binom{39}{13} = 0.182 \ldots$.

Again this result could have been anticipated by the following reasoning. North knows that four aces and 35 non-aces are divided among the remaining three players. His partner's hand can be selected in $\binom{39}{13}$ ways, of which $\binom{35}{13}$ lead to the event "no ace." (Cf. problems 3–5.)

(d) We conclude with an example in which the sample space contains infinitely many points. Suppose we shoot at a target until it is hit for the first time. The theory of the next chapter will lead us to attribute probability $(1 - p)p^{n-1}$ to the event that exactly n trials are required; here p (the probability of a miss) is a constant with $0 < p < 1$ (cf. also problem 7). In other words, our sample space consists of the points $1, 2, 3, \ldots$ with corresponding probabilities $(1 - p)$, $(1 - p)p$, $(1 - p)p^2, \ldots$. Assuming that the first trial results in failure (hypothesis H), what is the probability that more than three trials will be necessary? Here $Pr\{H\} = 1 - (1 - p) = p$. More than three trials are required only when the first results in failure. In our case, therefore, $A = AH$ and $Pr\{AH\} = (1 - p)p^3\{1 + p + p^2 + \ldots\} = p^3$. Hence $Pr\{A \mid H\} = p^2$. Among all cases which do not end with success at the first trial, those which are continued beyond the third trial should have an average frequency p^2.

Taking conditional probabilities of various events with respect to a particular hypothesis H amounts to choosing H as a new sample space; we have to multiply all probabilities by the constant factor $1/Pr\{H\}$ in order to reduce the total probability of the new sample space to unity. This formulation shows that *all general theorems on probabilities are valid also for conditional probabilities with respect to any particular hypothesis H.* As an example we mention the fundamental relation for the probability of the occurrence of either A or B or both. We have

(1.5) $\qquad Pr\{A \cup B \mid H\} = Pr\{A \mid H\} + Pr\{B \mid H\} - Pr\{AB \mid H\}.$

Similarly, all theorems of Chapter 4 concerning probabilities of the realization of m among N events carry over to conditional probabilities, but we shall not need them.

Formula (1.3) is often used in the form

(1.6) $$Pr\{AH\} = Pr\{A \mid H\} \cdot Pr\{H\}.$$

This is the so-called theorem on compound probabilities. To generalize it to three events A, B, C we first take $H = BC$ as hypothesis and then apply (1.6) once more; it follows that

(1.7) $$Pr\{ABC\} = Pr\{A \mid BC\} \cdot Pr\{B \mid C\} \cdot Pr\{C\}.$$

A further generalization to four or more events is straightforward.

We conclude with a simple formula which is frequently useful. Let $H_1, \cdots, H_n$ be a set of mutually exclusive events of which one necessarily occurs (that is, the union of $H_1, \cdots, H_n$ is the entire sample space). Then any event A can occur only in conjunction with some H_j, or in symbols,

(1.8) $$A = AH_1 \cup AH_2 \cup \cdots \cup AH_n.$$

Since the AH_j are mutually exclusive, their probabilities add. Applying (1.6) to $H = H_j$ and adding, we get

(1.9) $$Pr\{A\} = \Sigma Pr\{A \mid H_j\} \cdot Pr\{H_j\}.$$

This formula is useful because an evaluation of the conditional probabilities $Pr\{A \mid H_j\}$ is sometimes easier than a direct calculation of $Pr\{A\}$. Examples will be found in the next section.

2. Compound Experiments

The use of conditional probabilities greatly simplifies formulations. In applications, many experiments are described in terms of conditional probabilities (although the adjective "conditional" is usually omitted). We shall give a few examples which will reveal a general scheme more effectively than a direct description could.

Examples. (a) *Families*. We want to interpret the following statement. "The probability of a family with exactly k children is p_k (where $p_0 + p_1 + \ldots = 1$). For any family size all sex distributions have equal probabilities." Letting b stand for boy and g for girl, our sample space consists of the points 0 (no children), b, g, bb, bg, gb, gg, bbb, ... The second assumption in quotation marks can be stated more formally thus: if it is known that the family has exactly n children, each of the 2^n possible sex distributions has conditional probability 2^{-n}.

The probability of the hypothesis is p_n, and we see from (1.6) that the absolute probability of any arrangement of n letters b and g is $p_n \cdot 2^{-n}$.

Let A stand for the event "the family has boys but no girls." Its probability is obviously $Pr\{A\} = p_1 \cdot 2^{-1} + p_2 \cdot 2^{-2} + p_3 \cdot 2^{-3} + \ldots$. Incidentally, this is a special case of (1.9). The hypothesis H_j in this case is "family has j children." We now ask the question: if it is known that a family has no girls, what is the (conditional) probability that it has only one child? Here A is the hypothesis. Let H be the event "only one child." Then AH means "one child and no girl," and

$$(2.1) \quad Pr\{H \mid A\} = \frac{Pr\{AH\}}{Pr\{A\}} = \frac{p_1 2^{-1}}{p_1 2^{-1} + p_2 2^{-2} + p_3 2^{-3} + \ldots}.$$

(b) *Mixed Populations.* Suppose a human population consists of subpopulations or strata $H_1, H_2, \ldots$. These may be races, age groups, professions, etc. Let p_j be the probability that an individual chosen at random belongs to H_j. Saying "the probability that an individual in H_j is left-handed is q_j" is short for "the conditional probability of the event A (left-handedness) on the hypothesis that an individual belongs to H_j is q_j." The probability that an individual chosen at random is left-handed is $p_1 q_1 + p_2 q_2 + p_3 q_3 + \ldots$, which is again a special case of (1.9). Given that an individual is left-handed, the conditional probability of his belonging to stratum H_j is

$$(2.2) \quad Pr\{H_j \mid A\} = \frac{p_j q_j}{p_1 q_1 + p_2 q_2 + \ldots}.$$

(c) *Polya's Urn Scheme.* An urn contains b black and r red balls. Random drawings are made. The ball drawn is always replaced, and, in addition, c balls of the color drawn are added to the urn. Here we are given conditional probabilities only. *If* the first ball is black, the (conditional) probability of a black ball at the second drawing is $(b + c)/(b + c + r)$. The absolute probability of the sequence black, black is therefore, by (1.6),

$$(2.3) \quad \frac{b}{b + r} \cdot \frac{b + c}{b + c + r}.$$

If the first two drawings result in black, then the urn contains $b + r + 2c$ balls among which $b + 2c$ are black. The (conditional) probability of a black ball at the third trial becomes, again using (1.6),

$$(2.4) \quad \frac{b + 2c}{b + 2c + r}.$$

In this way we can calculate all probabilities. It is easily seen that any sequence of n drawings resulting in n_1 black and n_2 red balls ($n_1 + n_2 = n$) has the same probability as the event of extracting *first* n_1 black and *then* n_2 red balls, namely,

$$(2.5) \quad p_{n_1,n} = \frac{b(b+c)(b+2c) \cdots (b+n_1c-c) \cdot r(r+c) \cdots (r+n_2c-c)}{(b+r)(b+r+c)(b+r+2c) \cdots (b+r+nc-c)}.$$

This scheme [1] was devised for the analysis of phenomena like contagious diseases, where the occurrence of certain events increases their future probabilities. More general probability models of this kind will be taken up later on. [Polya's scheme is discussed in problems 15–19 and again in Chapter 6, problems 33–35; Chapter 9, problem 13; and Chapter 10, problem 9; cf. also Chapter 15, example (10.a), and Chapter 17, problems 5 and 6.]

(d) Die A has four red and two white faces, whereas die B has two red and four white faces. The game starts by flipping a coin once: if it falls heads, the game continues by throwing die A alone; if it falls tails, die B is to be used. Here again we know conditional probabilities only. For example, the conditional probability of the sequence (red, red, white), assuming heads at the first trial, is $(4 \cdot 4 \cdot 2) \cdot 6^{-3}$. For 3 throws of the die we have 16 sample points. Each of the points (H, R, R, R) and (T, W, W, W) has probability $4/27$; the 6 points (H, R, R, W), (H, R, W, R), (H, W, R, R), (T, W, W, R), (T, W, R, W), (T, R, W, W) have probability $2/27$ each; the 6 points obtained by interchanging W and R have probability $1/27$; finally, (H, W, W, W) and (T, R, R, R) have probability $1/54$ each.

Note that the probability of red at any trial is $1/2$. If it is not known which die is used and at the first 2 throws red is observed, then the conditional probability of red at the third trial is

$$(2.6) \quad \frac{Pr\{(H, R, R, R)\} + Pr\{(T, R, R, R)\}}{Pr\{(H, R, R)\} + Pr\{(T, R, R)\}} = \frac{3}{5}$$

(cf. problem 14).

(e) The following example is famous and illustrative, but somewhat artificial. Imagine a population of $N + 1$ urns, each containing N red and white balls; the urn number k contains k red and $N - k$ white balls ($k = 0, 1, 2, \cdots, N$). An urn is chosen at random and n random drawings are made from it, the ball drawn being replaced each time.

---

[1] F. Eggenberger and G. Polya, Über die Statistik verketteter Vorgänge, *Zeitschrift für Angewandte Mathematik und Mechanik*, vol. 3 (1923), pp. 279–289.

Suppose that all n balls turn out to be red (event A). We seek the (conditional) probability that the next drawing will yield a red ball also (event B). If the first choice falls on urn number k, then the probability of extracting in succession n red balls is $(k/N)^n$. Hence, by (1.9),

$$(2.7) \qquad Pr\{A\} = \frac{1^n + 2^n + \cdots + N^n}{N^n(N+1)}.$$

The event AB means that $n+1$ drawings yield red balls, and therefore

$$(2.8) \qquad Pr\{AB\} = Pr\{B\} = \frac{1^{n+1} + 2^{n+1} + \cdots + N^{n+1}}{N^{n+1}(N+1)}.$$

The required probability is $Pr\{B \mid A\} = Pr\{B\}/Pr\{A\}$.

The sums in (2.7) and (2.8) can be considered Riemann sums approximating integrals, so that when N is large

$$(2.9) \qquad N^{-1} \sum_{k=1}^{N} \left(\frac{k}{N}\right)^n \sim \int_0^1 x^n \, dx = \frac{1}{n+1}.$$

We have therefore for large N approximately

$$(2.10) \qquad Pr\{B \mid A\} \approx \frac{n+1}{n+2}.$$

This formula can be interpreted roughly as follows: if all compositions of an urn are equally probable, and if n trials yielded red balls, the probability of a red ball at the next trial is $(n+1)/(n+2)$. This is the so-called law of succession of Laplace (1812).

Before the ascendance of the modern theory, the notion of equal probabilities was often used as synonymous for "no advance knowledge." Laplace himself has illustrated the use of (2.10) by computing the probability that the sun will rise tomorrow, given that it has risen daily for 5000 years or $n = 1{,}826{,}213$ days. It is said that Laplace was ready to bet 1,826,214 to 1 in favor of regular habits of the sun, and we should be in a position to better the odds since regular service has followed for another century. A historical study would be necessary to render justice to Laplace and to understand his intentions. His successors, however, used similar arguments in routine work and recommended methods of this kind to physicists and engineers in cases where the formulas have no operational meaning. We would have to reject the method even if, for sake of argument, we were to concede that our universe was chosen at random from a collection in which all conceivable possibilities were equally likely. In fact, the assumed rising of the sun on February 5, 3123 B.C., is by no means more certain than that the sun will rise tomorrow. We believe in both for the same reasons.

Note on Bayes's Rule. In (2.1) and (2.2) we have calculated certain conditional probabilities directly from the definition. The beginner is advised always to do so and not to memorize the formula (2.12) which we shall now derive. It retraces in a general way what we did in special cases, but it is only a way of rewriting (1.3). We had a collection of events $H_1, H_2, \ldots$ which are mutually exclusive and exhaustive, that is, every sample point belongs to one, and only one, among the H_j. We were interested in

(2.11) $$Pr\{H_k \mid A\} = \frac{Pr\{AH_k\}}{Pr\{A\}}.$$

If (1.6) and (1.9) are introduced into (2.11), it takes the form

(2.12) $$Pr\{H_k \mid A\} = \frac{Pr\{A \mid H_k\}\, Pr\{H_k\}}{\sum_j Pr\{A \mid H_j\}\, Pr\{H_j\}}.$$

If the events H_k are called causes, then (2.12) becomes "Bayes's rule for the probability of causes." Mathematically, (2.12) is a special way of writing (1.3) and nothing more. The formula is useful in many statistical applications of the type described in the above examples (a–d). Unfortunately, Bayes's rule has been somewhat discredited by metaphysical applications of the type described in example (e). In routine practice this kind of argument can be dangerous. A quality-control engineer is concerned with one particular machine and not with an infinite population of machines from which one was chosen at random. He has been advised to use Bayes's rule on the grounds that it is logically acceptable and corresponds to our way of thinking. Plato used this type of argument to prove the existence of Atlantis, and philosophers used it to prove the absurdity of Newton's mechanics. In our case it overlooks the circumstance that the engineer desires success and that he will do better by estimating and minimizing the sources of various types of errors in prediction and guessing. The modern method of statistical tests and estimation is less intuitive but more realistic. It may be not only defended but also applied.

3. Statistical Independence

In the above examples the conditional probability $Pr\{A \mid H\}$ generally does not equal the absolute probability $Pr\{A\}$. Popularly speaking, the information as to whether H has occurred changes our way of betting on the event A. Only when $Pr\{A \mid H\} = Pr\{A\}$, this information does not permit any inference as to the occurrence of A. In this case we shall say that A is statistically independent of H. Now (1.6) shows that the condition $Pr\{A \mid H\} = Pr\{A\}$ can be written in the form

(3.1) $$Pr\{AH\} = Pr\{A\} \cdot Pr\{H\}.$$

This equation is symmetric in A and H, and shows that whenever A is statistically independent of H so is H of A. It is therefore preferable to start from the following symmetric

Definition 1. *Two events A and H are said to be statistically independent (or independent, for short) if equation* (3.1) *holds.* This definition is accepted also if $Pr\{H\} = 0$, in which case $Pr\{A \mid H\}$ is not defined.

Examples. (a) A card is chosen at random from a deck of playing cards. For reasons of symmetry we expect the events "spade" and "ace" to be independent. As a matter of fact, their probabilities are 1/4 and 1/13, and the probability of their simultaneous realization is 1/52.

(b) Two true dice are thrown. We verify that the events "ace with first die" and "even face with second" are independent; the probability of their simultaneous realization, $3/36 = 1/12$, is the product of their probabilities, namely 1/6 and 1/2.

(c) In a random permutation of the four letters (a, b, c, d) the events "a precedes b" and "c precedes d" are independent. This is intuitively clear and easily verified.

(d) *Sex Distribution.* We return to example (1.b) but now consider families with three children. We assume that each of the eight possibilities bbb, bbg, $\cdots$, ggg has probability 1/8. Let H be the event "the family has children of both sexes," and A the event "there is at most one girl." Then $Pr\{H\} = 6/8$, and $Pr\{A\} = 4/8$. The simultaneous realization of A and H means one of the possibilities bbg, bgb, gbb, and therefore $Pr\{AH\} = 3/8 = Pr\{A\} \cdot Pr\{H\}$. Thus in families with three children the two events are independent. Note that this is not the case for families with two or four children. This shows that it is not always obvious whether or not we have independence.

If H occurs, then the complementary event $\bar{H}$ does not occur, and vice versa. Statistical independence implies that no inference can be drawn from the occurrence of H to that of A; therefore statistical independence of A and H should mean the same as independence of A and $\bar{H}$ (and, because of symmetry, also of $\bar{A}$ and H, and of $\bar{A}$ and $\bar{H}$). This assertion is easily verified, using the relation $Pr\{\bar{H}\} = 1 - Pr\{H\}$. If (3.1) holds, then (since $A\bar{H} = A - AH$)

(3.2) $\quad Pr\{A\bar{H}\} = Pr\{A\} - Pr\{AH\} = Pr\{A\} - Pr\{A\} \cdot Pr\{H\}$

$$= Pr\{A\} \cdot Pr\{\bar{H}\},$$

as expected.

Suppose now that three events A, B, and C are pairwise independent so that

(3.3)
$$Pr\{AB\} = Pr\{A\} \cdot Pr\{B\}$$
$$Pr\{AC\} = Pr\{A\} \cdot Pr\{C\}$$
$$Pr\{BC\} = Pr\{B\} \cdot Pr\{C\}.$$

One might think that this always implies the independence of such pairs of events as AB and C. Unfortunately this is *not* necessarily so. We shall exhibit an example in which (3.3) is true but the simultaneous occurrence of A, B, and C is impossible, so that AB and C cannot be independent.

Example. (*e*) Two dice are thrown and three events are defined as follows. A means "odd face with first die"; B means "odd face with second die"; finally, C means "odd sum" (one face even, the other odd). If each of the 36 sample points has probability 1/36, then any two of the events are clearly independent. The probability of each is 1/2, and so is its conditional probability, assuming that *one* of the other two events has occurred. Nevertheless, the three events cannot occur simultaneously. The information that A but not B has occurred assures that C has occurred, and a similar statement holds for all other combinations.

It is desirable to reserve the term statistical independence for the case where no such inference is possible. For this it is necessary that (3.3) holds, but we must in addition assume that

(3.4)
$$Pr\{ABC\} = Pr\{A\}\,Pr\{B\}\,Pr\{C\}.$$

This equation insures that A and BC are independent and also that the same is true of B and AC, and C and AB. Furthermore, it can now be proved also that $A \cup B$ and C are independent. In fact, by the fundamental relation (6.4) of Chapter 1 we have

(3.5) $\quad Pr\{(A \cup B)C\} = Pr\{AC\} + Pr\{BC\} - Pr\{ABC\}.$

Now, applying (3.3) and (3.4) to the right side, we can factor out $Pr\{C\}$. The other factor is $Pr\{A\} + Pr\{B\} - Pr\{AB\} = Pr\{A \cup B\}$ so that

(3.6) $\quad Pr\{(A \cup B)C\} = Pr\{(A \cup B)\}\,Pr\{C\}.$

This makes it plausible that the conditions (3.3) and (3.4) together suffice to avoid embarrassment; any event expressible in terms of A and B will be independent of C.

In the general case of n events the following definition proves satisfactory.

Definition 2. *The events $A_1, A_2, \cdots, A_n$ are called mutually independent if for all combinations $1 \leq i < j < k < \cdots \leq n$ the multiplication rules*

$$Pr\{A_iA_j\} = Pr\{A_i\} \, Pr\{A_j\}$$

$$Pr\{A_iA_jA_k\} = Pr\{A_i\} \, Pr\{A_j\} \, Pr\{A_k\}$$

(3.7)
$$\cdots \cdots \cdots \cdots \cdots \cdots \cdots$$

$$Pr\{A_1A_2 \cdots A_n\} = Pr\{A_1\} \, Pr\{A_2\} \cdots Pr\{A_n\}$$

apply.

The first line stands for $\binom{n}{2}$ equations, the second for $\binom{n}{3}$, etc. We have, therefore, $\binom{n}{2} + \binom{n}{3} + \cdots + \binom{n}{n} = (1+1)^n - \binom{n}{1} - \binom{n}{0}$ $= 2^n - n - 1$ conditions which must be satisfied. On the other hand, the $\binom{n}{2}$ conditions stated in the first line suffice to insure *pairwise independence*. The whole system (3.7) looks like a complicated set of conditions, but it will soon become apparent that its validity is usually obvious and requires no checking. The distinction between mutual and pairwise independence is of theoretical rather than practical interest. Practical examples of pairwise independent events which are not mutually independent apparently do not exist. The possibility of such an occurrence was discovered by S. Bernstein.

4. Repeated Trials

The notion of statistical independence finally enables us to formulate analytically the intuitive concept of experiments "repeated under identical conditions."

Consider the sample space $\mathfrak{S}$ representing a certain conceptual experiment. Let the sample points be $E_1, E_2, \ldots$ and denote their probabilities by $p_1, p_2, \ldots$. The possible results of a succession of two similar experiments are the pairs (E_j, E_k) and they form a new sample space. In it probabilities can be assigned in many ways. However, if the experimentalist says that two measurements are performed under identical conditions, he implies independence: the first outcome should have no influence on the second. This means that the two events "first outcome is E_j" and "second outcome is E_k" should be statistically independent or that

(4.1) $$Pr\{E_j, E_k\} = p_j p_k.$$

This equation assigns a probability to every pair (E_j, E_k) of possible outcomes. Before we can use (4.1) as a definition of probabilities in the new sample space, we must show that the quantities $p_j p_k$ add to unity. Now, in the sum $\Sigma\Sigma p_j p_k$ each term appears once, and only once, so that $\Sigma\Sigma p_j p_k = (p_1 + p_2 + \ldots)(p_1 + p_2 + \ldots) = 1$. Hence (4.1) is acceptable as a definition of probabilities.

Let A and B be two arbitrary events in the original sample space $\mathfrak{S}$. We denote the event "A occurred at first trial and B at second" by (A, B). Suppose A contains the points $E_{a_1}, E_{a_2}, \ldots$ and B the points $E_{b_1}, E_{b_2}, \ldots$. Then (A, B) is the union of all pairs (E_{a_j}, E_{b_k}), and as before we see that

(4.2) $$Pr\{(A, B)\} = \Sigma\Sigma p_{a_j} p_{b_k} = (\Sigma p_{a_j})(\Sigma p_{b_k})$$
$$= Pr\{A\} Pr\{B\}.$$

Hence the events A and B are independent. We see that the definition (4.1) entails that all events at the second trial are independent of events at the first trial. For the purposes of probability theory this describes "identical experiments."

These considerations obviously also apply to a succession of r experiments and lead to the

Definition. Let $\mathfrak{S}$ be a sample space with sample points $E_1, E_2, \ldots$ and corresponding probabilities $p_1, p_2, \ldots$. By r independent trials corresponding to $\mathfrak{S}$ we mean the sample space whose points are the r-tuples $(E_{j_1}, E_{j_2}, \ldots, E_{j_r})$ to which the probabilities

(4.3) $$Pr\{(E_{j_1}, E_{j_2}, \ldots, E_{j_r})\} = p_{j_1} p_{j_2} \cdots p_{j_r}$$

are assigned.

In other words, each point of the new space is a sample of size r (with possible repetitions) of points of the original space, and probabilities are defined by the multiplication rule (4.3). The reader is reminded that (4.3) is *not* the only possible definition of probabilities. In other words, repeated trials are not necessarily independent. For example, in sampling without replacement we are concerned with *dependent* trials. Equation (4.3) defines independent trials or, in physical terms, trials repeated under identical conditions.

The argument which led to (4.2) shows more generally the truth of the following

Theorem. *Suppose that a system of events* $A_1, A_2, \cdots, A_r$ *is such that the jth trial alone decides whether or not* A_j *occurs; then the events* $A_1, \cdots, A_r$ *are mutually independent if the trials are independent, that is, if* (4.3) *holds.*

If $\mathfrak{S}$ contains a finite number, N, of points, then there are N^r sample points $(E_{j_1}, \cdots, E_{j_r})$. If each point of $\mathfrak{S}$ has probability $1/N$, then (4.3) assigns probability N^{-r} to each point $(E_{j_1}, \cdots, E_{j_r})$. The new approach is conceptually preferable to a formal assignment of equal probabilities because it applies to sample spaces with unequal probabilities and also to infinite sample spaces. It is indispensable for the general theory of probability where we consider even a single trial as the first in a potentially infinite sequence. We are then dealing only with infinite sequences $(E_{j_1}, E_{j_2}, \ldots)$ of possible outcomes, and in this new space probabilities are defined in a way consistent with (4.3). Unfortunately this leads beyond the theory of discrete sample spaces, to which the present volume is restricted. We have a more elementary theory but pay for it by the necessity of changing the sample space according to the number of trials.

In the preceding discussion we have considered only repetitions of the same experiment, but successions of unlike experiments can be treated in the same way. If we first toss a coin, then throw a die, we naturally assume that the two experiments are independent. This amounts to assigning probabilities by the product rule. Thus $Pr\{(\text{heads, ace})\} = 1/2 \cdot 1/6$, etc. In this particular case this is equivalent to assigning equal probabilities to all twelve sample points, but in general we must proceed as in (4.3). Let $\mathfrak{S}'$ and $\mathfrak{S}''$ be two sample spaces and denote their points by $E_1', E_2', \ldots$ and $E_1'', E_2'', \ldots$. Let the corresponding probabilities be $p_1', p_2', \ldots$ and $p_1'', p_2'', \ldots$. The succession of the two experiments is described by the space with points (E_j', E_k''). Saying that the two experiments are independent means defining probabilities by

(4.4) $$Pr\{(E_j', E_k'')\} = p_j' p_k''.$$

Examples. (a) *Permutations.* We have considered the $n!$ permutations of $a_1, a_2, \cdots, a_n$ as points of a sample space and attributed probability $1/n!$ to each. We may consider the *same sample space* as representing $n-1$ successive experiments as follows. Begin by writing down a_1. The first experiment consists in putting a_2 either before or after a_1. This done, we have three places for a_3 and the second experiment consists of a choice among them. This decides on the relative order of a_1, a_2, and a_3. Now we have four places to choose from for a_4. In general, when $a_1, \cdots, a_k$ are put into some relative order, we proceed with experiment number k, which consists in selecting one of the $k+1$ places for a_{k+1}. As an example, take $n=5$. The permutation $(a_4, a_2, a_1, a_5, a_3)$ is built up successively by choosing for $a_2, a_3, a_4,$ and a_5 the first, last, first, and fourth of the available places. In other words, we have a succession of $n-1$ experiments of which the kth can result in k different choices (sample points), each having probability $1/k$. The experiments are independent, that is, the probabilities are multiplicative. Each permutation of the n elements has probability $1/2 \cdot 1/3 \cdots 1/n$, in accordance with the original definition.

(b) *Sampling without Replacement.* Let the population be $(a_1, \cdots, a_n)$. In sampling without replacement each choice removes an element. After k steps there remain $n-k$ elements, and the next choice can be described by specifying the number ν of the place of the element chosen ($\nu = 1, 2, \cdots, n-k$). In this way the taking of a sample of size r without replacement becomes a succession of r experiments where the first has n possible results, the second $n-1$, the third $n-2$, etc. We attribute equal probabilities to all results of the individual experiments and postulate that the r experiments are independent. This amounts to attributing probability $1/(n)_r$ to each sample in accordance with our definition of random samples. [Note that for $n = 100$, $r = 3$, the sample (a_{13}, a_{40}, a_{81}) means choices number 13, 39, 79, respectively. We must say that at the third experiment the seventy-ninth element of the reduced population of $n-2$ was chosen, for with the original numbering the outcomes of the third experiment would depend on the first two choices.] We see that the notion of repeated independent experiments permits us to study sampling as a succession of individual operations.

4a. A Guide to Abstract Language

The notions with which probability theory deals occur also in other branches of mathematics. *Sample space* is simply an abstract space in which a probability measure is defined. The term *repeated trials refers to combinatorial product spaces* with congruent component spaces; a measure is defined on the product space and induces measures on the component spaces. *Independence of trials means product measure.* Saying that event A *depends only on trial number k* is an abbreviation for "A is a cylindrical set with base in the kth component space." The phrase "if it is known that A occurred" is a translation of "if $x \in A$," where x stands for a point in sample space. *Successive experiments* refer to product spaces with different component spaces, and "independent" again refers to product measure.

* 5. Applications to Genetics

The theory of heredity, originated by G. Mendel (1822–1884), provides instructive illustrations for the applicability of simple probability models. We shall restrict ourselves to indications concerning the most elementary problems. In describing the biological background, we shall necessarily oversimplify and concentrate on such facts as are pertinent to the mathematical treatment.

Heritable characters depend on special carriers, called *genes*. All cells of the body, except the reproductive cells or gametes, carry exact replicas of the same gene structure. The salient fact is that genes appear in pairs. The reader may picture the genes as a vast collection of beads on short pieces of string, the chromosomes. These appear in pairs so that the two genes of a pair occupy similar positions on two related chromosomes. In the simplest case each gene of a particular pair can assume two forms (alleles), A and a. Then three different pairs can be formed, and, with respect to this particular pair, the organism belongs to one of the three *genotypes* AA, Aa, aa (there is no distinction between Aa and aA). For example, peas carry a pair of genes such that A causes red, and a causes white, blossom color. The three genotypes are in this case distinguishable as red, pink, and white. Each pair of genes determines one heritable factor, but the majority of observable properties of organisms depend on several factors. For some characteristics (e.g., eye color, and left-handedness) the influence of one particular pair of genes is predominant, and in such cases the effects of Mendelian laws are readily observable. Other characteristics, such as height, can be understood as the cumulative effect of a very large number of genes [cf. Chapter 10, example (5.c)]. Here we shall study genotypes and inheritance for only one particular pair of genes with respect to which we have the three genotypes AA, Aa, aa. Frequently there are N different forms $A_1, \cdots, A_N$ for the two genes and, accordingly, $\binom{N+1}{2}$ genotypes $A_1A_1, A_1A_2, \cdots, A_NA_N$. The theory applies to this case with obvious modifications (cf. problem 21). The following calculations apply also to the case where A is *dominant* and a *recessive*. By this is meant that Aa-individuals have the same observable properties as AA, so that only the pure aa-type shows an observable influence of the a-gene. All shades of partial dominance

---

* Starred sections treat special topics and may be omitted at first reading.

appear in nature. Typical partially recessive properties are blue eyes, left-handedness, etc.

The reproductive cells, or gametes, are formed by a splitting process and receive *one* gene only. Organisms of the pure AA- and aa-genotypes (or homozygotes) produce therefore gametes of only one kind, but Aa-organisms (hybrids or heterozygotes) produce A- and a-gametes in equal numbers. New organisms are derived from two parental gametes from which they receive their genes. Therefore each pair includes a paternal and a maternal gene, and any gene can be traced back to one particular ancestor in any generation, however remote.

The genotypes of offspring depend on a chance process. At every occasion, each parental gene has probability $1/2$ to be transmitted, and the successive trials are independent. In other words, we conceive of the genotypes of n offspring as the result of n independent trials, each of which corresponds to the tossing of two coins. For example, the genotypes of descendants of an $Aa \times Aa$ pairing are AA, Aa, aa with respective probabilities $1/4$, $1/2$, $1/4$. An $AA \times aa$ union can have only Aa-offspring, etc.

Looking at the population as a whole, we conceive of the pairing of parents as the result of a second chance process. We shall investigate only the so-called *random mating*, which is defined by this condition: if r descendants in the first filial generation are chosen at random, then their parents form a random sample of size r, with possible repetitions, from the aggregate of all possible parental pairs. In other words, each descendant is to be regarded as the product of a random selection of parents, and all the selections are mutually independent. Random mating is an idealized model of the conditions in many natural populations and in many field experiments. However, if red peas are sown in one corner of the field and white peas in another, parents of like color will unite more often than under random mating. Preferential selectivity (such as blonde preferring blondes) violates the condition of random mating. Extreme non-random mating is represented by self-fertilizing plants and artificial inbreeding. Such assortative mating systems have been analyzed mathematically, but we shall restrict our attention mainly to random mating.

The genotype of an offspring is the result of four independent random choices. The genotypes of the two parents can be selected in $3 \cdot 3$ ways, their genes in $2 \cdot 2$ ways. However, we may combine two selections and describe the process as one of double selection thus: the paternal and maternal gene are each selected independently and at random from the population of all genes carried by males or females of the parental population.

Suppose that the three genotypes AA, Aa, aa occur among males and females in the same ratios, $u:2v:w$. We shall suppose $u + 2v + w = 1$ and call u, $2v$, w, the *genotype frequencies*. Put

(5.1) $$p = u + v, \quad q = v + w.$$

Clearly the numbers of A- and a-genes are as $p:q$, and since $p + q = 1$ we shall call p and q the *gene frequencies* of A and a. In each of the two selections an A-gene is selected with probability p, and, because of the assumed independence, the probability of an offspring being AA is p^2. The genotype Aa can occur in two ways, and its probability is therefore $2pq$. Thus, under random mating conditions *an offspring belongs to the genotypes AA, Aa, or aa with probabilities*

(5.2) $$u_1 = p^2, \quad 2v_1 = 2pq, \quad w_1 = q^2.$$

Examples. (a) All parents are Aa (heterozygotes). Then $u = w = 0$, $2v = 1$, and $p = q = 1/2$; (b) AA- and aa-parents are mixed in equal proportions. Then $u = w = 1/2$, $v = 0$, and again $p = q = 1/2$; (c) $u = w = 1/4$, $2v = 1/2$. Again $p = q = 1/2$. In all three cases we have for the filial generation $u_1 = 1/4$, $2v_1 = 1/2$, $w_1 = 1/4$.

For a better understanding of the implications of (5.2) let us fix the gene frequencies p and q ($p + q = 1$) and consider all systems of genotype frequencies u, $2v$, w for which $u + v = p$ and $v + w = q$. They all lead to the same probabilities (5.2) for the first filial generation. Among them there is the particular distribution

(5.3) $$u = p^2, \quad 2v = 2pq, \quad w = q^2.$$

If the frequencies u, v, w in the original generation stand in the particular relation (5.3) (as in example c), then we find for the genotype probabilities in the first filial generation $u_1 = u$, $v_1 = v$, and $w_1 = w$. Therefore we call genotype distributions of the form (5.3) *stable*. To every ratio $p:q$ there corresponds a stable distribution.

Equations (5.2) give the genotype probabilities for a randomly selected individual of the second generation. In a large population we must expect the actual genotype frequencies to be very close to the theoretical distribution.[2] Now, whatever the distribution $u:2v:w$ in the parental generation, equations (5.2) define a *stable* distribution; in it the genes A and a appear with frequencies [cf. (5.1)] $u_1 + v_1 = u + v = p$ and $v_1 + w_1 = v + w = q$. In other words, if the

---

[2] Without this our probability model would be void of operational meaning. The statement is made precise by the law of large numbers and the central limit theorem, which permits us to estimate the effect of chance fluctuations.

observed frequencies coincided exactly with the calculated probabilities, then the first filial generation would have a stable genotype distribution which would perpetuate itself without change in all succeeding generations. In practice, deviations will be observed, but for large populations we can say: *whatever the composition of the parent population may be, random mating will within one generation produce approximately a stable genotype distribution with unchanged gene frequencies.* From the second generation on, there is no tendency towards a systematic change: a steady state is reached with the first filial generation. This was first noticed by *G. H. Hardy*,[3] who thus resolved assumed difficulties in Mendelian laws. It follows in particular that under conditions of random mating the frequencies of the three genotypes must stand in the ratios $p^2: 2pq: q^2$. This can in turn be used to check the assumption of random mating.

Hardy also pointed out that emphasis must be put on the word "approximately." Because of chance fluctuations the actual genotype frequencies will never coincide exactly with the theoretical probabilities (5.2). Therefore, even with a stable distribution we must expect small changes from generation to generation. This leads us to the following picture. Starting from any parent population, random mating tends to establish the stable genotype distribution (5.3) within *one* generation. For a stable distribution there is no tendency towards a systematic change of any kind. However, chance fluctuations will change the gene frequencies p and q from generation to generation, and the genetic composition will slowly drift. There are no restoring forces seeking to reestablish original frequencies. On the contrary, our simplified model leads to the conclusion [cf. Chapter 15, example 2, X] that, whenever a population is bounded in size, one gene should ultimately die out, so that the population should eventually belong to one of the pure types, AA or aa. In nature this does not occur because of the creation of new genes by mutations, selections, and many other effects. These more complicated processes of evolution can be studied by more refined mathematical tools (Markov chains, diffusion theory).

Hardy's theorem is frequently interpreted to imply a strict stability for all times. It is a common fallacy to believe that the law of large

---

[3] G. H. Hardy, Mendelian Proportions in a Mixed Population, Letter to the Editor, *Science*, N.S., vol. 28 (1908), pp. 49–50. Anticipating the language of Chapters 9 and 15, we can describe the situation as follows. The frequencies of the three genotypes in the nth generation are three random variables whose expected values are given by (5.2) and do not depend on n. Their actual values will vary from generation to generation and form a stochastic process of the Markov type.

numbers acts as a force endowed with memory seeking a return to the original state, and many wrong conclusions have been drawn from this assumption. (The biological processes here considered are typical of the important class of Markov processes which will be studied in detail in Chapter 15.)

It should also be noted that Hardy's law does not apply to the distribution of two pairs of genes (e.g., eye color and left-handedness). With respect to two pairs we have to distinguish nine genotypes $AABB$, $AABb$, $\cdots$, $aabb$. There is still a tendency towards a stable distribution, but equilibrium is not reached in the first generation (cf. problem 25).

* 6. Sex-linked Characters

In the introduction to the preceding section it was mentioned that genes lie on chromosomes. These appear in pairs and are transmitted as units, so that all genes on a chromosome stick together.[4] Our scheme for the inheritance of genes therefore applies also to chromosomes as units. Sex is determined by two chromosomes; females are XX, males XY. The mother necessarily transmits an X-chromosome, and the sex of offspring depends on the chromosome transmitted by the father. Accordingly, male and female gametes are produced in equal numbers. The difference in birth rate for boys and girls is explained by variations in prenatal survival chances.

It has been said that both genes and chromosomes appear in pairs. There is an exception inasmuch as the genes situated on the X-chromosome have no corresponding gene on Y. Females have two X-chromosomes, and hence two of such X-linked genes; however, in males the X-genes appear as singles. Typical are two sex-linked genes causing colorblindness and haemophilia. With respect to each of them, females can still be classified into the three genotypes, AA, Aa, aa, but, having only *one* gene, males have only the two genotypes A and a. Note that a son always has the father's Y-chromosome so that a sex-linked character cannot be inherited from father to son. However, it can pass from father to daughter and from her to a grandson.

We now proceed to generalize the analysis of the preceding section. Assume again random mating and let the frequencies of the genotypes AA, Aa, aa in the *female* population be u, $2v$, w, respectively. As before put $p = u + v$, $q = v + w$. The frequencies of the two *male* genotypes A and a will be denoted by p' and q' ($p' + q' = 1$). Then p and p'

---

* Starred sections treat special topics and may be omitted at first reading.

[4] This picture is somewhat complicated by occasional breakings and recombinations of chromosomes, cf. Chapter 2, section 8, problem 11.

are the frequencies of the A-gene in the female and male populations, respectively. The probability for a female descendant to be of genotype AA, Aa, aa will be denoted by u_1, $2v_1$, w_1; the analogous probabilities for the male types A and a are p_1', q_1'. Now a male offspring receives his X-chromosome from the female parent, and hence

(6.1) $$p_1' = p, \quad q_1' = q.$$

For the three female genotypes we find, as in section 5,

(6.2) $$u_1 = pp', \quad 2v_1 = pq' + qp', \quad w_1 = qq'.$$

Hence

(6.3) $$p_1 = u_1 + v_1 = \frac{p + p'}{2}, \quad q_1 = v_1 + w_1 = \frac{q + q'}{2}.$$

We can interpret these formulas as follows. Among the male descendants the genes A and a appear approximately with the frequencies p, q of the maternal population; the gene frequencies among female descendants are approximately p_1 and q_1, or half-way between those of the paternal and maternal populations. We discern a tendency towards equalization of the gene frequencies. In fact, from (6.1) and (6.3) we get

(6.4) $$p_1' - p_1 = \frac{p - p'}{2}, \quad q_1' - q_1 = \frac{q - q'}{2}.$$

This means that random mating will in one generation reduce approximately by one-half the differences between gene frequencies among males and females. However, it will not eliminate the differences, and a tendency towards further reduction will subsist. In contrast to Hardy's law, we have here no stable situation after one generation. We can pursue the systematic component of the changes from generation to generation by neglecting chance fluctuations and identifying the theoretical probabilities (6.2) and (6.3) with corresponding actual frequencies in the first filial generation.[5] For the second generation we obtain by the same process

(6.5) $$p_2 = \frac{p_1 + p_1'}{2} = \frac{3p}{4} + \frac{p'}{4}, \quad q_2 = \frac{q_1 + q_1'}{2} = \frac{3q}{4} + \frac{q'}{4}$$

and, of course, $p_2' = p_1$, $q_2' = q_1$. A few more trials will lead to the

---

[5] In the terminology introduced in footnote 3 we can interpret p_n and q_n as the expected values of the gene frequencies in the nth female generation. With this interpretation the formulas for p_n and q_n are no longer approximations but exact.

general expression for the probabilities p_n and q_n among females of the nth descendant generation. Put

(6.6) $$\alpha = \frac{2p + p'}{3}, \quad \beta = \frac{2q + q'}{3}.$$

(Note that $\alpha + \beta = 1$.) Then

(6.7) $$p_n = \frac{p_{n-1} + p'_{n-1}}{2} = \alpha + (-1)^n \frac{p - p'}{3 \cdot 2^n},$$

$$q_n = \frac{q_{n-1} + q'_{n-1}}{2} = \beta + (-1)^n \frac{q - q'}{3 \cdot 2^n},$$

and $p_n' = p_{n-1}$, $q_n' = q_{n-1}$. Hence

(6.8) $$p_n \to \alpha, \quad p_n' \to \alpha, \quad q_n \to \beta, \quad q_n' \to \beta.$$

The genotype frequencies in the female population are given by (6.2) or

(6.9) $$u_n = p_{n-1}p'_{n-1}, \quad 2v_n = p_{n-1}q'_{n-1} + q_{n-1}p'_{n-1},$$

$$w_n = q_{n-1}q'_{n-1}.$$

Hence

(6.10) $$u_n \to \alpha^2, \quad 2v_n \to 2\alpha\beta, \quad w_n \to \beta^2.$$

These formulas show that there is a strong systematic tendency, from generation to generation, towards a state where the genotypes A and a appear among males with frequencies α and β, and the female genotypes AA, Aa, aa have probabilities α^2, $2\alpha\beta$, β^2, respectively. The convergence is very fast, as indicated by (6.7). In practice, equilibrium will be reached after three or four generations. To be sure, small chance fluctuations will be superimposed on the described changes, but these represent the prevailing systematic tendency.

Our main conclusion is that under random mating we can expect the sex-linked genotypes A and a among males, and AA, Aa, aa among females to occur approximately with the frequencies α, β, α^2, $2\alpha\beta$, β^2, respectively, where $\alpha + \beta = 1$.

Application. Many sex-linked genes, like colorblindness, are *recessive* and cause defects. Let a be such a gene. Then all a-males and all aa-females show the defect. Females of Aa-type may transmit the defect to their offspring, but are not themselves affected. Hence we expect that a *recessive sex-linked defect which occurs among males with frequency α occurs among females with frequency α^2*. If one man in 100 is colorblind, one woman in 10,000 should be affected.

*7. Selection

As a typical example of the influence of selection we shall investigate the case where aa-individuals cannot multiply. This happens when the a-gene is recessive and lethal, so that aa-individuals are born but cannot survive. Another case occurs when artificial interference by breeding or laws prohibits mating of aa-individuals.

Assume random mating among AA- and Aa-individuals, but no mating of aa-types. Let the frequencies with which the genotypes AA, Aa, aa appear in the *total* population be $u, 2v, w$. The corresponding frequencies for *parents* are then

$$(7.1) \qquad u^* = \frac{u}{1-w}, \qquad 2v^* = \frac{2v}{1-w}, \qquad w^* = 0.$$

We can proceed as in section 5, but have to use the quantities (7.1) instead of $u, 2v, w$. Hence, (5.1) is to be replaced by

$$(7.2) \qquad p = \frac{u+v}{1-w}, \qquad q = \frac{v}{1-w}.$$

The probabilities of the three genotypes in the first filial generation are again given by (5.2) or $u_1 = p^2$, $2v_1 = 2pq$, $w_1 = q^2$.

As before, in order to investigate the systematic changes from generation to generation, we have to replace u, v, w by u_1, v_1, w_1 and thus obtain probabilities u_2, v_2, w_2 for the second descendant generation, etc. In general we get from (7.2)

$$(7.3) \qquad p_n = \frac{u_n + v_n}{1 - w_n}, \qquad q_n = \frac{v_n}{1 - w_n}$$

and

$$(7.4) \qquad u_{n+1} = p_n^2, \qquad 2v_{n+1} = 2p_n q_n, \qquad w_{n+1} = q_n^2.$$

A comparison of (7.3) and (7.4) shows that

$$(7.5) \qquad p_{n+1} = \frac{u_{n+1} + v_{n+1}}{1 - w_{n+1}} = \frac{p_n}{1 - q_n^2} = \frac{1}{1 + q_n}$$

and similarly

$$(7.6) \qquad q_{n+1} = \frac{v_{n+1}}{1 - w_{n+1}} = \frac{q_n}{1 + q_n}.$$

From (7.6) we can calculate q_n explicitly. In fact

$$(7.7) \qquad \frac{1}{q_{n+1}} = 1 + \frac{1}{q_n}$$

* Starred sections treat special topics and may be omitted at first reading.

whence successively

(7.8) $\quad \dfrac{1}{q_1} = 1 + \dfrac{1}{q}, \quad \dfrac{1}{q_2} = 2 + \dfrac{1}{q}, \quad \dfrac{1}{q_3} = 3 + \dfrac{1}{q}, \quad \cdots, \quad \dfrac{1}{q_n} = n + \dfrac{1}{q}$

or

(7.9) $\quad\quad\quad\quad q_n = \dfrac{q}{1 + nq}, \quad w_{n+1} = \left(\dfrac{q}{1 + qn}\right)^2.$

We see that the unproductive (or undesirable) genotype gradually drops out, but the process is extremely slow. For $q = 0.1$ it takes ten generations to reduce the frequency of a-genes by one-half; this reduces the frequency of the aa-type approximately from 1 to $\frac{1}{4}$ per cent. (If a is sex-linked, the elimination proceeds much faster as shown in problem 23; for a generalized selection scheme cf. problem 24.[6])

8. Problems for Solution

1. Three dice are rolled. If no two show the same face, what is the probability that at least one is an ace?

2. Given that a throw with 10 dice produced at least one ace, what is the probability p of two or more aces?

3. *Bridge.* In a bridge party West has no ace. What probability should be attributed to the event of his partner having two or more aces? [Cf. example (1.c).]

4. *Bridge.* North and South have ten trumps between them (trumps being cards of a specified suit). Find the probability that all three remaining trumps are in the same hand (either East or West has no trumps).

5. *Continuation.* If it is known that the king of trumps is included among the three, what is the probability that he is "unguarded" (that is, one player has the king, the other the remaining two trumps)?

6. In a bolt factory machines A, B, C manufacture, respectively, 25, 35, and 40 per cent of the total. Of their output 5, 4, and 2 per cent are defective bolts. A bolt is drawn at random from the produce and is found defective. What are the probabilities that it was manufactured by machines A, B, C?

7. In example (1.d) suppose that an even number n of shots was fired. What is the probability that $n = 2$?

8. Suppose that 5 men out of 100 and 25 women out of 10,000 are colorblind. A colorblind person is chosen at random. What is the probability of his being male? (Assume males and females to be in equal numbers.)

---

9. Let [7] the probability p_n that a family has exactly n children be αp^n when $n \geq 1$, and $p_0 = 1 - \alpha p(1 + p + p^2 + \ldots)$. Suppose that all sex distributions of n children have the same probability. Show that for $k \geq 1$ the probability

---

[6] For a further analysis of various eugenic effects (which are frequently different from the ideas of enthusiastic proponents of sterilization laws) cf. G. Dahlberg, *Mathematical Methods for Population Genetics*, New York and Basel, 1948.

[7] According to A. J. Lotka, American family statistics satisfies our hypothesis with $p = 0.7358$. Cf. Théorie analytique des associations biologiques II, *Actualités scientifique et industrielles*, no. 780, Paris, 1939.

that a family contains exactly k boys is

$$\frac{2\alpha p^k}{(2-p)^{k+1}}.$$

10. *Continuation.* Given that a family includes at least one boy, what is the probability that there are two or more?

11. Let the events $A_1, A_2, \cdots, A_n$ be independent and $Pr\{A_k\} = p_k$. Find the probability p that none of the events occur.

12. *Continuation.* Show that always $p < e^{-\Sigma p_k}$.

13. *Continuation.* From Bonferoni's inequality (6.7 of Chapter 4) deduce that the probability of k or more of the events $A_1, \cdots, A_n$ occurring simultaneously is less than $(p_1 + \cdots + p_n)^k / k!$.

14. *To example* (2d). If the die turns up a red face n times in succession, the probability that it is die A is

$$\frac{1}{1+(1/2)^n}.$$

15. *To Polya's urn scheme, example* (2.c). Given that the second ball was black, what is the probability that the first was black?

16. *To Polya's urn scheme, example* (2.c). Show by induction that the probability of a black ball at any trial is $b/(b+r)$.

17. *Continuation.* Prove by induction: for any $m < n$ the probabilities that the mth and the nth drawings produce the combinations (b, b) or (b, r) are

$$\frac{b(b+c)}{(b+r)(b+r+c)} \qquad \frac{br}{(b+r)(b+r+c)}.$$

Generalize to more than two drawings.

18. *Time symmetry of Polya's scheme.* Let A and B stand either for black or red (so that AB can be any of the four combinations). Show that the probability of A at the nth drawing, given that the mth drawing yields B, is the same as the probability of A at the mth drawing when the nth drawing yields B.

19. In the Polya scheme let $p_k(n)$ be the probability of k black balls in the first n drawings. Prove the recurrence relation

$$p_k(n+1) = p_k(n)\frac{r+(n-k)c}{b+r+nc} + p_{k-1}(n)\frac{b+(k-1)c}{b+r+nc}$$

where $p_{-1}(n)$ is to be interpreted as 0. Use this relation for a new proof of (2.5).

Applications in Biology

20. Under random mating less than half the population belongs to genotype Aa.

21. Generalize the results of section 5 to the case where each gene can have any of the forms $A_1, A_2, \cdots, A_k$, so that there are $\binom{k+1}{2}$ genotypes instead of three (multiple alleles).

22. *Brother-sister mating.* Two parents are selected at random from a population in which the genotypes AA, Aa, aa occur with frequencies u, $2v$, w. This process is repeated in their progeny. Find the probabilities that both parents of the first, second, third filial generation belong to AA [cf. Chapter 16, example (4.b)].

23. Selection. Let a be a recessive sex-linked gene, and suppose that a selection process makes mating of a-males impossible. If the genotypes AA, Aa, aa appear among females with frequencies u, $2v$, w, show that for female descendants of the first generation $u_1 = u + v$, $2v_1 = v + w$, $w_1 = 0$ and hence $p_1 = p + q/2$, $q_1 = q/2$. That is to say, the frequency of the a-gene among females is reduced to one-half.

24. The selection problem of section 7 can be generalized by assuming that only the fraction λ ($0 < \lambda \leq 1$) of the aa-class is eliminated. Show that

$$p = \frac{u+v}{1-\lambda w}, \quad q = \frac{v + (1-\lambda)w}{1-\lambda w}.$$

More generally, (7.3) is to be replaced by

$$p_{n+1} = \frac{p_n}{1 - \lambda q_n^2}, \quad q_{n+1} = q_n \frac{1 - \lambda q_n}{1 - \lambda q_n^2}.$$

(The general solution of these equations appears to be unknown.)

25. Consider simultaneously two pairs of genes with possible forms (A, a) and (B, b), respectively. Any person transmits to each descendant one gene of each pair, and we shall suppose that each of the four possible combinations has probability $1/4$. (This is the case if the genes are on separate chromosomes; otherwise there is strong dependence.) There are then nine genotypes, and we assume that their frequencies in the parent population are U_{AABB}, U_{aaBB}, U_{AAbb}, U_{aabb}, $2U_{AaBB}$, $2U_{Aabb}$, $2U_{AABb}$, $2U_{aaBb}$, $4U_{AaBb}$. Put $p_{AB} = U_{AABB} + U_{AABb} + U_{AaBB} + U_{AaBb}$, $p_{Ab} = U_{AAbb} + U_{Aabb} + U_{AABb} + U_{AaBb}$, $p_{aB} = U_{aaBB} + U_{aaBb} + U_{AaBB} + U_{AaBb}$, $p_{ab} = U_{aabb} + U_{Aabb} + U_{aaBb} + U_{AaBb}$. Compute the corresponding quantities for the first descendant generation. Show that for it $p_{AB}^{(1)} = p_{AB} - \delta$, $p_{Ab}^{(1)} = p_{Ab} + \delta$, $p_{aB}^{(1)} = p_{aB} + \delta$, $p_{ab}^{(1)} = p_{ab} - \delta$ with $2\delta = p_{AB}p_{ab} - p_{Ab}p_{aB}$. The stable distribution is given by

$$p_{AB} - 2\delta = p_{Ab} + 2\delta, \text{ etc.}$$

(Notice that Hardy's law does *not* apply: the composition changes from generation to generation.)

26. Assume that the genotype frequencies in a population are $u = p^2$, $2v = 2pq$, $w = q^2$. Given that a man is of genotype Aa, the probability that his brother is of the same genotype is $(1 + pq)/2$.

Note: The following problems are on family relations and give a meaning to the notion of degree of relationship, according to which brothers are as close to each other as grandfather and grandson. Each problem is a continuation of the preceding one. Random mating and the notations of section 5 are assumed. We are here concerned with a special case of Markov chains (cf. Chapter 15). Matrix algebra simplifies the writing.

27. Number the genotypes AA, Aa, aa by 1, 2, 3, respectively, and let p_{ik} ($i, k = 1, 2, 3$) be the conditional probability that an offspring be of genotype k if it is known that the male (or female) parent is of genotype i. Compute the nine probabilities p_{ik}, assuming that the probabilities for the other parent to be of genotype 1, 2, 3 are p^2, $2pq$, q^2, respectively.

28. Show that p_{ik} is also the conditional probability that the parent is of genotype k if it is known that a specified offspring is of genotype i.

29. Prove that the conditional probability of a grandson (grandfather) to be of genotype k if it is known that the grandfather (grandson) is of genotype i is given by

$$p_{ik}^{(2)} = p_{i1}p_{1k} + p_{i2}p_{2k} + p_{i3}p_{3k}$$

[The matrix $(p_{ik}^{(2)})$ is the square of the matrix (p_{ik}).]

30. Show that $p_{ik}^{(2)}$ is also the conditional probability that a man is of genotype k if it is known that a specified brother (not twin) is of genotype i.

31. Show that the conditional probability of a man to be of genotype k when it is known that a specified great-grandfather (or great-grandson, or uncle, or nephew) is of genotype i is given by

$$p_{ik}^{(3)} = p_{i1}^{(2)}p_{1k} + p_{i2}^{(2)}p_{2k} + p_{i3}^{(2)}p_{3k} = p_{i1}p_{1k}^{(2)} + p_{i2}p_{2k}^{(2)} + p_{i3}p_{3k}^{(2)}.$$

[The matrix $(p_{ik}^{(3)})$ is the third power of the matrix (p_{ik}). This procedure gives a precise meaning to the notion of the degree of family relationship.]

32. More generally, define probabilities $p_{ik}^{(n)}$ that a descendant of nth generation is of genotype k if a specified ancestor was of genotype i. Prove by induction that the $p_{ik}^{(n)}$ are given by the elements of the following matrix

$$\begin{pmatrix} p^2 + pq/2^{n-1} & 2pq + q(q-p)/2^{n-1} & q^2 - q^2/2^{n-1} \\ p^2 + p(q-p)/2^n & 2pq + (1 - 4pq)/2^n & q^2 + q(p-q)/2^n \\ p^2 - p^2/2^{n-1} & 2pq + p(p-q)/2^{n-1} & q^2 + pq/2^{n-1} \end{pmatrix}$$

(This shows that the influence of an ancestor decreases from generation to generation by the factor $\frac{1}{2}$.)

CHAPTER 6

THE BINOMIAL AND THE POISSON DISTRIBUTIONS

1. Bernoulli Trials [1]

Repeated independent trials are called Bernoulli trials if there are only two possible outcomes for each trial and their probabilities remain the same throughout the trials. It is usual to denote the two probabilities by p and q, and to refer to the outcome with probability p as *"success,"* S, and to the other as *"failure,"* F. Clearly, p and q must be non-negative, and

(1.1) $$p + q = 1.$$

The sample space of each individual trial is formed by the two points S and F. The sample space of n Bernoulli trials contains 2^n points or successions of n symbols S and F, each point representing one possible outcome of the compound experiment. Since the trials are independent, the probabilities multiply. In other words, *the probability of any specified sequence is the product obtained on replacing the symbols S and F by p and q, respectively.* Thus $Pr\{(SSFSF \cdots FFS)\} = ppqpq \cdots qqp$.

Examples. The most familiar example of Bernoulli trials is provided by successive tosses of a true or symmetric coin; here $p = q = 1/2$. If the coin is unbalanced, we still assume that the successive tosses are independent so that we have a model of Bernoulli trials in which the probability p for success can have an arbitrary value. If four faces of a "good" die are red and two black, then we have only two distinguishable outcomes, to which for obvious reasons we attribute probabilities $p = 2/3$ and $q = 1/3$. Often there are several possible outcomes, but we have no interest in distinguishing among them and prefer to describe any result as a simple alternative. Thus with good dice the distinction between ace (S) and non-ace (F) leads to Bernoulli trials with $p = 1/6$, whereas distinguishing between even or odd leads to Bernoulli trials with $p = 1/2$. If the die is unbalanced, the succes-

[1] James Bernoulli (1654–1705). His main work, the *Ars Conjectandi*, was published in 1713.

sive throws still form Bernoulli trials, but the corresponding probabilities p are different. A royal flush in poker or double ace in rolling 2 dice may represent success; calling all other outcomes failure, we have Bernoulli trials with $p = 1/649,740$ and $p = 1/36$ respectively. Reductions of this type are usual in statistical applications. For example, washers produced in mass production may vary continuously in thickness, but, on inspection, they are classified as conforming (S) or defective (F) according as their thickness is, or is not, within prescribed limits.

The Bernoulli scheme of trials is a theoretical model, and only experience can show whether it is suitable for the description of specified physical experiments. Our knowledge that successive tossings of a coin conform to the Bernoulli scheme is derived from experimental evidence. The man in the street, and also the philosopher K. Marbe,[2] believe that after a run of seventeen heads tail becomes more probable. This argument has nothing to do with imperfections of physical coins; it endows nature with memory, or, in our terminology, it denies the statistical independence of successive trials. Marbe's theory cannot be refuted by logic but has been rejected because of lack of empirical support.

In sampling practice, industrial quality control, etc., the scheme of Bernoulli trials provides an ideal standard even though it can never be fully attained. Thus, in the above example of the production of washers, there are many reasons why the output cannot conform to the Bernoulli scheme. The machines are subject to changes, and hence the probabilities do not remain constant; there is a persistence in the action of machines, and therefore long runs of deviations of like kind are more probable than they would be if the trials were truly independent. From the point of view of quality control, however, it is desirable that the process conform to the Bernoulli scheme, and it is an important discovery that, within certain limits, production can be made to behave in that way. The purpose of continuous control is then to discover at an early stage flagrant departures from the ideal scheme and to use them as an indication of impending trouble.

2. The Binomial Distribution

Frequently we are interested only in the total number of successes produced in a succession of n Bernoulli trials but not in their order. The number of successes can be 0, 1, $\cdots$, n, and our first problem is

[2] *Die Gleichförmigkeit in der Welt*, Munich, 1916. There exists a huge critical literature on Marbe's theory.

to determine the corresponding probabilities. Now the event "n trials result in k successes and $n - k$ failures" can happen in as many ways as k letters S can be distributed among n places. In other words, our event contains $\binom{n}{k}$ points, and, by definition, each point has the probability $p^k q^{n-k}$. This proves the

Theorem. *Let $b(k; n, p)$ be the probability that n Bernoulli trials with probabilities p for success and $q = 1 - p$ for failure result in k successes and $n - k$ failures ($0 \leq k \leq n$). Then*

$$(2.1) \qquad b(k; n, p) = \binom{n}{k} p^k q^{n-k}.$$

In particular, the probability of no success is q^n, and the probability of at least one success is $1 - q^n$.

The number of successes in n Bernoulli trials will later be considered as a special random variable (cf. Chapter 9). In the general terminology, the function (2.1) is the "distribution" of this random variable, and we shall refer to it as the *binomial distribution*. The attribute "binomial" refers to the fact that (2.1) represents the kth term of the binomial expansion of $(p + q)^n$. This remark shows also that $b(0; n, p) + b(1; n, p) + \cdots + b(n; n, p) = (p + q)^n = 1$, as is required by the notion of probability. Formula (2.1) is sometimes called Newton's theorem.

Examples. (*a*) The probabilities of $0, 1, \cdots, 6$ heads in 6 tosses of a symmetric coin are $1/64$, $6/64$, $15/64$, $20/64$, $15/64$, $6/64$, $1/64$, respectively. These are also the probabilities of having $0, 1, \cdots, 6$ odd digits among 6 random digits. Again, the same figures can, in a rough approximation, be interpreted as probabilities that among 6 children $0, 1, \cdots, 6$ will be girls.

(*b*) In 10 throws of an ideal die the probability of exactly one ace is $10(1/6)(5/6)^9 = 0.323011\ldots$. The probability of at least one ace is $1 - (5/6)^{10} = 0.838494\ldots$. The probability of more than one ace is the difference, or $0.515483\ldots$.

(*c*) *Weldon's Dice Data.* Let an experiment consist in throwing 12 dice and let us count fives and sixes as "success." If the dice are perfect, the probability of success is $p = 1/3$, and the number of successes should follow the binomial distribution $b(k; 12, 1/3)$. Table 1 gives these probabilities, together with the corresponding observed average frequencies in 26,306 actual experiments. The agreement looks good, but for such extensive data it is really very bad. Statisticians usually judge closeness of fit by the chi-square criterion. According to it, deviations as large as those observed would happen with true dice only four times out of 10,000. It is, therefore, reasonable to assume that the dice were biased. A bias with probability of success $p = 0.3377$ fits the observations.[3]

TABLE 1

WELDON'S DICE DATA

k	$b(k; 12, 1/3)$	Observed Frequency	$b(k; 12, 0.3377)$
0	0.007 707	0.007 033	0.007 123
1	.046 244	.043 678	.043 584
2	.127 171	.124 116	.122 225
3	.211 952	.208 127	.207 736
4	.238 446	.232 418	.238 324
5	.190 757	.197 445	.194 429
6	.111 275	.116 589	.115 660
7	.047 689	.050 597	.050 549
8	.014 903	.015 320	.016 109
9	.003 312	.003 991	.003 650
10	.000 497	.000 532	.000 558
11	.000 045	.000 152	.000 052
12	.000 002	.000 000	.000 002

(d) We have encountered several special cases of the binomial distribution. For example, the probability of the digit 7 appearing among n random digits exactly k times is given by (2.1) with $p = 1/10$. Table 1 of Chapter 3 gives the binomial distribution with $n = 100$, $p = 1/10$, together with actual counts of the occurrence of the digit 7. In Chapter 4, section 4, we applied the binomial distribution to a card-guessing problem, and the columns b_n of Table 1 give the terms of the binomial distributions with $n = 3, 4, 5, 6, 10$ and $p = 1/n$. In the problem of random distributions of objects in cells we found formula (3.1) of Chapter 3, which is again a special case of the binomial distribution for $p = 1/n$.

(e) If the probability of success is 0.01, how many trials are necessary in order for the probability of at least one success to be $1/2$ or more? Here we seek the smallest integer n for which $1 - (0.99)^n \geq 1/2$, or $-n \log (0.99) \geq \log 2$; therefore $n \geq 70$.

[3] R. A. Fisher, *Statistical Methods for Research Workers*, Edinburgh-London, 1932, p. 66, or T. C. Fry, *Probability and Its Engineering Uses*, New York, 1928, pp. 303ff.

(f) *Power-supply Problems.* Typical of many applications is the following problem. A motor generates power to be used intermittently by n workers. To get a crude idea of the load to be expected we imagine that at any given time each worker has the same probability p of requiring a unit of power. If they work independently, the probability of exactly k workers requiring power at the same time should be $b(k; n, p)$. If, on the average, a worker uses power for 12 minutes per hour, one would put $p = 1/5$. The binomial distribution is used in practice to compute the probability of undesirable overloads or, what amounts to the same thing, to find an optimum number n of workers per motor. A finer analysis of the situation would require knowledge of the frequency of overloads, their average duration, etc. These problems necessitate the consideration of the process with regard to time dependence and lead to the theory of stochastic ($=$ random) processes.

(g) *Sampling.* If a population consists of a red and b black elements, then random sampling with replacement conforms to the Bernoulli scheme, and the probability that a sample of size n will contain exactly k red elements is given by the binomial distribution (2.1) with $p = a/(a + b)$. In sampling without replacement the corresponding probability is given by the hypergeometric distribution

$$(2.2) \qquad \frac{\binom{a}{k}\binom{b}{n-k}}{\binom{a+b}{n}}$$

[cf. (5.1) of Chapter 2]. When the population size $a + b$ is large compared to n, the probability (2.2) is close to $b(k; n, a/(a + b))$, so that the binomial distribution applies *approximately* also to sampling without replacement from large populations (cf. problem 42 of section 8 of Chapter 2). This fact is used in many ways. For example, in industrial quality control, lots containing N items each are subjected to sampling inspection with the aim of eliminating all lots containing an excess of defective items. The probability of a sample of size n containing k defective items is given by (2.2), but this formula is cumbersome and is usually replaced by the binomial distribution, even though sampling without replacement is used.

(h) *Banach's Match-box Problem.*[4] A certain mathematician always carries two match boxes; every time he wants a match, he selects a box at random. Inevitably a moment occurs when, for the first time, he

---

[4] Communicated by H. Steinhaus.

finds a box empty. At that moment the other box may contain $r = 0, 1, 2, \ldots$, matches, and we wish to find the corresponding probabilities u_r. Assume that initially each box contained N matches. The event "when first box is found empty, the second contains r matches" means that out of the first $N + (N - r)$ matches drawn, N are from the first box. This corresponds to N successes in $2N - r$ trials, and therefore

$$(2.3) \qquad u_r = \binom{2N - r}{N} \frac{1}{2^{2N-r}}.$$

[Cf. Table 2. The discussion of the problem is continued in example (3.f) of Chapter 9.]

TABLE 2
Probabilities (2.3)

r	u_r	U_r	r	u_r	U_r
0	0.079 589	0.079 589	15	0.023 171	0.917 941
1	.079 589	.159 178	16	.019 081	.937 022
2	.078 785	.237 963	17	.015 447	.952 469
3	.077 177	.315 140	18	.012 283	.964 752
4	.074 790	.389 931	19	.009 587	.974 338
5	.071 674	.461 605	20	.007 338	.981 676
6	.067 902	.529 506	21	.005 504	.987 180
7	.063 568	.593 073	22	.004 041	.991 220
8	.058 783	.651 855	23	.002 901	.994 121
9	.053 671	.705 527	24	.002 034	.996 155
10	.048 363	.753 890	25	.001 392	.997 547
11	.042 989	.796 879	26	.000 928	.998 475
12	.037 676	.834 555	27	.000 602	.999 077
13	.032 538	.867 094	28	.000 379	.999 456
14	.027 676	.894 770	29	.000 232	.999 688

u_r is the probability that, at the moment the first match box is found empty, the second contains exactly r matches, assuming that initially each box contained 50 matches. $U_r = u_0 + u_1 + \cdots + u_r$ is the corresponding probability of having not more than r matches.

3. The Central Term

From (2.1) we see that

$$(3.1) \qquad \frac{b(k; n, p)}{b(k - 1; n, p)} = \frac{(n - k + 1)p}{kq} = 1 + \frac{(n + 1)p - k}{kq}.$$

This ratio is greater than or less than one according as k is less than or greater than $(n+1)p$. Accordingly, the term $b(k; n, p)$ is greater than the preceding one if $k < (n+1)p$ and is smaller if $k > (n+1)p$. If $(n+1)p = m$ happens to be an integer, then $b(m; n, p) = b(m-1; n, p)$. In general, *if m is the largest integer not exceeding $(n+1)p$, then as k goes from 0 to n the terms $b(k; n, p)$ first increase and then decrease, reaching their greatest value for $k = m$.* If $(n+1)p$ is an integer, there are two equal terms. We shall call $b(m; n, p)$ the *central term*. Often m is called "the most probable number of successes," but it must be remembered that for large values of n all terms $b(k; n, p)$ are small so that even the most probable number of successes has only a small probability. In 100 tossings of a true coin the most probable number of heads is 50, but its probability is only $0.079589\ldots$.

In the next chapter we shall find that the central term $b(m; n, p)$ is approximately $(2\pi npq)^{-\frac{1}{2}}$. There we shall also deduce simple approximations to other terms of the binomial distribution and to their sums. These approximations enable us to evaluate probabilities whose direct computation is cumbersome, e.g., the probability that in 1000 tossings of a coin the number of heads is between 485 and 508.

4. The Poisson Approximation [5]

In many applications we deal with Bernoulli trials where, comparatively speaking, n is large and p is small, whereas the product

$$(4.1) \qquad \lambda = np$$

is of moderate magnitude. In such cases it is convenient to use an approximation formula to $b(k; n, p)$ which is due to Poisson and which we proceed to derive. We have $b(0; n, p) = (1-p)^n$ or, substituting from (4.1),

$$(4.2) \qquad b(0; n, p) = \left(1 - \frac{\lambda}{n}\right)^n.$$

Passing to logarithms and using the Taylor expansion [formula (6.9) of Chapter 2] we find

$$(4.3) \qquad \log b(0; n, p) = n \log\left(1 - \frac{\lambda}{n}\right) = -\lambda - \frac{\lambda^2}{2n} - \ldots,$$

so that for large n,

$$(4.4) \qquad b(0; n, p) \approx e^{-\lambda}.$$

---

[5] Siméon D. Poisson (1781–1840). His book, *Recherches sur la probabilité des jugements en matière criminelle et en matière civile, précédées des règles générales du calcul des probabilités*, appeared in 1837.

Furthermore

(4.5) $$b(1; n, p) = \frac{np}{q} b(0; n, p) = \frac{\lambda}{1 - \lambda/n} b(0; n, p),$$

and therefore

(4.6) $$b(1; n, p) \approx \lambda e^{-\lambda}.$$

Similarly

(4.7) $$b(2; n, p) = \frac{n(n-1)p^2}{2q^2} b(0; n, p) \approx \frac{\lambda^2}{2} e^{-\lambda}.$$

Continuing in this way, it is seen that for sufficiently large n and any fixed λ

(4.8) $$b(k; n, p) \approx \frac{\lambda^k}{k!} e^{-\lambda}.$$

This is the famous *Poisson approximation to the binomial distribution*. For any fixed λ the error committed in (4.8) tends to zero with increasing n. We shall see that the error is of the order of magnitude λ^2/n so that (4.8) is of practical use if λ^2 is small as compared to n. Before discussing the error term we proceed to illustrate the use of (4.8) and to calculate the error involved in it in a few special cases. It is convenient to have a symbol for the right-hand member in (4.8), and we shall put

(4.9) $$p(k; \lambda) = e^{-\lambda} \frac{\lambda^k}{k!}.$$

With this notation $p(k; \lambda)$ should be an approximation to $b(k; n, \lambda/n)$ when n is sufficiently large.

Examples. (a) The entries p_m of the last column of Table 1 in Chapter 4 give the values $p(m; 1)$. In the preceding columns b_m stands for $b(m; N, 1/N)$. The table enables us to compare the Poisson distribution $p(m; 1)$ with the binomial distributions with $p = 1/n$ and $n = 3, 4, 5, 6, 10$. It will be seen that the agreement is surprisingly good despite the small values of n.

(b) Table 3 compares $p(k; 1)$ to the binomial distribution with $n = 100$, $p = 1/100$. It shows the approximation to be satisfactory for many purposes. As an example take the occurrence of the combination (7, 7) among 100 pairs of random digits, which should have the binomial distribution $b(k; 100, 1/100)$. The last column of Table 3 gives the actual counts in 100 batches of 100 random digits each recorded in Table 2 of Chapter 2. To obtain relative frequencies all entries of the last column should be divided by 100. These frequencies agree reasonably with the theoretical probabilities. (As judged by the χ^2-criterion, chance fluctuations should, in about 75 out of 100 similar cases, produce larger deviations of observed frequencies from the theoretical probabilities.)

TABLE 3

An Example of the Poisson Approximation

k	$b(k; 100, 1/100)$	$p(k; 1)$	N_k
0	0.366 032	0.367 879	41
1	.369 730	.367 879	34
2	.184 865	.183 940	16
3	.060 999	.061 313	8
4	.014 942	.015 328	0
5	.002 898	.003 066	1
6	.000 463	.000 511	0
7	.000 063	.000 073	0
8	.000 007	.000 009	0
9	.000 001	.000 001	0

The first columns illustrate the Poisson approximation to the binomial distribution. The last column records the number of batches of 100 pairs of random digits each from Table 2 of Chapter 2 in which the combination (7, 7) appears exactly k times.

(c) *Birthdays.* What is the probability, p_k, that in a company of 500 people exactly k will have birthdays on New Year's day? If the 500 people are chosen at random, we may apply the scheme of 500 Bernoulli trials with probability of success $p = 1/365$. Then $p_0 = (364/365)^{500} = 0.2537\ldots$ For the Poisson approximation we put $\lambda = 500/365 = 1.3699\ldots$ Then $p(0; \lambda) = 0.2541$, which involves an error only in the fourth decimal place. For $k = 1, 2, \ldots$ the correct values of p_k as calculated from the binomial formula are $p_1 = 0.3484\ldots$, $p_2 = 0.2388\ldots$, $p_3 = 0.1089\ldots$, $p_4 = 0.0372\ldots$, $p_5 = 0.0101\ldots$, $p_6 = 0.0023\ldots$. The corresponding Poisson approximations are $p(1; \lambda) = 0.3481\ldots$, $p(2; \lambda) = 0.2385\ldots$, $p(3; \lambda) = 0.1089\ldots$, $p(4; \lambda) = 0.0373\ldots$, $p(5; \lambda) = 0.0102\ldots$, $p(6; \lambda) = 0.0023\ldots$. All errors are in the fourth decimal place.

(d) *Defective Items.* Suppose that screws are produced under statistical quality control so that it is legitimate to apply the Bernoulli scheme of trials. If the probability of a screw being defective is $p = 0.015$, then the probability that a box of 100 screws contains no defective is $(0.985)^{100} = 0.22061$. The corresponding Poisson approximation is $e^{-1.5} = 0.22313\ldots$, which should be close enough for most practical purposes. We now ask: how many screws should a box contain in order that the probability of finding at least 100 conforming screws be 0.8 or better? If $100 + x$ is the required number, then x is a small integer. To apply the Poisson approximation for $n = 100 + x$ trials we should put $\lambda = np$, but np is approximately $100p = 1.5$. We then require the smallest integer x for which

$$(4.10) \qquad e^{-1.5}\left\{1 + \frac{1.5}{1} + \cdots + \frac{(1.5)^x}{x!}\right\} \geq 0.8.$$

In tables [6] we find that for $x = 1$ the left side is approximately 0.56, while for $x = 2$ it is 0.809. Thus the Poisson approximation would lead to the conclusion that 102 screws are required. Since 0.809 is dangerously near the given threshold of 0.8, the number 103 is safer. Actually the probability of finding at least 100 conforming screws in a box of 102 is

$$(0.985)^{102} + \binom{102}{1}(0.985)^{101}(0.015) + \binom{102}{2}(0.985)^{100}(0.015)^2$$
$$= 0.8022\ldots.$$

(e) *Centenarians.* At birth any particular person has a small chance of living 100 years, and in a large community the number of yearly births is large. Owing to wars, epidemics, etc., different lives are not statistically independent, but as a first approximation we may compare n births to n Bernoulli trials with death after 100 years as success. In a stable community, where neither size nor mortality rate changes appreciably, it is reasonable to expect that the frequency of years in which exactly k centenarians die is approximately $p(k; \lambda)$, with λ depending on the size and health of the community. Records of Switzerland confirm this conclusion.[7]

(f) *Misprints, Raisins, etc.* If in printing a book there is a constant probability of any letter being misprinted, and if the conditions of printing remain unchanged, then we have as many Bernoulli trials as there are letters. The frequency of pages containing exactly k misprints will then be approximately $p(k; \lambda)$, where λ is a characteristic of the printer. Occasional fatigue of the printer, difficult passages, etc., will increase the chances of errors and may produce clusters of mis-

---

[6] E. C. Molina, *Poisson's Exponential Binomial Limit*, New York, 1942. (These are tables giving $p(k; \lambda)$ and $p(k; \lambda) + p(k + 1; \lambda) + \ldots$ for k ranging from 0 to 100.)

[7] E. J. Gumbel, *Les centenaires*, Aktuárske Vědy, Prague, vol. 7 (1937), pp. 1–8.

prints. Thus the Poisson formula may be used to discover radical departures from uniformity or from the state of statistical control. A similar argument applies in many cases. For example, if many raisins are distributed in the dough, we should expect that thorough mixing will result in the frequency of loaves with exactly k raisins to be approximately $p(k; \lambda)$ with λ a measure of the density of raisins in the dough.

(g) *Poisson Approximation to the Hypergeometric Distribution.* In the sampling problem (g) of section 2 we have used the binomial distribution as an approximation to the hypergeometric distribution (2.2). Now if the number a of red elements is small as compared to the number b of black elements, then the probability $p = a/(a + b)$ of "success" is small, and we can in turn approximate $b(k; n, p)$ by the Poisson expressions $p(k; \lambda)$. In this case the ratio $a/(a + b)$ is approximately the same as a/b, and we conclude therefore: *if a is small as compared to b and $\lambda = na/b$ is of moderate magnitude, then the hypergeometric distribution (2.2) can be approximated by the Poisson expressions $p(k; \lambda)$.* It is easy to verify this result directly by the method used to derive (4.8).

As a specific example suppose that out of the crop of a grain field seeds are taken for the planting of a new field of equal size. A particular plant may be mother to $k = 0, 1, 2, \ldots$ new plants, and we require the corresponding probabilities. If there are n plants in all, we may consider the seeds as a random sample of size n from the population of all seeds. Suppose that the given plant has a seeds, and that there are $a + b$ seeds in all. Then formula (2.2) applies. If the number of plants is large, then a will be small as compared to b, and the probability that our plant is mother to exactly k plants is, approximately, $p(k; \lambda)$ with $\lambda = na/b$.

Estimate of the Error. In practice we are not so much interested in the absolute error $|b(k; n, \lambda/n) - p(k; \lambda)|$ as in the relative error

(4.11) $$\Delta = \frac{b(k; n, \lambda/n)}{p(k; \lambda)} - 1.$$

Now, using (2.1) and (4.9), it is readily seen that

(4.12) $$\frac{b(k; n, \lambda/n)}{p(k; \lambda)} = \left(1 - \frac{1}{n}\right)\left(1 - \frac{2}{n}\right) \cdots \left(1 - \frac{k-1}{n}\right) e^\lambda \left(1 - \frac{\lambda}{n}\right)^{n-k}.$$

To estimate the product to the right we pass to logarithms and use the fact that all terms in the Taylor expansion [cf. formula (6.9) of Chapter 2] of $\log(1 - x)$ are negative. It follows that for $x > 0$

(4.13) $$\log(1-x) < -x - \frac{x^2}{2} < -x,$$

and also

(4.14) $$\log(1-x) > -x - \frac{x^2}{2(1-x)} > \frac{-x}{1-x}.$$

Hence

$$\log \frac{b(k; n, \lambda/n)}{p(k; \lambda)} < -\frac{1 + 2 + \cdots + k - 1}{n} + \lambda - (n-k)\left(\frac{\lambda}{n} + \frac{\lambda^2}{2n^2}\right)$$

(4.15) $$= -\frac{k(k-1)}{2n} - \frac{\lambda^2}{2n} + \frac{k\lambda}{n} + \frac{k\lambda^2}{2n^2}$$

$$= -\frac{(k-\lambda)^2}{2n} + \frac{k}{2n} + \frac{k\lambda^2}{2n^2}.$$

On the other hand,

$$\log \frac{b(k; n, \lambda/n)}{p(k; \lambda)}$$

(4.16) $$> -\frac{1 + 2 + \cdots + (k-1)}{n - k + 1} + \lambda - n\left(\frac{\lambda}{n} + \frac{\lambda^2}{2n(n-\lambda)}\right) + \frac{k\lambda}{n}$$

$$\geq -\frac{(k-\lambda)^2}{2(n-k+1)} + \frac{(k-\lambda-1)\lambda^2}{2(n-\lambda)(n-k+1)} - \frac{k(k-1)\lambda}{n(n-k+1)}.$$

It follows that *if λ is restricted to a finite interval the relative error* (4.11) *tends uniformly to zero as* $(k - \lambda)^2/n \to 0$. The relative error (4.11) can be appreciable only if $(k - \lambda)^2/n$ is not small. However, $p(k; \lambda)$ is largest when k is near λ, and when $k > 2\lambda$ the terms $p(k; \lambda)$ decrease faster than a geometric series with ratio 1/2. Also, it follows from (4.15) that for $k > 2\lambda + 1$ the Poisson approximation overestimates $b(k; n, \lambda/n)$. Hence, if $(k - \lambda)^2/n$ is not small, both $p(k; \lambda)$ and $b(k; n, \lambda/n)$ are negligible, and we conclude that *in every finite λ-interval as $n \to \infty$ the difference $b(k; n, \lambda/n) - p(k; \lambda)$ tends to zero uniformly for all k.*

5. The Poisson Distribution

In the preceding section we have used the Poisson expression (4.9) merely as a convenient approximation to the binomial distribution in the case of large n and small p. In connection with the matching problem and the occupancy problem of Chapter 4 (sections 4 and 5) we have studied quite different probability distributions, which also led to the Poisson expressions $p(k; \lambda)$ as a limiting form. We have here a special case of the remarkable fact that there exist a few distributions of great universality which occur in a surprisingly great variety of problems. The three principal distributions, with ramifications throughout probability theory, are the binomial distribution, the normal distribution (to be introduced in the following chapter), and the *Poisson distribution*

$$p(k;\lambda) = e^{-\lambda}\frac{\lambda^k}{k!}, \tag{5.1}$$

which we shall now consider on its own merits.

We note first that on adding the equations (5.1) for $k = 0, 1, 2, \ldots$ we get on the right side $e^{-\lambda}$ times the Taylor series for e^λ. Hence for any fixed λ the quantities $p(k; \lambda)$ add to unity, and therefore it is possible to conceive of an ideal experiment in which $p(k; \lambda)$ is the probability of exactly k successes. We shall now indicate why many physical experiments and statistical observations actually lead to such an interpretation of (5.1). The examples of the next section will illustrate the wide range and the importance of various applications of (5.1). The true nature of the Poisson distribution will become apparent only in connection with the theory of stochastic processes (cf. Chapter 17, where a new approach to the Poisson distribution is given).

Consider a sequence of random events occurring in time, such as radioactive disintegrations or incoming calls at a telephone exchange. Each event is represented by a point on the time axis, and we are concerned with chance distributions of points. There exist many different types of such distributions, but their study belongs to the domain of continuous probabilities which we have postponed to the second volume. Here we shall be content to show that the simplest physical assumptions lead to $p(k; \lambda)$ as the probability of finding exactly k points (events) within a fixed interval of specified length. Our methods are necessarily crude, and we shall return to the same problem with more adequate methods in Chapter 17.

The physical assumptions which we want to express mathematically are that the conditions of the experiment remain constant in time, and that non-overlapping time intervals are statistically independent in the sense that information concerning the number of events in one interval reveals nothing about the other. The theory of probabilities in a continuum makes it possible to express these statements directly, but being restricted to discrete probabilities, we have to use an approximate finite model and pass to the limit.

Imagine the unit time interval divided into a great number n of intervals, each of length $1/n$. Either a particular subinterval is empty, or it contains at least one of our random points (or events), and we agree to call the two possibilities failure and success, respectively. The probability p_n of success must be the same for all n subintervals, since they have the same length. The assumed independence of non-overlapping intervals then implies that we have n Bernoulli trials, and the probability of exactly k successes is given by $b(k; n, p_n)$. Now the number of successes is not necessarily the same as the number of random points, since a subinterval may contain several random points. However, it is natural to introduce the additional assumption that the probability of two or more random points during a very short time interval is in the limit negligible.[8] In this case the probability of finding exactly k random points in the unit time interval is given by the limit of $b(k; n, p_n)$ as $n \to \infty$. Now when we divide each subinterval into two parts of equal length, we find that $p_n = 2p_{2n} - p_{2n}^2$; this equation states that success in an interval of length $1/n$ means either success in the left half, or success in the right half, or in both. It follows that $p_n < 2p_{2n}$, and this suggests that np_n increases monotonically (which can be proved rigorously). If $np_n \to \lambda$, then $b(k; n, p_n) \sim b(k; n, \lambda/n) \to p(k; \lambda)$, and we find (5.1) as the probability that there is a total of k random points contained in our unit interval. The assumption $np_n \to \infty$ leads to no sensible result, as it would imply infinitely many random points even in the smallest interval.

If, instead of the unit interval, we take an arbitrary interval of length t and again use a subdivision into intervals of length $1/n$, then we have Bernoulli trials with the same probability p_n of success, but the number of trials is the integer nearest to nt rather than n. The passage to the limit is the same, but we get λt instead of λ. This leads us to consider

$$(5.2) \qquad p(k; \lambda t) = e^{-\lambda t} \frac{(\lambda t)^k}{k!}$$

as the probability of finding exactly k points in a fixed interval of length t. In particular, the probability of no point in an interval of length t is

$$(5.3) \qquad p(0; \lambda t) = e^{-\lambda t}.$$

and the probability of one or more points is therefore $1 - e^{-\lambda t}$.

The parameter λ is a physical constant which determines the density of points on the t-axis. The larger λ, the smaller is the probability (5.3) of finding no point. Suppose that a physical experiment is repeated a great number N of times and that each time we count the

---

[8] This assumption is implicit in the intuitive picture of isolated random points. However, it is necessary to exclude the possibility of our events appearing in doublets. For example, if the events are automobile accidents, then the probability of two events within a short time is negligible in comparison with the probability of one event. On the other hand, an accident is likely to involve two cars, and if the events mean "a car smashed" then they are likely to appear in pairs and our assumption does not apply.

number of events in an interval of fixed length t. Let N_k be the number of times that exactly k events are observed. Then

(5.4) $$N_0 + N_1 + N_2 + \ldots = N.$$

The total number of points observed in the N experiments is

(5.5) $$N_1 + 2N_2 + 3N_3 + \ldots = T,$$

and T/N is the average. Now if N is large, we expect that

(5.6) $$N_k \approx Np(k; \lambda t)$$

(this lies at the root of all applications of probability and will be justified and made more precise by the law of large numbers in Chapter 10). Substituting from (5.6) into (5.5), we find

$$T \approx N\{p(1; \lambda t) + 2p(2; \lambda t) + 3p(3; \lambda t) + \ldots\}$$

(5.7) $$= Ne^{-\lambda t}\lambda t \left\{1 + \frac{\lambda t}{1} + \frac{(\lambda t)^2}{2!} + \ldots\right\} = N\lambda t$$

and hence

(5.8) $$\lambda t \approx \frac{T}{N}.$$

This relation gives us a means of estimating λ from observations and of comparing theory with experiments. The examples of the next section will illustrate this point.

Spatial Distributions. We have considered the distribution of random events or points along the t-axis, but the same argument applies to the distribution of points in plane or space. Instead of intervals of length t we have domains of area or volume t, and the fundamental assumption is that the probability of finding k points in any specified domain depends only on the area or volume of the domain, but not on its shape. Otherwise we have the same assumptions as before: (1) if t is small, the probability of finding more than one point in a domain of volume t is small as compared to t; (2) non-overlapping domains are mutually independent. To find the probability that a domain of volume t contains exactly k random points we subdivide it into n subdomains and approximate the required probability by the probability of k successes in n trials. This means neglecting the possibility of finding more than one point in the same subdomain, but our assumption (1) implies that the error tends to zero as $n \to \infty$. In the limit we get again the Poisson distribution (5.2). Stars in space, raisins in cake, weed seeds among grass seeds, flaws in materials, animal litters in fields are distributed in accordance with the Poisson law. For numerical examples cf. section 6, examples (b) and (e).

6. Observations Fitting the Poisson Distribution [9]

(a) *Radioactive Disintegrations.* A radioactive substance emits α-particles, and the number of particles reaching a given portion of space during time t is the best-known example of random events obeying the Poisson law. Of course, the substance continues to decay, and in the long run the density of α-particles will decline. However, with radium it takes years before a decrease of matter can be detected; for relatively short periods the conditions must be considered constant, and we have an ideal realization of the hypotheses which led to the Poisson distribution.

In a famous experiment [10] a radioactive substance was observed during $N = 2608$ time intervals of 7.5 seconds each; the number of particles reaching a counter was obtained for each period. Table 4 records the number N_k of periods with exactly k particles. The total number of particles is $T = \Sigma k N_k = 10{,}094$, the average $T/N = 3.870$. The theoretical values $Np(k; 3.870)$ are seen to be rather close to the observed numbers N_k. To judge the closeness of fit, an estimate of the probable magnitude of chance fluctuations is required. Statisticians judge the closeness of fit by the χ^2-criterion. Measuring by this

TABLE 4

EXAMPLE (a): RADIOACTIVE DISINTEGRATIONS

k	N_k	$Np(k; 3.870)$
0	57	54.399
1	203	210.523
2	383	407.361
3	525	525.496
4	532	508.418
5	408	393.515
6	273	253.817
7	139	140.325
8	45	67.882
9	27	29.189
$k \geq 10$	16	17.075
Total	2608	2608.000

[9] The Poisson distribution has become known as the law of small numbers or of rare events. These are misnomers which proved detrimental to the realization of the fundamental role of the Poisson distribution. The following examples will show how misleading the two names are.

[10] Rutherford, Chadwick, and Ellis, *Radiations from Radioactive Substances*, Cambridge, 1920, p. 172. Table 4 and the χ^2-estimate of the text are taken from H. Cramér, *Mathematical Methods of Statistics*, Upsala and Princeton, 1945, p. 436.

standard one should expect that under ideal conditions about 17 out of 100 comparable cases show worse agreement than exhibited in Table 4.

(b) *Flying-bomb Hits on London.* As an example of a spatial distribution of random points consider the statistics of flying-bomb hits in the south of London during World War II. The entire area is divided into $N = 576$ small areas of $t = \frac{1}{4}$ square kilometers each, and Table 5 records the number N_k of areas with exactly k hits.[11] The total number of hits is $T = \Sigma k N_k = 537$, the average $\lambda t = T/N = 0.9323 \ldots$. The fit of the Poisson distribution is surprisingly good: as judged by the χ^2-criterion, under ideal conditions some 88 per cent of comparable observations should show a worse agreement. It is interesting to note that most people believed in a tendency of the points of impact to cluster. In this case we would have a higher frequency of areas with either many hits or no hit, and a deficiency in the intermediate classes. Table 5 indicates perfect randomness and homogeneity of the area; we have here an instructive illustration of the established fact that to the untrained eye randomness appears as regularity or tendency to cluster.

TABLE 5

EXAMPLE (b): FLYING-BOMB HITS ON LONDON

k	0	1	2	3	4	5 and over
N_k	229	211	93	35	7	1
$p(k; 0.9323)$	226.74	211.39	98.54	30.62	7.14	1.57

(c) *Chromosome Interchanges in Cells.* Irradiation by X-rays produces certain processes in organic cells which we call chromosome interchanges. As long as radiation continues, the probability of such interchanges remains constant, and, according to theory, the numbers N_k of cells with exactly k interchanges should follow a Poisson distribution. The theory is also able to predict the dependence of the parameter λ on the intensity of radiation, the temperature, etc., but we shall not enter into these details. Table 6 records the result of eleven different series of experiments.[12] These are arranged according

[11] The figures are taken from R. D. Clarke, An Application of the Poisson Distribution, *Journal of the Institute of Actuaries*, vol. 72 (1946), p. 48.

[12] D. G. Catcheside, D. E. Lea, and J. M. Thoday, Types of Chromosome Structural Change Induced by the Irradiation of *Tradescantia* Microspores, *Journal of Genetics*, vol. 47 (1945–46), pp. 113–136. Our table is Table IX of this paper, except that the χ^2-levels were recomputed, using a single degree of freedom.

TABLE 6

EXAMPLE (c): CHROMOSOME INTERCHANGES INDUCED BY X-RAY IRRADIATION

Experiment Number		Cells with k Interchanges				Total N	χ^2-Level in Per Cent
		0	1	2	≥ 3		
1	Observed N_k	753	266	49	5	1073	95
	$Np(k; 0.35508)$	752.3	267.1	47.4	6.2		
2	Observed N_k	434	195	44	9	682	85
	$Np(k; 0.45601)$	432.3	197.1	44.9	7.7		
3	Observed N_k	280	75	12	1	368	65
	$Np(k; 0.27717)$	278.9	77.3	10.7	1.1		
4	Observed N_k	2278	273	15	0	2566	65
	$Np(k; 0.11808)$	2280.2	269.2	15.9	0.7		
5	Observed N_k	593	143	20	3	759	45
	$Np(k; 0.25296)$	589.4	149.1	18.8	1.7		
6	Observed N_k	639	141	13	0	793	45
	$Np(k; 0.21059)$	642.4	135.3	14.2	1.1		
7	Observed N_k	359	109	13	1	482	40
	$Np(k; 0.28631)$	362.0	103.6	14.9	1.5		
8	Observed N_k	493	176	26	2	697	35
	$Np(k; 0.33572)$	498.2	167.3	28.1	3.4		
9	Observed N_k	793	339	62	5	1199	20
	$Np(k; 0.39867)$	804.8	320.8	64.0	9.4		
10	Observed N_k	579	254	47	3	883	20
	$Np(k; 0.40544)$	588.7	238.7	48.4	7.2		
11	Observed N_k	444	252	59	1	756	5
	$Np(k; 0.49339)$	461.6	227.7	56.2	10.5		

to goodness of fit. The last column indicates the approximate percentage of ideal cases in which chance fluctuations would produce a worse agreement (as judged by the χ^2-standard). It is difficult to imagine a better agreement between theory and observation.

(d) *Connections to Wrong Number.* Table 7 shows statistics of telephone connections to a wrong number.[13] A total of $N = 267$

TABLE 7

EXAMPLE (d): CONNECTIONS TO WRONG NUMBER

k	N_k	$Np(k; 8.74)$
0–2	1	2.05
3	5	4.76
4	11	10.39
5	14	18.16
6	22	26.45
7	43	33.03
8	31	36.09
9	40	35.04
10	35	30.63
11	20	24.34
12	18	17.72
13	12	11.92
14	7	7.44
15	6	4.33
≥ 16	2	4.65
	267	267.00

numbers was observed; N_k indicates how many numbers had exactly k wrong connections. The Poisson distribution $p(k; 8.74)$ shows again an excellent fit. (As judged by the χ^2-criterion the deviations are near the median value.) In Thorndike's paper the reader will find other telephone statistics following the Poisson law. In some cases (as with party lines, calls from groups of coinboxes, etc.) there is an obvious interdependence among the events, and the Poisson distribution no longer fits.

(e) *Bacteria and Blood Counts.* Figure 3 reproduces a photograph of a Petri plate with bacterial colonies, which are visible under the microscope as dark spots.

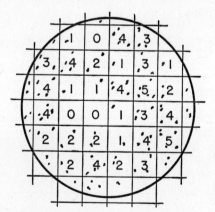

FIGURE 3. *Bacteria on a Petri Plate.*

[13] The observations are taken from F. Thorndike, Applications of Poisson's Probability Summation, *The Bell System Technical Journal*, vol. 5 (1926), pp. 604–624. The paper contains a graphical analysis of 32 different statistics.

TABLE 8
Example (e): Counts of Bacteria

k	0	1	2	3	4	5	6	7	χ^2-Level
Observed N_k	5	19	26	26	21	13	8		97
Poisson theor.	6.1	18.0	26.7	26.4	19.6	11.7	9.5		
Observed N_k	26	40	38	17	7				66
Poisson theor.	27.5	42.2	32.5	16.7	9.1				
Observed N_k	59	86	49	30	20				26
Poisson theor.	55.6	82.2	60.8	30.0	15.4				
Observed N_k	83	134	135	101	40	16	7		63
Poisson theor.	75.0	144.5	139.4	89.7	43.3	16.7	7.4		
Observed N_k	8	16	18	15	9	7			97
Poisson theor.	6.8	16.2	19.2	15.1	9.0	6.7			
Observed N_k	7	11	11	11	7	8			53
Poisson theor.	3.9	10.4	13.7	12.0	7.9	7.1			
Observed N_k	3	7	14	21	20	19	7	9	85
Poisson theor.	2.1	8.2	15.8	20.2	19.5	15	9.6	9.6	
Observed N_k	60	80	45	16	9				78
Poisson theor.	62.6	75.8	45.8	18.5	7.3				

The last entry in each row includes the figures for higher classes and should be labeled "k or more."

The plate is divided into small squares. Table 8 reproduces the observed numbers of squares with exactly k dark spots in eight experiments with as many different kinds of bacteria.[14] The last column again indicates how many per cent of similar experiments conducted under ideal conditions must be expected to show a worse χ^2-agreement. We have here a representative of an important practical application of the Poisson distribution to spatial distributions of random points.

[14] The table is taken from J. Neyman, *Lectures and Conferences on Mathematical Statistics* (mimeographed), Dept. of Agriculture, Washington, 1938. The original (by T. Matuszewski, J. Supinska, and J. Neyman) appeared, together with related material, in *Zentralblatt für Bakteriologie, Parasitenkunde und Infektionskrankheiten*, II Abt., vol. 95 (1936).

7. The Multinomial Distribution

The binomial distribution can easily be generalized to the case of n repeated independent trials where each trial can have one of several outcomes. Denote the possible outcomes of each trial by $E_1, \cdots, E_r$, and suppose that the probability of the realization of E_i in each trial is $p_i (i = 1, \cdots, r)$. For $r = 2$ we have Bernoulli trials; in general, the numbers p_i are subject only to the condition

$$(7.1) \qquad p_1 + \cdots + p_r = 1, \qquad p_i \geq 0.$$

The result of n trials is a succession like $E_3 E_1 E_2 \ldots$. *The probability that in n trials E_1 occurs k_1 times, E_2 occurs k_2 times, etc., is*

$$(7.2) \qquad \frac{n!}{k_1! k_2! \cdots k_r!} p_1^{k_1} p_2^{k_2} p_3^{k_3} \cdots p_r^{k_r};$$

here the k_i are arbitrary non-negative integers subject to the obvious condition

$$(7.3) \qquad k_1 + k_2 + \cdots + k_r = n.$$

If $r = 2$, then (7.2) reduces to the binomial distribution with $p_1 = p$, $p_2 = q$, $k_1 = k$, $k_2 = n - k$. The proof in the general case proceeds along the same lines, starting with formula (4.7) of Chapter 2.

Formula (7.2) is called the *multinomial distribution* because the right-hand member is the general term of the *multinomial* expansion of $(p_1 + \cdots + p_r)^n$. Its main application is to *sampling with replacement* when the individuals are classified into more than two categories (e.g., according to professions).

Examples. (a) In rolling 12 dice, what is the probability of getting each face twice? Here $E_1, \cdots, E_6$ represent the six faces, all k_i equal 2, and all $p_i = 1/6$. Therefore, the answer is $(12!)(2)^{-6}(6)^{-12} = 0.0034\ldots$.

(b) The distribution of 800 random digits corresponds to a multinomial distribution with $r = 10$, $n = 800$, $p_1 = \cdots = p_{10} = 1/10$. Among the 800 first decimals [15] of e the frequencies of occurrence of the digits $0, 1, \cdots, 9$ are: 74, 73, 83, 86, 79, 67, 78, 82, 84, 94. The corresponding frequencies for π are 74, 92, 83, 79, 80, 73, 77, 75, 76, 91. Of course, there is no logical reason why such expansions should have properties of random sampling digits. The fact is that the usual statistical tests would never raise the suspicion that the above figures were not obtained by performing 800 independent trials.

---

[15] The decimals of e are taken from *Intermédiaire des recherches mathématiques*, vol. 2 (1946), p. 112; those of π from *Mathematical Tables and Other Aids to Computation*, vol. 2 (1947), p. 245, and vol. 3 (1948), p. 19.

(c) *Multiple Bernoulli Trials.* Two sequences of Bernoulli trials with probabilities of success and failure p_1, q_1, and p_2, q_2, respectively, may be considered one compound experiment with four possible outcomes in each trial, namely, the combinations (S, S), (S, F), (F, S), (F, F). The assumption that the two original sequences are independent is translated into the statement that the probabilities of the four outcomes are p_1p_2, p_1q_2, q_1p_2, q_1q_2, respectively. If k_1, k_2, k_3, k_4 are four integers adding to n, the probability that in n trials SS will appear k_1 times, SF k_2 times, etc., is

(7.4) $$\frac{n!}{k_1!k_2!k_3!k_4!} p_1^{k_1+k_2} q_1^{k_3+k_4} p_2^{k_1+k_3} q_2^{k_2+k_4}.$$

A special case occurs in *sampling inspection.* An item is conforming or defective with probabilities p and q. It may or may not be inspected with corresponding probabilities p' and q'. The decision of whether an item is inspected is made without knowledge of its quality, so that we have independent trials. [Cf. problems 30 and 31, and example (1.c) of Chapter 9.]

8. Problems for Solution

1. Assuming all sex distributions to be equally probable, what proportion of families with exactly six children should be expected to have three boys and three girls?

2. A bridge player had no ace in three consecutive hands. Has he reason to complain of ill luck?

3. How long has a series of random digits to be in order for the probability of the digit 7 appearing to be at least 9/10?

4. How many independent bridge dealings are required in order for the probability of a preassigned player having four aces at least once to be 1/2 or better? Solve again for some player instead of a given one.

5. If the probability of hitting a target is 1/5 and ten shots are fired independently, what is the probability of the target being hit at least twice?

6. In problem 5, find the conditional probability of the target being hit at least twice, assuming that at least one hit is scored.

7. Find the probability that a hand of 13 bridge cards selected at random contains exactly 2 red cards and compare it with the corresponding probability in Bernoulli trials with $p = 1/2$. (For a description of bridge cards cf. footnote 1, Chapter 1.)

8. What is the probability that the birthdays of six people fall in 2 calendar months leaving exactly 10 months free? (Assume independence and that all months are equally probable.)

9. In rolling 6 true dice, find the probability of obtaining (*a*) at least one, (*b*) exactly one, (*c*) exactly two, aces. Compare with the Poisson approximations.

10. If there are on the average 1 per cent left-handers, estimate the chances of having at least four left-handers among 200 people.

11. A book of 500 pages contains 500 misprints. Estimate the chances that a page contains at least three misprints.

12. Colorblindness appears in 1 per cent of the people in a certain population. How large must a random sample (with replacements) be if the probability of its containing a colorblind person is to be 0.95 or more?

13. In the preceding exercise, what is the probability that a sample of 100 will contain (a) no, (b) two or more, colorblind people?

14. Estimate the number of raisins which a cookie should contain on the average if it is desired that the probability of a cookie to contain at least one raisin be 0.99 or more.

15. The probability of a royal flush in poker is $p = 1/649{,}740$. How large has n to be to render the probability of no royal flush in n games smaller than $1/e \approx 1/3$? (*Note:* No calculations are necessary for the solution.)

---

16. Two people toss a true coin n times each. Find the probability that they will score the same number of heads.

17. In a sequence of Bernoulli trials with probability p for success, find the probability that a successes will occur before b failures. (*Note:* The issue is decided after at most $a + b - 1$ trials. This problem played a role in the classical theory of games in connection with the question of how to divide the pot when the game is interrupted at a moment when one player lacks a points to victory, the other b points.)

18. Show that for the largest term of the multinomial distribution we have $|k_j - np_j| \leq r$ for all j.

19. *Inequalities for the "tails" of the binomial distribution.* Using (3.1), prove

$$(8.1) \qquad b(k; n, p) + b(k+1; n, p) + \cdots + b(n; n, p) \leq \frac{n}{k - np} b(k; n, p)$$

for $k > np + 1$, and

$$(8.2) \qquad b(k; n, p) + b(k-1; n, p) + \cdots + b(0; n, p) \leq \frac{n}{np - k} b(k; n, p)$$

for $k < np$.

(This theorem will be used in Chapter 7, section 2.)

20. *Further inequalities.* As in section 3 let m be the central index. Prove for $k > (n+1)p$

$$(8.3) \qquad \frac{b(k; n, p)}{b(k-1; n, p)} < e^{-\frac{k - (n+1)p}{n}}$$

and hence, by multiplication,

$$(8.4) \qquad b(k; n, p) < 2e^{-\frac{\{k + \frac{1}{2} - (n+1)p\}^2}{2n}} b(m; n, p).$$

21. Without further computations show that (8.4) holds also for $k < np - 1$.

22. Verify the identity

$$(8.5) \qquad \sum_{\nu=0}^{k} b(\nu; n_1, p) b(k - \nu; n_2, p) = b(k; n_1 + n_2, p)$$

and interpret it probabilistically.

Hint: Use formula (5.7) of Chapter 2. Equation (8.5) is a special case of *convolutions*, to be taken up in Chapter 11; another example is (8.6).

23. Verify the identity

(8.6) $$\sum_{\nu=0}^{k} p(\nu; \lambda_1)p(k - \nu; \lambda_2) = p(k; \lambda_1 + \lambda_2).$$

24. Let

(8.7) $$B(k; n, p) = \sum_{\nu=0}^{k} b(\nu; n, p)$$

be the probability of at most k successes in n trials. Then

(8.8) $$B(k; n + 1, p) = B(k; n, p) - pb(k; n, p),$$

$$B(k + 1; n + 1, p) = B(k; n, p) + qb(k + 1; n, p).$$

Verify this (a) from the definition, (b) analytically.

25. With the same notation

(8.9) $$B(k; n, p) = (n - k)\binom{n}{k}\int_0^q t^{n-k-1}(1 - t)^k\, dt$$

and

(8.10) $$1 - B(k; n, p) = n\binom{n-1}{k}\int_0^p t^k(1 - t)^{n-k-1}\, dt.$$

Hint: Integrate by parts or differentiate both sides with respect to p. Deduce one formula from the other.

Note: The integral in (8.10) is the *incomplete beta function*. Tables of $1 - B(k; n, p)$ to 7 decimals for k and n up to 50 and $p = 0.01, 0.02, 0.03, \ldots$ are given in K. Pearson, *Tables of the Incomplete Beta Function*, London (Biometrika Office), 1934.

26. Deduce the binomial distribution (a) by induction, (b) from the general summation formula (3.1) of Chapter 4.

27. Prove $\Sigma kb(k; n, p) = np$, and $\Sigma k^2 b(k; n, p) = n^2 p^2 + npq$.

28. Prove $\Sigma k^2 p(k; \lambda) = \lambda^2 + \lambda$.

29. *Multiple Poisson distribution.* When n is large and $np_j = \lambda_j$ is moderate, the multinomial distribution (7.2) can be approximated by

$$e^{-(\lambda_1 + \cdots + \lambda_r)} \frac{\lambda_1^{k_1}\lambda_2^{k_2} \cdots \lambda_r^{k_r}}{k_1! k_2! \cdots k_r!}.$$

Prove also that the terms of this distribution add to unity.

30. *Multiple Bernoulli trials.* In example (7.c) find the conditional probabilities p and q of (S, F) and (F, S), respectively, assuming that one of these combinations has occurred. Show that $p > 1/2$ or $p < 1/2$, according as $p_1 > p_2$ or $p_2 > p_1$.

31. *Continuation.*[16] If in n pairs of trials exactly m resulted in one of the combinations (S, F) or (F, S), show that the probability that (S, F) has occurred exactly k times is $b(k; m, p)$.

---

[16] A. Wald, Sequential Tests of Statistical Hypotheses, *Annals of Mathematical Statistics*, vol. 16 (1945), p. 166. Wald uses the results given above to devise a practical method of comparing two empirically given sequences of trials (say, the output of two machines), with a view of selecting the one with the greater probability of success. Using the above result, he reduces this problem to the simpler one of finding whether in a sequence of Bernoulli trials the frequency of success falls significantly short of 1/2 or exceeds 1/2.

32. Combination of the binomial and Poisson distributions. Suppose that the probability of an insect laying r eggs is $p(r; \lambda)$ and that the probability of an egg developing is p. Assuming mutual independence of the eggs, show that the probability of a total of k survivors is given by the Poisson distribution with parameter λp.

Note: Another example for the same situation: the probability of k chromosome breakages is $p(k; \lambda)$, and the probability of a breakage healing is p.

33. Polya's scheme of contagion. The urn scheme [example (2.c) of Chapter 5] permits the following obvious generalization. Successive (dependent trials) can each result only in S or F. In the first trial the probabilities of S and F are p and q. If the first n trials ($n = 1, 2, \ldots$) resulted in k successes and $n - k$ failures, the conditional probabilities of S and F in the $(n + 1)$st trial become $(p + k\gamma)/(1 + n\gamma)$ and $(q + (n - k)\gamma)/(1 + n\gamma)$. Here $\gamma > -1$ is a constant replacing $c/(b + r)$ in the original scheme. Show that *the probability of exactly k successes in the first n trials is given by*

$$(8.11) \quad \pi(k; n)$$

$$= \binom{n}{k} \frac{p(p + \gamma)(p + 2\gamma) \cdots (p + k\gamma - \gamma)q(q + \gamma)(q + 2\gamma) \cdots (q + n\gamma - k\gamma - \gamma)}{1(1 + \gamma)(1 + 2\gamma) \cdots (1 + n\gamma - \gamma)}.$$

34. Continuation: Limiting form of Polya's distribution. If $n \to \infty$, $p \to 0$, $\gamma \to 0$ so that $np \to \lambda$, $n\gamma \to \rho^{-1}$, then

$$(8.12) \quad \pi(k; n) \to \binom{\lambda\rho + k - 1}{k} \left(\frac{\rho}{1 + \rho}\right)^{\lambda\rho} \left(\frac{1}{1 + \rho}\right)^k.$$

35. Continuation. As $\rho \to \infty$ the last expression tends to $p(k; \lambda)$.

CHAPTER 7

THE NORMAL APPROXIMATION TO THE BINOMIAL DISTRIBUTION

1. The Normal Distribution

In order to avoid later interruptions we pause here to introduce two functions of great importance.

Definition. The function

$$(1.1) \qquad \phi(x) = \frac{1}{(2\pi)^{1/2}} e^{-x^2/2}$$

is called the normal density function; its integral

$$(1.2) \qquad \Phi(x) = \frac{1}{(2\pi)^{1/2}} \int_{-\infty}^{x} e^{-y^2/2} \, dy$$

is the normal distribution function.

The graph of $\phi(x)$ is the symmetric, bell-shaped curve shown in Figure 4. Note that different units are used along the two axes. The maximum of $\phi(x)$ is $(2\pi)^{-1/2} = 0.399$, approximately, so that in an ordinary Cartesian system the curve $y = \phi(x)$ would be much flatter.

Lemma 1. The domain bounded by the graph of $\phi(x)$ and the x-axis has unit area, that is,

$$(1.3) \qquad \int_{-\infty}^{+\infty} \phi(x) \, dx = 1.$$

Proof. We have

$$(1.4) \qquad \left\{ \int_{-\infty}^{+\infty} \phi(x) \, dx \right\}^2 = \int_{-\infty}^{+\infty} \int_{-\infty}^{+\infty} \phi(x)\phi(y) \, dx \, dy$$

$$= \frac{1}{2\pi} \int_{-\infty}^{+\infty} \int_{-\infty}^{+\infty} e^{-(x^2+y^2)/2} \, dx \, dy.$$

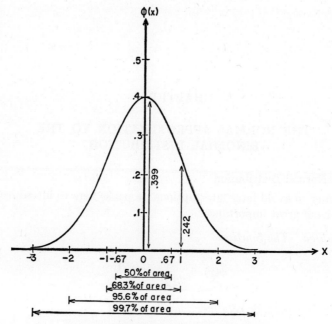

FIGURE 4. *The Normal Density Function.*

This double integral can be expressed in polar coordinates thus:

$$(1.5) \quad \frac{1}{2\pi} \int_0^{2\pi} d\theta \int_0^{\infty} e^{-r^2/2} r \, dr = \int_0^{\infty} e^{-r^2/2} r \, dr = \left. -e^{-r^2/2} \right|_0^{\infty} = 1$$

which proves the assertion.

It follows from the definition and the lemma that *the function $\Phi(x)$ increases steadily from 0 to 1.* Its graph (Figure 5) is an S-shaped curve with

$$(1.6) \quad \Phi(-x) = 1 - \Phi(x).$$

Table 1 gives the values [1] of $\Phi(x)$ for positive x, and from (1.6) we get $\Phi(-x)$.

For many purposes it is convenient to have an elementary estimate of the "tail," $1 - \Phi(x)$, for large x. Such an estimate is given by

---

[1] For larger tables cf. *Tables of Probability Functions*, Vol. 2, National Bureau of Standards, New York, 1942. There $\phi(x)$ and $\Phi(x) - \Phi(-x)$ are given to 15 decimals for x from 0 to 1 in steps of 0.0001 and for $x > 1$ in steps of 0.001.

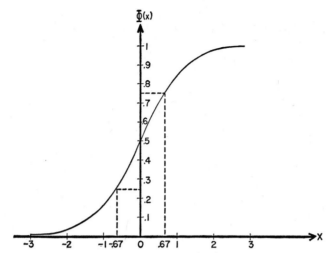

FIGURE 5. *The Normal Distribution Function.*

Lemma 2.[2] As $x \to \infty$

$$(1.7) \qquad 1 - \Phi(x) \sim \frac{1}{(2\pi)^{1/2} x} e^{-x^2/2};$$

more precisely, for every $x > 0$ the double inequality

$$(1.8) \qquad \frac{1}{(2\pi)^{1/2}} e^{-x^2/2} \left\{ \frac{1}{x} - \frac{1}{x^3} \right\} < 1 - \Phi(x) < \frac{1}{(2\pi)^{1/2}} e^{-x^2/2} \cdot \frac{1}{x}$$

holds (cf. problem 1).

Proof. By differentiation we may verify that

$$(1.9) \qquad \frac{1}{(2\pi)^{1/2}} e^{-x^2/2} \frac{1}{x} = \frac{1}{(2\pi)^{1/2}} \int_x^\infty e^{-y^2/2} \left\{ 1 + \frac{1}{y^2} \right\} dy.$$

The integrand on the right side is greater than the integrand of

$$(1.10) \qquad 1 - \Phi(x) = \frac{1}{(2\pi)^{1/2}} \int_x^\infty e^{-y^2/2} \, dy,$$

which proves the second inequality in (1.8). The first inequality follows in the same way, using as new integrand $e^{-y^2/2}\{1 - 3/y^4\}$ which is smaller than $e^{-y^2/2}$.

---

[2] Here and in the sequel the sign $\sim$ is used to indicate that the *ratio* of the two sides tends to one.

TABLE 1
The Normal Distribution

t	$\phi(t)$	$\Phi(t)$
0.0	0.398 942	0.500 000
0.1	.396 952	.539 828
0.2	.391 043	.579 260
0.3	.381 388	.617 911
0.4	.368 270	.655 422
0.5	.352 065	.691 462
0.6	.333 225	.725 747
0.7	.312 254	.758 036
0.8	.289 692	.788 145
0.9	.266 085	.815 940
1.0	.241 971	.841 345
1.1	.217 852	.864 334
1.2	.194 186	.884 930
1.3	.171 369	.903 200
1.4	.149 727	.919 243
1.5	.129 518	.933 193
1.6	.110 921	.945 201
1.7	.094 049	.955 435
1.8	.078 950	.964 070
1.9	.065 616	.971 283
2.0	.053 991	.977 250
2.1	.043 984	.982 136
2.2	.035 475	.986 097
2.3	.028 327	.989 276
2.4	.022 395	.991 802
2.5	.017 528	.993 790
2.6	.013 583	.995 339
2.7	.010 421	.996 533
2.8	.007 915	.997 445
2.9	.005 953	.998 134
3.0	.004 432	.998 650
3.1	.003 267	.999 032
3.2	.002 384	.999 313
3.3	.001 723	.999 517
3.4	.001 232	.999 663
3.5	.000 873	.999 767
3.6	.000 612	.999 841
3.7	.000 425	.999 892
3.8	.000 292	.999 928
3.9	.000 199	.999 952
4.0	.000 134	.999 968
4.1	.000 089	.999 979
4.2	.000 059	.999 987
4.3	.000 039	.999 991
4.4	.000 025	.999 995
4.5	.000 016	.999 997

Note on Terminology. The term *distribution function* is used in the mathematical literature for any never-decreasing function $F(x)$ which tends to 0 as $x \to -\infty$, and to 1 as $x \to \infty$. Statisticians currently prefer the term *cumulative distribution function*, but the adjective "cumulative" is redundant. A *density function* is a non-negative function $f(x)$ whose integral, extended over the entire x-axis, is unity. The integral from $-\infty$ to x of any density function is a distribution function. The older term *frequency function* is a synonym for density function.

The normal distribution function is often called the *Gaussian distribution*, but it was used in probability theory earlier by DeMoivre and Laplace. If the origin and the unit of measurement are changed, then $\Phi(x)$ is transformed into $\Phi((x-a)/b)$; this function is called the normal distribution function with mean a and variance b^2 (or standard deviation $|b|$). The function $2\Phi(x 2^{1/2}) - 1$ is often called *error function*.

2. The DeMoivre-Laplace Limit Theorem

Let S_n stand for the number of successes in n Bernoulli trials with probability p for success. Then $b(k; n, p)$ is the probability of the event that $S_n = k$. In practice we are usually interested in *the probability of the event that the number of successes lies between preassigned limits α and β*. If α and β are integers and $\alpha < \beta$, then this event is defined by the inequality $\alpha \leq S_n \leq \beta$, and its probability is

(2.1) $\quad Pr\{\alpha \leq S_n \leq \beta\}$
$$= b(\alpha; n, p) + b(\alpha+1; n, p) + \ldots + b(\beta; n, p).$$

This sum may involve many terms, and a direct evaluation is usually impractical. Fortunately, whenever n is large, the normal distribution function can be used to derive simple approximations to the probability (2.1). This discovery is due to DeMoivre [3] and Laplace.[4] We shall see that its importance goes far beyond the domain of numerical calculations.

Our first aim is to derive an asymptotic formula for the individual terms

(2.2) $$b(k; n, p) = \frac{n!}{k!(n-k)!} p^k q^{n-k}.$$

The probability p will be kept fixed, but we shall let $n \to \infty$. As the number n of trials increases, we must expect that also the numbers of successes and failures will increase, and we are therefore interested mainly in such combinations of n and k, for which

(2.3) $\quad\quad\quad\quad n \to \infty, \quad k \to \infty, \quad n - k \to \infty.$

---

[3] Abraham DeMoivre (1667–1754). His *The Doctrine of Chances* appeared in 1718.

[4] Pierre S. Laplace (1749–1827). His *Théorie analytique des probabilités* appeared in 1812.

We may then express the factorials in (2.2) by means of Stirling's formula [5]

(2.4) $$r! \sim (2\pi)^{1/2} r^{r+1/2} e^{-r}$$

as $r \to \infty$. We get [6]

(2.5) $$b(k; n, p) \sim \left\{\frac{n}{2\pi k(n-k)}\right\}^{1/2} \left(\frac{np}{k}\right)^k \left(\frac{nq}{n-k}\right)^{n-k}$$

The last two factors on the right equal unity for $k = np$, and their product decreases as $|k - np|$ increases. It is, therefore, natural to replace k by the new variable

(2.6) $$\delta_k = k - np.$$

[From Chapter 6, section 3, we know that the number of successes for which $b(k; n, p)$ is greatest lies in the interval $(np - q, np + p)$. Accordingly, δ_k is approximately the deviation of the number k of successes from its most probable value.] With the notation (2.6) we have $k = np + \delta_k$ and $n - k = nq - \delta_k$, so that (2.5) becomes

(2.7) $$b(k; n, p)$$
$$\sim \left\{\frac{n}{2\pi(np+\delta_k)(nq-\delta_k)}\right\}^{1/2} \frac{1}{(1+\delta_k/np)^{np+\delta_k}(1-\delta_k/nq)^{nq-\delta_k}}.$$

To evaluate the last fraction we pass to logarithms. In the interval $|\delta_k| \leq npq$ we may use Taylor's expansion [Chapter 2, formula (6.9)], and find for the logarithm of the denominator

$$(np + \delta_k) \log(1 + \delta_k/np) + (nq - \delta_k) \log(1 - \delta_k/nq)$$

(2.8) $$= (np + \delta_k)\left(\frac{\delta_k}{np} - \frac{\delta_k^2}{2n^2 p^2} + \frac{\delta_k^3}{3n^3 p^3} - + \cdots\right)$$

$$- (nq - \delta_k)\left(\frac{\delta_k}{nq} + \frac{\delta_k^2}{2n^2 q^2} + \frac{\delta_k^3}{3n^3 q^3} + \cdots\right).$$

---

[5] It will be recalled that in Chapter 2 we did not complete the proof of Stirling's formula but showed only that $r! \sim Cr^{r+1/2}e^{-r}$, where C is a positive constant. In the text it is assumed that $C = (2\pi)^{1/2}$. If we want to *prove* this fact, then the factor $(2\pi)^{1/2}$ in equations (2.5), (2.7), and (2.11) must be replaced by C. In this case a factor $(2\pi)^{1/2}/C$ must be inserted on the right sides in (2.14), (2.17), and (2.20). To show that this factor really equals 1 it suffices to choose x_β and $-x_\alpha$ very large. The right side in the modified equation (2.20) is then arbitrarily near to $C/(2\pi)^{1/2}$, while the left side is near 1 (cf. Chapter 6, problem 19).

[6] The symbol $\sim$ indicates that the ratio of the two sides tends to unity.

Reordering the terms according to powers of δ_k, we get

$$(2.9) \qquad \frac{\delta_k^2}{2n}\left(\frac{1}{p}+\frac{1}{q}\right) - \frac{\delta_k^3}{6n^2}\left(\frac{1}{p^2}-\frac{1}{q^2}\right) + \cdots.$$

We now suppose that k increases with n in such a manner that

$$(2.10) \qquad \frac{\delta_k^3}{n^2} \to 0.$$

[In this case also $\delta_k/n \to 0$ so that (2.3) holds and the expansion (2.8) is justified.] It follows from (2.10) that the term within braces in (2.7) is asymptotically equivalent to $(2\pi npq)^{-\frac{1}{2}}$. The logarithm of the denominator in (2.7) is given by (2.9), but in view of (2.10) all terms except the first one may be neglected; this first term equals $\delta_k^2/2npq$. Combining these results, we have

$$(2.11) \qquad b(k; n, p) \sim (2\pi npq)^{-\frac{1}{2}} e^{-\delta_k^2/2npq}.$$

This is the desired asymptotic formula. We simplify it by the use of a more convenient notation. Put

$$(2.12) \qquad h = \frac{1}{(npq)^{\frac{1}{2}}}$$

and define a function x_k of the variable k by

$$(2.13) \qquad x_k = (k - np)h = \frac{\delta_k}{(npq)^{\frac{1}{2}}}.$$

In terms of these quantities and the normal density function $\phi(x)$ we can rewrite (2.11) in the form

$$(2.14) \qquad b(k; n, p) \sim h\phi(x_k).$$

This formula has been derived under the sole condition (2.10), which can be rewritten as
$$(2.15) \qquad hx_k^3 \to 0.$$

We have thus the

Theorem. *If n and k vary in such a way that $(k - np)^3/n^2 \to 0$, then the asymptotic formula (2.14) holds.*

The error committed in replacing $b(k; n, p)$ by $h\phi(x_k)$ decreases with increasing npq. Figure 6 illustrates the theorem in the case $n = 10$, $p = 0.2$ where npq is only 1.6. It is seen that even in this extremely unfavorable case the approximation is surprisingly good.[7]

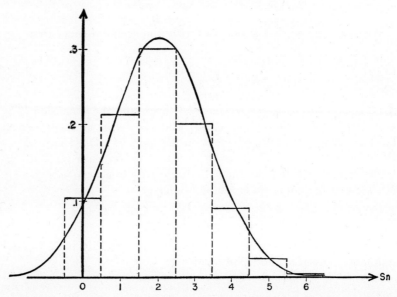

FIGURE 6. *The Normal Approximation to the Binomial Distribution.* The step function gives the probabilities $b(k; 10, 1/5)$ of k successes in 10 Bernoulli trials with $p = 1/5$. The continuous curve gives for each integer k the corresponding normal approximation.

Our theorem leads directly to simple approximations for the sum (2.1). If

(2.16) $$hx_\alpha^3 \to 0 \quad \text{and} \quad hx_\beta^3 \to 0,$$

then (2.14) holds uniformly for all terms in (2.1), and therefore

(2.17) $$Pr\{\alpha \leq S_n \leq \beta\} \sim h\{\phi(x_\alpha) + \phi(x_{\alpha+1}) + \ldots + \phi(x_\beta)\}.$$

The right side is a Riemann sum approximating an integral. The points $x_\alpha, x_{\alpha+1}, \cdots, x_\beta$ are uniformly spaced at distance h, and the contribution $h\phi(x_k)$ is asymptotically equivalent to the area under the curve $y = \phi(x)$ between $x_k - h/2$ and $x_k + h/2$. In other words,

---

[7] The values of $b(k; 10, 0.2)$ for $k = 0, 1, \cdots, 6$ are 0.1074, 0.2684, 0.3020, 0.2013, 0.0880, 0.0264, 0.0055. The corresponding approximations $h\phi(x_k)$ are 0.0904, 0.2307, 0.3154, 0.2307, 0.0904, 0.0189, 0.0021.

it is claimed that for all k of the interval $\alpha \leq k \leq \beta$ uniformly

$$(2.18) \qquad h\phi(x_k) \sim \int_{x_k-h/2}^{x_k+h/2} \phi(x)\,dx = \Phi(x_{k+\frac{1}{2}}) - \Phi(x_{k-\frac{1}{2}}).$$

To verify (2.18) it suffices to note that the definition (1.1) of $\phi(x)$ implies that in the interval $x_k - h/2 \leq x \leq x_k + h/2$ we have the double inequality

$$(2.19) \qquad \phi(x_k) e^{-|x_k|\frac{h}{2} - \frac{h^2}{8}} \leq \phi(x) \leq \phi(x_k) e^{|x_k|\frac{h}{2}}.$$

Since $x_k h \to 0$, the assertion (2.18) follows. If we add (2.18) for $k = \alpha, \alpha + 1, \cdots, \beta$, then all intermediate terms on the right cancel, and only $\Phi(x_{\beta+\frac{1}{2}}) - \Phi(x_{\alpha-\frac{1}{2}})$ remains. We have thus proved the

DeMoivre-Laplace limit theorem. If α and β vary so that $hx_\alpha^3 \to 0$ and $hx_\beta^3 \to 0$, then

$$(2.20) \qquad Pr\{\alpha \leq S_n \leq \beta\} \sim \Phi(x_{\beta+\frac{1}{2}}) - \Phi(x_{\alpha-\frac{1}{2}}),$$

where $h = (npq)^{-\frac{1}{2}}$ and $x_t = (t - np)h$. In words, the percentage difference between the two sides in (2.20) tends to zero together with hx_β^3 and hx_α^3.

In particular, (2.20) holds if α and β are restricted to values for which x_α and x_β remain within a fixed interval. [The case where α and β are so large that the condition (2.16) is not satisfied will be discussed in section 5 and in problem 14.]

In statistical applications (2.20) is usually used for values α and β for which $|x_\alpha|$ and $|x_\beta|$ do not exceed 3 or 4. In theoretical applications it is often necessary to use (2.20) for intervals (α, β) which are far off the central part of the binomial distribution and for which both x_α and x_β are large. In such cases both sides of (2.20) are small, and it becomes important to know that their *ratio* is near unity so that not only their absolute, but also their percentage, difference tends to zero.

Examples. (a) Let $p = 1/2$, $n = 200$, $\alpha = 95$, $\beta = 105$. In this case, $x = Pr\{95 \leq S_n \leq 105\}$ may be interpreted as the probability that in 200 tossings of a coin the number of heads deviates from 100 by at most 5. A cumbersome direct computation [8] shows that $x = 0.56325 \ldots$. To find the right side in (2.20) we calculate $h = (50)^{-\frac{1}{2}} = 0.141421 \ldots$, and hence $-x_{\alpha-\frac{1}{2}} = x_{\beta+\frac{1}{2}} = (5.5)h = 0.7778 \ldots$. From tables we get $\Phi(x_{\beta+\frac{1}{2}}) - \Phi(x_{\alpha-\frac{1}{2}}) = 0.56331 \ldots$.

---

[8] J. V. Uspensky, *Introduction to Mathematical Probability*, New York, 1937, p. 131. Uspensky uses the normal approximation $\Phi(x_\beta) - \Phi(x_\alpha)$, instead of $\Phi(x_{\beta+\frac{1}{2}}) - \Phi(x_{\alpha-\frac{1}{2}})$ and gets 0.56050 ... with an error 0.0027 ... instead of our 0.00006.

The error is 0.00006. This is less than should be expected in general.

(b) Let $p = 1/10$, $n = 500$, $\alpha = 50$, $\beta = 55$. The correct value is $Pr\{50 \leq S_n \leq 55\} = 0.317573 \ldots$. Now $h = (45)^{-1/2} = 0.1490712 \ldots$, and we get the approximation $\Phi(5.5h) - \Phi(-0.5h) = 0.3235 \ldots$. The error is about 2 per cent.

(c) Let $n = 100$, $p = 0.3$. Table 2 shows in a typical example (for relatively small n) how the normal approximation deteriorates as the interval (α, β) moves away from the central term.

TABLE 2

COMPARISON OF THE BINOMIAL DISTRIBUTION FOR $n = 100$, $p = 0.3$ AND THE NORMAL APPROXIMATION

Number of Successes	Probability	Normal Approximation	Percentage Error
$9 \leq S_n \leq 11$	0.000 006	0.000 03	+400
$12 \leq S_n \leq 14$	.000 15	.000 33	+100
$15 \leq S_n \leq 17$	.002 01	.002 83	+40
$18 \leq S_n \leq 20$	.014 30	.015 99	+12
$21 \leq S_n \leq 23$	.059 07	.058 95	0
$24 \leq S_n \leq 26$	.148 87	.144 47	−3
$27 \leq S_n \leq 29$	.237 94	.234 05	−2
$31 \leq S_n \leq 33$	.230 13	.234 05	+2
$34 \leq S_n \leq 36$	.140 86	.144 47	+3
$37 \leq S_n \leq 39$	.058 89	.058 95	0
$40 \leq S_n \leq 42$	.017 02	.015 99	−6
$43 \leq S_n \leq 45$	.003 43	.002 83	−18
$46 \leq S_n \leq 48$	.000 49	.000 33	−33
$49 \leq S_n \leq 51$	.000 05	.000 03	−40

The limit theorem (2.20) takes on a simpler form if, instead of S_n, we introduce the *reduced number of successes* defined by

$$(2.21) \qquad S_n^* = \frac{S_n - np}{(npq)^{1/2}}.$$

This amounts to measuring the deviations of S_n from np in units of $(npq)^{1/2}$. The quantity np *is called the mean, and* $(npq)^{1/2}$ *the standard deviation of* S_n; this terminology is suggested by the theory of random variables (cf. Chapter 9). The inequality $\alpha \leq S_n \leq \beta$ is the same as $x_\alpha \leq S_n^* \leq x_\beta$, and (2.18) states that for arbitrary fixed $x_\alpha < x_\beta$

$$(2.22) \qquad Pr\{x_\alpha \leq S_n^* \leq x_\beta\} \sim \Phi\left(x_\beta + \frac{h}{2}\right) - \Phi\left(x_\alpha - \frac{h}{2}\right),$$

where $h = (npq)^{-1/2}$. Now $h \to 0$ as $n \to \infty$, and therefore the right side tends to $\Phi(x_\beta) - \Phi(x_\alpha)$. Thus we have the following

Corollary to the Limit Theorem. For every fixed $a < b$

(2.23) $$Pr\{a \leq S_n{}^* \leq b\} \to \Phi(b) - \Phi(a).$$

This is a weakened version of (2.20) but represents the traditional form of Laplace's limit theorem. The dropping of $h/2$ in (2.22) introduces an error which tends to zero as $n \to \infty$ but has a considerable influence when npq is of moderate magnitude (as is the case in the three examples (a)-(c)).

The main fact revealed by (2.23) is that for large n the probability on the left is practically independent of p. This permits us to compare fluctuations in different series of Bernoulli trials simply by referring to our standard units.

Examples. (d) Let us find a number a such that, for large n, the inequality $|S_n{}^*| > a$ has a probability near $1/2$. For this it is necessary that $\Phi(a) - \Phi(-a) = 1/2$ or $\Phi(a) = 3/4$. From tables of the normal distribution we find that $a = 0.6745$, and hence the two inequalities

(2.24) $$|S_n - np| < 0.6745(npq)^{1/2} \quad \text{and} \quad |S_n - np| > 0.6745(npq)^{1/2}$$

are about equally probable. In particular, the probability is about $1/2$ that in n tossings of a coin the number of heads lies within the limits $n/2 \pm 0.337 n^{1/2}$, and, similarly, that in n throws of a die the number of aces lies within the interval $n/6 \pm 0.251 n^{1/2}$. The probability of S_n lying within the limits $np \pm 2(npq)^{1/2}$ is about $\Phi(2) - \Phi(-2) = 0.9545 \ldots$, and for $np \pm 3(npq)^{1/2}$ the probability is $0.9973 \ldots$.

(e) *A Competition Problem.* This example illustrates practical applications of formula (2.23). Two competing railroads operate one train each between Chicago and Los Angeles, which leave and arrive simultaneously and have comparable equipment. We suppose that n passengers select trains independently and at random so that the number of passengers in each train is the outcome of n Bernoulli trials with $p = 1/2$. If a train carries $s < n$ seats, then there is a positive probability $f(s)$ that more than s passengers will turn up, in which case not all patrons can be accommodated. Using the approximation (2.23), we find

(2.25) $$f(s) \approx 1 - \Phi\left(\frac{2s - n}{n^{1/2}}\right).$$

If s is so large that $f(s) < 0.01$, then the number of seats will be sufficient in 99 out of 100 cases. More generally, the company may decide on an arbitrary risk level α and determine s so that $f(s) < \alpha$. For that purpose it suffices to put

$$(2.26) \qquad s \geq \tfrac{1}{2}(n + t_\alpha n^{1/2}),$$

where t_α is the root of the equation $\alpha = 1 - \Phi(t_\alpha)$, which can be found from tables. For example, if $n = 1000$ and $\alpha = 0.01$, then $t_\alpha \approx 2.33$ and $s = 537$ seats should suffice. If both railroads accept the risk level $\alpha = 0.01$, the two trains will carry a total of 1074 seats of which 74 will be empty. The loss due to competition (or chance fluctuations) is remarkably small. In the same way, 514 seats should suffice in about 80 per cent of all cases, and 549 seats in 999 out of 1000 cases.

Similar considerations apply in other competitive supply problems. For example, if m movies compete for the same n patrons, each movie will put for its probability of success $p = 1/m$, and (2.26) is to be replaced by $s \geq (1/m)[n + t_\alpha n^{1/2}(m-1)^{1/2}]$. The total number of empty seats under this system is $ms - n \approx t_\alpha n^{1/2}(m-1)^{1/2}$. For $\alpha = 0.01$, $n = 1000$, and $m = 2, 3, 4$, this number is about 74, 126, and 147, respectively. The loss of efficiency because of competition is remarkably small.

Theorem (2.23) is historically the first limit theorem of probability. From a modern point of view it is only an exceedingly special case of the *central limit theorem*, to which we shall return in Chapter 10 but whose general derivation must be postponed to the second volume. Statisticians use (2.23) as an approximation even where npq is relatively small, and in such cases an estimate of the error is desired. In Uspensky's book (cited in the last footnote) the reader will find an excellent bound for the error in the case of intervals (a, b) near the origin. The derivation requires Fourier analysis. Serge Bernstein devoted a series of papers to the investigation of the error term in the general case and discussed how the definition of x_t should be modified in order to improve the convergence in (2.20). His papers are written in Russian and are difficult to obtain. A simplified derivation with an improvement of his results is, however, available in English.[9]

Note on Optional Stopping. It is essential to note that our limit and *approximation theorems are valid only if the number n of trials is fixed in advance independently of the outcome of the trials*. If a gambler has the privilege of stopping at a moment favorable to him, his ultimate gain cannot be judged from the normal approximation, for now the duration of the game depends on chance. For *every fixed n* it is very improbable that S_n^* is large. However, in the long run, even the

[9] W. Feller, On the Normal Approximation to the Binomial Distribution, *Annals of Mathematical Statistics*, vol. 16 (1945), pp. 319–329.

most improbable thing is bound to happen, and we shall see that in a continued game S_n^* is practically certain to have a sequence of maxima of the order of magnitude $(\log \log n)^{1/2}$ (this is the law of the iterated logarithm of Chapter 8, section 5).

3. The Law of Large Numbers

On several occasions we have mentioned that our *intuitive notion of probability* is based on the following assumption. If in n identical trials A occurs ν times, and if n is very large, then ν/n should be near the probability p of A. Clearly, a formal mathematical theory can never refer directly to real life, but it should at least provide theoretical counterparts to the phenomena which it tries to explain. Accordingly, we require that the vague introductory remark be made precise in the form of a theorem. For this purpose we translate "identical trials" as "Bernoulli trials" with probability p for success. Then—so we expect—S_n/n should be near p. The Laplace limit theorem gives precise meaning to this statement. For every $\epsilon > 0$ the event $\left|\dfrac{S_n}{n} - p\right| < \epsilon$ is the same as $|S_n - np| < \epsilon n$ or $|S_n^*| < \epsilon \left(\dfrac{n}{pq}\right)^{1/2}$. Hence, for large n,

$$(3.1) \quad Pr\left\{\left|\frac{S_n}{n} - p\right| < \epsilon\right\} \approx \Phi\left(\epsilon\left(\frac{n}{pq}\right)^{1/2}\right) - \Phi\left(-\epsilon\left(\frac{n}{pq}\right)^{1/2}\right),$$

and the right side is near unity. Thus we have one form of *the law of large numbers which asserts that*

$$(3.2) \quad Pr\left\{\left|\frac{S_n}{n} - p\right| < \epsilon\right\} \to 1.$$

In words: as n increases, the probability that the average number of successes deviates from p by more than any preassigned ϵ tends to zero. Its practical value depends entirely on the more precise form (3.1). In fact, even the use of (3.1) depends on the assurance that it is not only a limit relation, but that even for a given n the right side is a reasonable approximation to the left side. (Cf. the remarks concerning the error terms at the conclusion of the preceding section.)

Examples. (*a*) *Random Digits.* According to Chapter 2, Table 1, the digit 7 appeared among 10,000 digits 968 times. In an ideal sequence of 10,000 Bernoulli trials with $p = 0.1$ the standard deviation is 30, and the probability of a deviation $\left|S_n - \dfrac{n}{10}\right| > 32$ is about

$$1 - \Phi\left(\frac{32}{30}\right) + \Phi\left(-\frac{32}{30}\right) = 2\left\{1 - \Phi\left(\frac{32}{30}\right)\right\} = 0.386 \ldots.$$ This can be interpreted roughly by saying that in four out of ten ideal cases the deviation should be larger than the one observed.

(b) *Random Digits, Continued.* In example (2.b) of Chapter 2, we considered an event with $p = 0.3024$. In $n = 1200$ trials this event had an average frequency of 0.3142. The deviation from p is $\epsilon = 0.0118$. In this case $(pq)^{\frac{1}{2}} = 0.4593$ and $\epsilon(n/pq)^{\frac{1}{2}} \approx 0.890 \ldots.$ Hence the probability of $\left|\dfrac{S_n}{n} - p\right| > \epsilon$ is in this case about 0.37.... This indicates that in about 37 per cent of all cases the average number of successes should deviate from p by more than it does in our material.

(c) *Sampling.* A fraction p of a certain population are smokers. Suppose that p is unknown and that random sampling with replacement is to be used to determine p. It is desired to find p with an error not exceeding 0.005. How large should the sample size n be? If p' is the fraction of smokers in the sample, we desire that $|p' - p| < 0.005$. However, no sample size can give absolute assurance that $|p' - p| < 0.005$; it is conceivable that the sample contains only smokers. Since absolute certainty is unattainable, we settle for an arbitrary *confidence level* α, say, $\alpha = 0.95$, and require that $|p' - p| < 0.005$ with probability 0.95 or better. Note that np' is the number of successes in n trials, and hence

$$Pr\{|p' - p| < 0.005\} = Pr\left\{\left|\frac{S_n}{n} - p\right| < 0.005\right\}.$$

This means that n should be selected large enough to make the left side of (3.1), with $\epsilon = 0.005$, greater than 0.95. For the present purposes the normal approximation is sufficient. The root x of $\Phi(x) - \Phi(-x) = 0.95$ is $x = 1.96\ldots$, and hence we should have $0.005(n/pq)^{\frac{1}{2}} \geq 1.96$. Thus we are led to the inequality $n \geq 392^2 pq$ or $n \geq 160{,}000 pq$, approximately. This inequality depends on the unknown p, but pq never exceeds $1/4$, and hence the sample size $n = 40{,}000$ would be safe under all circumstances; with it the odds are about 20 to 1 that $|p' - p| < 0.005$. If it is known in advance that $p \leq 1/10$, then $pq \leq 9/100$, and a sample size of 15,000 should suffice, etc.

The theorem (3.2) is again only a special case of the general (weak) law of large numbers (cf. Chapter 10). A stronger and much more interesting theorem is the *strong law of large numbers*, to be proved in Chapter 8, section 4.

4. Relation to the Poisson Approximation

The error of the normal approximation will be small if npq is large. On the other hand, if n is large and p small, the terms $b(k; n, p)$ were found to be near the Poisson probabilities $p(k; \lambda)$ with $\lambda = np$. If λ is small, then only the Poisson approximation can be used. However, if λ is large, we can use either the normal or the Poisson approximation. This implies that for large values of λ it must be possible to approximate the Poisson distribution by the normal distribution, and in Chapter 10, example (1.c) we shall see that this is indeed so (cf. also problem 9). Here we shall be content to illustrate the point by a numerical and a practical example.

Examples. (a) Consider the Poisson distribution $p(k; 100)$. We can take it as an approximation, say, to the binomial distribution with $n = 100{,}000{,}000$ and $p = 1/1{,}000{,}000$. Then also $npq \approx 100$; this quantity, even though not large, suffices for the normal distribution to give reasonable approximations at least for the central sector of the binomial distribution. The Poisson distribution $p(k; 100)$ agrees with $b(k; 10^8, 10^{-6})$ to many decimals, and we can compare it with the normal approximation to the latter. Put, for brevity, $P(a, b) = p(a; 100) + p(a+1; 100) + \cdots + p(b; 100)$, so that $P(a, b)$ stands for $Pr\{a \leq S_n \leq b\}$ and should be approximated by $\Phi\left(\dfrac{b - 99.5}{10}\right) - \Phi\left(\dfrac{a - 100.5}{10}\right)$. The following sample gives an idea of the degree of approximation.

	Correct Values	Normal Approximation
$P(85, 90)$	0.113 84	0.110 49
$P(90, 95)$	.184 85	.179 50
$P(95, 105)$	.417 63	.417 68
$P(90, 110)$	.706 52	.706 28
$P(110, 115)$	.107 38	.110 49
$P(115, 120)$	.053 23	.053 35

(b) *A Telephone Trunking Problem.* The following problem is, with some simplifications, taken from actual practice.[10] A telephone exchange A is to serve 2000 subscribers in a nearby exchange B. It would be too expensive and extravagant to install 2000 trunklines from A to B. It will suffice to make the number N of lines so large that,

---

[10] E. C. Molina, Probability in Engineering, *Electrical Engineering*, vol. 54 (1935), pp. 423–427, or *Bell Telephone System Technical Publications Monograph* B-854. There the problem is treated by the Poisson method given in the text, which is preferable from the engineer's viewpoint.

under ordinary conditions, only one out of every hundred calls will fail to find an idle trunkline immediately at its disposal. Suppose that during the busy hour of the day on the average each subscriber requires a trunkline to B for 2 minutes. At a fixed moment of the busy hour we can reasonably compare the situation to a set of 2000 trials with a probability $p = 1/30$ in each that a line will be required. Under ordinary conditions these trials can be assumed to be independent (although this is not true when events like unexpected showers or earthquakes cause many people to call for taxicabs or the local newspaper; the theory no longer applies, and the trunks will be "jammed"). We have, then, 2000 Bernoulli trials with $p = 1/30$, and the smallest number N is required such that the probability of more than N "successes" will be smaller than 0.01; in symbols $Pr\{S_{2000} \geq N\} < 0.01$.

For the *Poisson approximation* we should take $\lambda = 200/3 \approx 66.67$. From the tables we find that the probability of 87 or more successes is about 0.0097, whereas the probability of 86 or more successes is about 0.013. This would indicate that *87 trunklines should suffice*. For the *normal approximation* we first find from tables the root x of $1 - \Phi(x) = 0.01$, which is $x = 2.327$. Then it is required that $(N - \frac{1}{2} - np)/(npq)^{\frac{1}{2}} \geq 2.327$. Since $n = 2000, p = 1/30$ this means $N \geq 67.17 + (2.327)(8.027) \approx 85.8$. Hence the normal approximation would indicate that *86 trunklines should suffice*.

For practical purposes the two solutions agree. The method yields further practical results. Conceivably, the installation might be cheaper if the 2000 subscribers were divided into two groups of 1000 each, and two separate groups of trunklines from A to B were installed. Using the above method, we find that actually some 10 additional trunklines would be required so that the first arrangement is more favorable.

5. Large Deviations [11]

Frequently we desire an estimate of the probability that the reduced number of successes S_n^*[cf. (2.21)] exceeds a given number x. Hence the upper limit of the interval is infinity, and it requires a special argument to show that our limit theorem (2.20) still applies.

Theorem. *If* $n \to \infty$ *and* x *varies as a function of* n *in such a way that* $x \to \infty$ *but* $x^3 h \to 0$, *then*

(5.1) $$Pr\{S_n^* > x\} \sim 1 - \Phi(x).$$

[11] The theorem is of general interest but will be used in this book only for the proof of the law of the iterated logarithm, Chapter 8.

In view of (1.7) *this is equivalent to*

$$(5.2) \qquad Pr\{S_n^* > x\} \sim \frac{1}{(2\pi)^{1/2} x} e^{-x^2/2}.$$

Proof. Choose in (2.20) the integers α and β so that x lies between x_α and $x_{\alpha+1}$, and that $x_\beta \approx x + \log x$. Then $x_\beta^3 h \to 0$ and (2.20) holds. Hence

$$(5.3) \qquad Pr\{\alpha < S_n < \beta\} \sim \{1 - \Phi(x_\alpha)\} - \{1 - \Phi(x_\beta)\}.$$

However, from (1.7) and the fact that $x_\beta \approx x_\alpha + \log x_\alpha$ it is readily seen that $1 - \Phi(x_\beta)$ is of smaller order of magnitude than $1 - \Phi(x_\alpha)$, while $1 - \Phi(x_\alpha) \sim 1 - \Phi(x)$. Hence

$$(5.4) \qquad Pr\{\alpha < S_n < \beta\} \sim 1 - \Phi(x).$$

On the other hand, from (2.14) and formula (8.1) of Chapter 6 we have

$$(5.5) \qquad Pr\{S_n \geq \beta\} \leq \frac{n}{\beta - np} b(\beta; n, p) \sim \frac{nh^2}{x_\beta} \phi(x_\beta).$$

Now $nh^2 = 1/pq$ is a constant, and

$$(5.6) \qquad \frac{1}{x_\beta} \phi(x_\beta) \sim 1 - \Phi(x_\beta).$$

We saw that the right side tends to zero faster than $1 - \Phi(x)$, which means that $Pr\{S_n \geq \beta\}$ is of smaller order of magnitude than $1 - \Phi(x)$. Combining this result with (5.4), we see then that

$$(5.7) \qquad Pr\{S_n > \alpha\} \sim 1 - \Phi(x),$$

and this is our theorem.

Further limit theorems for large deviations are given in problems 12–17.

6. Problems for Solution

1. Generalizing (1.8), prove that

$$(6.1) \qquad 1 - \Phi(x) \sim \frac{1}{(2\pi)^{1/2}} e^{-x^2/2} \left\{ \frac{1}{x} - \frac{1}{x^3} + \frac{1 \cdot 3}{x^5} - \frac{1 \cdot 3 \cdot 5}{x^7} + - \cdots + (-1)^k \frac{1 \cdot 3 \cdots (2k-1)}{x^{2k+1}} \right\}$$

and that for $x > 0$ the right side *overestimates* $1 - \Phi(x)$ if k is even, and *underestimates* if k is odd.

2. For every constant $a > 0$

(6.2) $$\left\{1 - \Phi\left(x + \frac{a}{x}\right)\right\} \div \left\{1 - \Phi(x)\right\} \to e^{-a}$$

as $x \to \infty$.

3. Prove the inequality

(6.3) $$he^{-h^2/8}e^{-x^2/2} \leq \int_{x-h/2}^{x+h/2} e^{-t^2/2}\,dt \leq he^{h^2(x^2 - \frac{1}{3} + h^2/80)/8}e^{-x^2/2}$$
if $h < 4.2^{1/2}$

[This improves (2.19).]

4. Find an approximation to the probability that the number of aces obtained in 12,000 rollings of a die is between 1900 and 2150.

5. Find a number k such that the probability is about 0.5 that the number of heads obtained in 1000 tossings of a coin will be between 440 and k.

6. A sample is taken in order to find the fraction f of females in a population. Find a sample size such that the probability of a sampling error less than 0.005 will be 0.99 or greater.

7. In 10,000 tossings, a coin fell heads 5400 times. Is it reasonable to assume that the coin is skew?

8. Find an approximation to the maximal term of the trinomial distribution

$$\frac{n!}{k!r!(n-k-r)!} p_1^k p_2^r (1 - p_1 - p_2)^{n-k-r}.$$

9. *Normal approximation to the Poisson distribution.* Using Stirling's formula, show that, if $\lambda \to \infty$, then for every fixed $\alpha < \beta$

(6.4) $$\sum_{\lambda + \alpha\lambda^{1/2} < k < \lambda + \beta\lambda^{1/2}} p(k; \lambda) \to \Phi(\beta) - \Phi(\alpha).$$

[Cf. Chapter 10, example (1.*c*).]

10. *Normal approximation to the hypergeometric distribution.* Let n, m, k be positive integers and suppose that

(6.5) $$\frac{r}{n+m} \to t, \quad \frac{n}{n+m} \to p, \quad \frac{m}{n+m} \to q, \quad h\{k - rp\} \to x$$

where $1/h = \{(n+m)pqt(1-t)\}^{1/2}$. Prove that

(6.6) $$\frac{\binom{n}{k}\binom{m}{r-k}}{\binom{n+m}{r}} \sim h\phi(x).$$

Hint: Use the normal approximation to the binomial distribution rather than Stirling's formula.

11. *Normal distribution and combinatorial runs.*[12] In Chapter 3, problem 15,

---

[12] A. Wald and J. Wolfowitz, On a Test Whether Two Samples Are from the Same Population, *Annals of Mathematical Statistics*, vol. 11 (1940), pp. 147–162. For more general results, see A. M. Mood, The Distribution Theory of Runs, *ibid.*, pp. 367–392.

we found that in an arrangement of n red and m black things the probability of having exactly k runs of red things is

$$\text{(6.7)} \qquad \pi_k = \binom{n-1}{k-1}\binom{m+1}{k} \div \binom{n+m}{n}.$$

Let $n \to \infty$, $m \to \infty$ so that (6.5) holds. For fixed $\alpha < \beta$ the probability that the number of red runs lies between $npq + \alpha(pqn)^{1/2}$ and $npq + \beta(pqn)^{1/2}$ tends to $\Phi(\beta) - \Phi(\alpha)$.

In the following problems $h^2 = npq$ and S_n^* is the reduced number of successes defined in (2.21). Finally

$$\text{(6.8)} \qquad F_n(x) = Pr\{S_n^* > x\}.$$

12. If x varies as a function of n so that $x^{3+a}h \to 0$ but $x \to \infty$, then [13]

$$\text{(6.9)} \qquad \frac{F_n(x)}{1 - \Phi(x)} = 1 + o(x^a),$$

where $o(x^a)$ stands for terms which are of smaller order of magnitude than x^a.

13. If $x^3 h \to 0$, $x \to \infty$, then [14] for any constant $a > 0$

$$\text{(6.10)} \qquad \frac{F_n(x) - F_n(x + a/x)}{F_n(x)} \to 1 - e^{-a}.$$

In words, the conditional probability of $x < S_n^* < x + a/x$, given that $S_n^* > x$, tends to $1 - e^{-a}$. [*Hint:* Use (5.2).]

14. *Probabilities of large deviations.* Starting with (2.5), prove the following theorem. If $n \to \infty$, and k varies so that $(k - np)/n \to 0$, then

$$\text{(6.11)} \qquad b(k; n, p) \sim \frac{h}{(2\pi)^{1/2}} e^{-\frac{x^2}{2} - f(x)}$$

where $x = (k - np)h$ and

$$\text{(6.12)} \qquad f(x) = \sum_{\nu=3}^{\infty} \frac{p^{\nu-1} - (-q)^{\nu-1}}{\nu(\nu-1)} h^{\nu-2} x^\nu.$$

Note: If $x^3 h \to 0$, then $f(x) \to 0$, and (6.11) reduces to (2.14). If x is of the order of magnitude of $h^{-1/3}$ but negligible as compared to $h^{-1/2}$, then

$$\text{(6.13)} \qquad f(x) \approx \frac{p-q}{6} x^3 h.$$

If x is of the order of magnitude of $h^{-1/2}$, then

$$\text{(6.14)} \qquad f(x) \approx \frac{p-q}{6} x^3 h + \frac{p^3 + q^3}{12} x^4 h^2,$$

etc.

---

[13] N. Smirnov, Über Wahrscheinlichkeiten grosser Abweichungen (in Russian, German summary), *Receuil Mathématique* [*Sbornik*] *Moscou*, vol. 40 (1933), pp. 443–454.

[14] A. Khintchine, Über einen neuen Grenzwertsatz der Wahrscheinlichkeitsrechnung, *Mathematische Annalen*, vol. 101 (1929), pp. 745–752. Cf. also problem 16.

15. *Continuation.* Prove that if $x \to \infty$, $xh \to 0$,

(6.15) $$f\left(x + \frac{a}{x}\right) - f(x) \to 0$$

and hence

(6.16) $$F_n(x) \sim e^{f(x)}\{1 - \Phi(x)\}.$$

16. Deduce (6.10) from (6.16), assuming only $xh \to 0$.

17. If $p > q$, then for large x

$$Pr\{S_n > x\} < Pr\{S_n < -x\}.$$

(*Hint:* Use problem 14.)

*CHAPTER 8

UNLIMITED SEQUENCES OF BERNOULLI TRIALS

This chapter discusses certain properties of randomness and the important law of the iterated logarithm for Bernoulli trials. The theory is of general interest, but the material covered in subsequent chapters is not connected with it. A different type of limit theorem for Bernoulli trials will be discussed in Chapter 12, section 5.

1. Infinite Sequences of Trials

In the preceding chapter we have dealt with probabilities connected with n Bernoulli trials and have studied their asymptotic behavior as $n \to \infty$. We turn now to a more general type of problem where the events themselves cannot be defined in a finite sample space.

Example. *A Problem in Runs.* Let α and β be two positive integers, and consider a potentially unlimited sequence of Bernoulli trials, such as tossing a coin or throwing dice. Suppose that Paul bets Peter that a run of α consecutive successes will occur before a run of β consecutive failures. It has an obvious intuitive meaning to speak of the probability of the event that Paul wins, but it must be remembered that in the mathematical theory the term event stands for "aggregate of sample points" and is meaningless unless an appropriate sample space has been defined. The model of a finite number, n, of Bernoulli trials is insufficient for our present purpose, but the difficulty is solved by a simple passage to the limit. In n trials Peter wins or loses, or the game remains undecided. Let the corresponding probabilities be x_n, y_n, z_n ($x_n + y_n + z_n = 1$). As the number n of trials increases, the probability z_n of a tie can only decrease, while both x_n and y_n necessarily increase. Hence $x = \lim x_n$, $y = \lim y_n$, and $z = \lim z_n$ exist. Nobody would hesitate to call them the probabilities of Peter's ultimate gain or loss or of a tie. However, the corresponding three events are defined only in the sample space of infinite sequences of trials, and this space is not discrete.

The example was introduced for illustration only, and the numerical values of x_n, y_n, z_n are not our immediate concern. We shall return to their calculation in

* Starred chapters treat special topics and may be omitted at first reading.

Chapter 13, example (2.b). The limits x, y, z may be obtained by a simpler method which is applicable to more general cases. We indicate it here because of its importance and intrinsic interest.

Let A be the event that *a run of α consecutive successes occurs before a run of β consecutive failures*. Then A means Peter's winning and $x = Pr\{A\}$. If u and v are the conditional probabilities of A under the hypotheses, respectively, that the first trial results in success or failure, then $x = pu + qv$ [Chapter 5, (1.9)]. Suppose first that the first trial results in success. In this case the event A can occur in $\alpha - 1$ mutually exclusive ways: (1) The following $\alpha - 1$ trials result in successes; the probability for this is $p^{\alpha-1}$. (2) The first failure occurs at the νth trial where $2 \leq \nu < \alpha$. Let this event be H_ν. Then $Pr\{H_\nu\} = p^{\nu-2}q$, and $Pr\{A \mid H_\nu\} = v$. Hence (using once more the formula for compound probabilities)

$$(1.1) \qquad u = p^{\alpha-1} + qv(1 + p + \cdots p^{\alpha-2}) = p^{\alpha-1} + v(1 - p^{\alpha-1}).$$

In the case of the first trial resulting in failure a similar argument leads to

$$(1.2) \qquad v = pu(1 + q + \cdots + q^{\beta-2}) = u(1 - q^{\beta-1}).$$

We have thus two simple linear equations for the two unknowns u and v, and find for $x = pu + qv$

$$(1.3) \qquad x = p^{\alpha-1} \frac{1 - q^\beta}{p^{\alpha-1} + q^{\beta-1} - p^{\alpha-1}q^{\beta-1}}.$$

To obtain y we have only to interchange p and q, and α and β. Thus

$$(1.4) \qquad y = q^{\beta-1} \frac{1 - p^\alpha}{p^{\alpha-1} + q^{\beta-1} - p^{\alpha-1}q^{\beta-1}}.$$

Since $x + y = 1$, we have $z = 0$: *the probability of a tie is zero.*

For example, in tossing a coin ($p = 1/2$) the probability that a run of two heads appears before a run of three tails is 0.7; for two consecutive heads before four consecutive tails the probability is 5/6, for three consecutive heads before four consecutive tails 15/22. In rolling dice there is probability 0.1753 that two consecutive aces will appear before five consecutive non-aces, etc.

In the present volume we are confined to the theory of discrete sample spaces, and this means a considerable loss of mathematical elegance. The general theory considers n Bernoulli trials only as the beginning of an infinite sequence of trials. A sample point is then represented by an infinite sequence of letters S and F, and the sample space is the aggregate of all such sequences. A finite sequence, like *SSFS*, stands for the aggregate of all points with this beginning, that is, for the compound event that in an infinite sequence of trials the first four result in S, S, F, S, respectively. In the infinite sample space the game of our example can be interpreted without a limiting process. Take any point, that is, a sequence *SSFSFF* .... In it a run of α consecutive S's may or may not occur. If it does, it may or may not be preceded by a run of β consecutive F's. In this way we get a classification of all sample points into three classes, representing the events "Peter wins," "Peter loses," "no decision." Their probabilities are the numbers x, y, z, computed above. The only trouble with this sample space is that it is not discrete, and we have not yet defined probabilities in general sample spaces.

Note that we are discussing a question of terminology rather than a genuine difficulty. In our example there was no question about the proper definition or interpretation of the number x. The trouble is only that for consistency we must either decide to refer to the number x as "the limit of the probability x_n that Peter wins in n trials" or else talk of the event "that Peter wins," which means referring to a nondiscrete sample space. We propose to do both. For simplicity of language we shall refer to events even when they are defined in the infinite sample space; for precision, the theorems will also be formulated in terms of finite sample spaces and passages to the limit. The events to be studied in this chapter share the following salient feature of our example. The event "Peter wins," although defined in an infinite space, is the union of the events "Peter wins at the nth trial" ($n = 1, 2, \ldots$), each of which depends only on a finite number of trials. The required probability x is the limit of a monotonic sequence of probabilities x_n which depend only on finitely many trials. We require no theory going beyond the model of n Bernoulli trials; we merely take the liberty of simplifying clumsy expressions [1] by calling certain numbers probabilities instead of using the term "limits of probabilities."

2. Systems of Gambling

The painful experience of many gamblers has taught us the lesson that no system of betting is successful in improving the gambler's chances. If the theory of probability is true to life, this experience must correspond to a provable statement.

For orientation let us consider a potentially unlimited sequence of Bernoulli trials and suppose that at each trial the bettor has the free choice of whether or not to bet. A "system" consists in fixed rules selecting those trials on which the player is to bet. For example, the bettor may make up his mind to bet at every seventh trial or to wait

---

[1] For the reader familiar with general measure theory the situation may be described as follows. We consider only events which either depend on a finite number of trials or are limits of *monotonic* sequences of such events. We calculate the obvious limits of probabilities and clearly require no measure theory for that purpose. However, only general measure theory shows that our limits are independent of the particular passage to the limit and are completely additive.

as long as necessary for seven heads to occur between two bets. He may bet only after head runs of length 13, or bet for the first time after the first head, for the second time after the first run of two consecutive heads, and generally, for the kth time, just after k heads have appeared in succession. In the latter case he would bet less and less frequently. Another possible system consists in betting whenever the accumulated number of heads exceeds the accumulated number of tails (in which case the average player would bet on tails). We need not consider the stakes at the individual trials; we want to show that no "system" changes the bettor's situation and that he can achieve the same result by betting every time. It goes without saying that this statement can be proved only for systems in the ordinary meaning where the bettor does not know the future (the existence or non-existence of genuine prescience is not our concern). It must also be admitted that the rule "go home after losing three times" does change the situation, but we shall rule out such uninteresting systems.

We define a system as a set of fixed rules which for every trial uniquely determines whether or not the bettor is to bet; at the kth trial the decision may depend on the outcomes of the first $k-1$ trials, but not on the outcome of trials number k, $k+1$, $k+2$, $\ldots$; finally the rules must be such as to ensure an indefinite continuation of the game. The last condition means the following. Since the set of rules is fixed, the event "in n trials the bettor bets more than r times" is well defined and its probability calculable. It is required that for every r, as $n \to \infty$, this probability tends to 1.

We now formulate our fundamental theorem to the effect that *under any system the successive bets form a sequence of Bernoulli trials with unchanged probability for success*. With an appropriate change of phrasing this theorem holds for all kinds of independent trials; the successive bets form in each case an exact replica of the original trials, so that no system can affect the bettor's fortunes. The importance of this statement was first recognized by von Mises, who introduced the impossibility of a successful gambling system as a fundamental axiom. The present formulation and proof follow Doob.[2] For simplicity we assume that $p = 1/2$.

Let A_k be the event "first bet occurs at the kth trial." Our definition of system requires that as $n \to \infty$ the probability tends to one that the first bet has occurred before the nth trial. This means that $Pr\{A_1\} + Pr\{A_2\} + \cdots + Pr\{A_n\} \to 1$, or

(2.1) $$\Sigma Pr\{A_k\} = 1.$$

[2] J. L. Doob, Note on Probability, *Annals of Mathematics*, vol. 37 (1936), pp. 363–367.

8.2] SYSTEMS OF GAMBLING

Next, let B_k be the event "head at kth trial." Then the event B "when first bet is made the trial results in heads" is the union of the events A_1B_1, A_2B_2, A_3B_3, ... which are mutually exclusive. Now A_k depends only on the outcome of the first $k-1$ trials, and B_k only on the trial number k. Hence A_k and B_k are independent and $Pr\{A_kB_k\} = Pr\{A_k\}Pr\{B_k\} = \frac{1}{2}Pr\{A_k\}$. Thus $Pr\{B\} = \Sigma Pr\{A_kB_k\} = \frac{1}{2}\Sigma Pr\{A_k\} = 1/2$. This shows that under this system the probability of heads at the first bet is $1/2$, and the same statement holds for all subsequent bets.

It remains to show that the bets are statistically independent. This means that the probability that the coin falls heads at both the first and the second bet should be $1/4$ (and similarly for all other combinations and for the subsequent trials). To verify this statement let A_k' be the event that the second bet occurs at the kth trial. Let E represent the event "heads at the first two bets"; it is the union of all events $A_jB_jA_k'B_k$ (if $j < k$, then A_j and A_k' are mutually exclusive and $A_jA_k' = 0$). Therefore

$$(2.2) \qquad Pr\{E\} = \sum_{j=1}^{\infty} \sum_{k=j+1}^{\infty} Pr\{A_jB_jA_k'B_k\}.$$

As before, we see that for fixed j and $k > j$, the event B_k (heads at kth trial) is independent of the event $A_jB_jA_k'$ (which depends only on the outcomes of the first $k-1$ trials). Hence

$$(2.3) \qquad Pr\{E\} = \frac{1}{2} \sum_{j=1}^{\infty} \sum_{k=j+1}^{\infty} Pr\{A_jB_jA_k'\}$$

$$= \frac{1}{2} \sum_{j=1}^{\infty} Pr\{A_jB_j\} \sum_{k=j+1}^{\infty} Pr\{A_k' \mid A_jB_j\}$$

[cf. Chapter 5, (1.9)]. Now, whenever the first bet occurs and whatever its outcome, the game is sure to continue, that is, the second bet occurs sooner or later. This means that for given A_jB_j with $Pr\{A_jB_j\} > 0$ the conditional probabilities that the second bet occurs at the kth trial must add to unity. The second series in (2.3) is therefore unity, and we have already seen that $\Sigma Pr\{A_jB_j\} = 1/2$. Hence $Pr\{E\} = 1/4$ as contended. A similar argument holds for any combination of trials.

Note that the situation is different when the player is permitted to vary arbitrarily the amounts which he puts down. Under such conditions there exist advantageous and disadvantageous strategies, and the game depends on the strategy. We shall return to this point in Chapter 14, section 2.

3. The Borel-Cantelli Lemmas

Two simple lemmas concerning infinite sequences of trials are used so frequently that they deserve special attention. We formulate them for Bernoulli trials, but they apply to more general cases. Lemma 1 is used in section 4; lemma 2 in sections 4 and 5.

We refer again to an infinite sequence of Bernoulli trials. Let $A_1, A_2, \ldots$ be an infinite sequence of events each of which depends only on a finite number of trials; in other words, we suppose that there exists an integer n_k such that A_k is an event in the sample space of the first n_k Bernoulli trials. Put

$$(3.1) \qquad a_k = Pr\{A_k\}.$$

(For example, A_k may stand for the event that trial number $2k$ concludes a run of at least k consecutive successes. Then $n_k = 2k$ and $a_k = p^k$.)

For every particular infinite sequence of letters S and F it is possible to establish whether it belongs to $0, 1, 2, \ldots$ or infinitely many among the $\{A_k\}$. This means that we can speak of the event U_r, that an unending sequence of trials produces more than r among the events $\{A_k\}$, and also of the event U_∞, that infinitely many among the $\{A_k\}$ occur. The event U_r is defined only in the infinite sample space, and its probability is the limit of $Pr\{U_{n,r}\}$, the probability that n trials produce more than r among the events $\{A_k\}$. Finally, $Pr\{U_\infty\} = \lim Pr\{U_r\}$; this limit exists since $Pr\{U_r\}$ can only decrease as r increases.

Lemma 1. *If Σa_k converges, then with probability one only finitely many events A_k occur.* More precisely, it is claimed that for r sufficiently large, $Pr\{U_r\} < \epsilon$ or: *to every $\epsilon > 0$ it is possible to find an integer r such that the probability that n trials produce one or more among the events $A_{r+1}, A_{r+2}, \ldots$ is less than ϵ for all n.*

Proof. Determine r so that $a_{r+1} + a_{r+2} + \ldots < \epsilon$; this is possible since Σa_k converges. Without loss of generality we may suppose that the A_k are ordered in such a way that $n_1 \leq n_2 \leq n_3 \leq \ldots$. Let N be the last subscript for which $n_N \leq n$. Then $A_1, \cdots, A_N$ are defined in the space of n trials, and the lemma asserts that the probability that one or more among the events $A_{r+1}, A_{r+2}, \cdots, A_N$ occur is less than ϵ. This is true, since by the fundamental inequality (6.6) of Chapter 1, we have for this probability

$$(3.2) \qquad Pr\{A_{r+1} \cup A_{r+2} \cup \cdots \cup A_N\} \leq a_{r+1} + a_{r+2} + \cdots + a_N \leq \epsilon,$$

as contended.

A satisfactory converse to the lemma is known only for the special case of mutually independent A_k. This situation occurs when the trials are divided into non-overlapping blocks and A_k depends only on the trials in the kth block (for example, A_k may be the event that the kth thousand of trials produces more than 600 successes).

Lemma 2. *If the events A_k are mutually independent, and if Σa_k diverges, then with probability one infinitely many A_k occur.* In other words, it is claimed that for every r the probability that n trials produce more than r among the events $\{A_k\}$ tends to 1 as $n \to \infty$.

Proof. As in the proof of lemma 1 let $A_1, A_2, \cdots, A_N$ be the events defined in the sample space of n trials. The probability that none of them occurs is, because of the assumed independence, $(1-a_1)(1-a_2)\cdots(1-a_N)$. Now $1 - x < e^{-x}$ for $0 < x < 1$, and hence $(1-a_1)(1-a_2)\cdots(1-a_N) < e^{-(a_1+a_2+\cdots+a_N)}$; with increasing N the last quantity tends to zero. We have thus proved that with probability one at least one among the $\{A_k\}$ occurs.

Next, divide the sequence $\{A_k\}$ into two subsequences $\{A_k'\}$ and $\{A_k''\}$ so that both series $\Sigma Pr\{A_k'\}$ and $\Sigma Pr\{A_k''\}$ diverge. Applying our result to these subsequences we find that, with probability one, at least one A_k' and one A_k'' occur. Therefore there is probability one that at least two among the $\{A_k\}$ occur. Applying, in turn, this statement to the sequences $\{A_k'\}$ and $\{A_k''\}$ we find that at least four among the $\{A_k\}$ are bound to occur, etc.

Example. What is the probability that in a sequence of Bernoulli trials the pattern *SFS* appears infinitely often? Let A_k be the event that the trials number k, $k+1$, and $k+2$ produce the sequence *SFS*. The events A_k are obviously not mutually independent, but the sequence $A_1, A_4, A_7, A_{10}, \ldots$ contains only mutually independent events (since no two depend on the outcome of the same trials). Since $a_k = p^2q$ is independent of k, the series $a_1 + a_4 + a_7 + \cdots$ diverges, and hence with probability one the pattern *SFS* occurs infinitely often. A similar argument obviously applies for arbitrary patterns.

4. The Strong Law of Large Numbers

The intuitive notion of probability is based on the expectation that the following is true: if the number of successes in the first n trials of a sequence of Bernoulli trials is $\mathbf{S}_n$, then

(4.1) $$\frac{\mathbf{S}_n}{n} \to p.$$

In the abstract theory this cannot be true for *every* sequence of trials; in fact, our sample space contains a point representing the conceptual possibility of an infinite sequence of uninterrupted successes, and for it $S_n/n = 1$. However, it is demonstrable that (4.1) holds with probability one, so that the cases where (4.1) does not hold form a negligible exception.

Note that we deal with a statement which is much stronger than the weak law of large numbers [Chapter 7, (3.2)]. The latter says that for every sufficiently large *fixed* n the average S_n/n is likely to be near p, but it does not say that S_n/n is bound to stay near p if the number of trials is increased. It leaves open the possibility that in n additional trials at least one of the events $S_{n+1}/(n+1) < p - \epsilon$, or $S_{n+2}/(n+2) < p - \epsilon, \ldots$, or $S_{2n}/2n < p - \epsilon$, occurs; the probability of this is the sum of a large number of probabilities of which we know only that they are individually small. We shall now prove that with probability one $S_n/n - p$ becomes *and remains* small.

Strong Law of Large Numbers. *For every $\epsilon > 0$ we have probability one that only finitely many of the events*

$$(4.2) \qquad \left| \frac{S_n}{n} - p \right| > \epsilon$$

occur. This implies that (4.1) holds with probability one. In terms of finite sample spaces, it is asserted that to every $\epsilon > 0$, $\delta > 0$ there corresponds an r such that for all ν the probability of the simultaneous realization of the ν inequalities

$$(4.3) \qquad \left| \frac{S_{r+k}}{r+k} - p \right| < \epsilon, \qquad k = 1, 2, \cdots, \nu$$

is greater than $1 - \delta$.

Proof. We shall prove a much stronger statement. Let A_k be the event

$$(4.4) \qquad \left| \frac{S_k - kp}{(kpq)^{1/2}} \right| > 2(\log k)^{1/2}.$$

Then, by formula (5.2) of Chapter 7

$$(4.5) \qquad Pr\{A_k\} \sim \frac{1}{2(2\pi)^{1/2}(\log k)^{1/2}} e^{-2 \log k} < \frac{1}{k^2},$$

and hence $\Sigma Pr\{A_k\}$ converges. Thus lemma 1 of the preceding section ensures that *with probability one only finitely many inequalities* (4.4) *hold*. On the other hand, (4.2) implies

$$\left| \frac{S_n - np}{(npq)^{1/2}} \right| > \frac{\epsilon}{(pq)^{1/2}} \cdot n^{1/2} \tag{4.6}$$

and for large n the right side is larger than $2(\log n)^{1/2}$. Hence, the realization of infinitely many inequalities (4.2) implies the realization of infinitely many A_k and has therefore probability zero.

The strong law of large numbers was first formulated by Cantelli (1917), after Borel and Hausdorff had discussed certain special cases. Like the weak law, it is only a very special case of a general theorem on random variables. Taken in conjunction with our theorem on the impossibility of gambling systems, the law of large numbers implies the existence of the limit (4.1) not only for the original sequence of trials but also for all subsequences obtained in accordance with the rules of section 2. *Thus the two theorems together describe the fundamental properties of randomness which are inherent in the intuitive notion of probability and whose importance was stressed with special emphasis by von Mises.*

5. The Law of the Iterated Logarithm

As in Chapter 7 let us again introduce the reduced number of successes in n trials

$$S_n^* = \frac{S_n - np}{(npq)^{1/2}}. \tag{5.1}$$

The Laplace limit theorem asserts that $Pr\{S_n^* > x\} \sim 1 - \Phi(x)$. Thus, for every particular value of n it is improbable to have a large S_n^*, but it is intuitively clear that in a prolonged sequence of trials S_n^* will sooner or later take on arbitrarily large values. Moderate values of S_n^* are most probable, but the maxima will slowly increase. How fast? In the course of the proof of the strong law of large numbers we have seen that with probability one

$$S_n^* < 2(\log n)^{1/2} \tag{5.2}$$

for all sufficiently large n: this gives us an upper bound for the fluctuations of S_n^*. More information is contained in the following remarkable theorem discovered by Khintchine.[3]

---

[3] A. Khintchine, Über einen Satz der Wahrscheinlichkeitsrechnung, *Fundamenta Mathematicae*, vol. 6 (1924), pp. 9–20. The discovery was preceded by partial results due to other authors. The present proof is arranged so as to permit straightforward generalization to more general random variables.

Theorem. *With probability one we have*

(5.3) $$\limsup_{n\to\infty} \frac{S_n^*}{(2 \log \log n)^{1/2}} = 1.$$

This means: if $\lambda > 1$, then there is probability one that only finitely many of the events

(5.4) $$S_n > np + \lambda(2npq \log \log n)^{1/2}$$

occur; if $\lambda < 1$, then there is probability one that (5.4) holds for infinitely many n.

For reasons of symmetry equation (5.3) implies that

(5.3a) $$\liminf_{n\to\infty} \frac{S_n^*}{(2 \log \log n)^{1/2}} = -1.$$

Proof. We start with two preliminary remarks concerning Bernoulli trials.

(1) There exists a constant $c > 0$ which depends on p, but not on n, such that

(5.5) $$Pr\{S_n > np\} > c$$

for all n. In fact, an inspection of the binomial distribution shows that the left side in (5.5) is never zero, and the Laplace limit theorem shows that it tends to $\frac{1}{2}$ as $n \to \infty$. Accordingly, the left side is bounded away from zero, as asserted.

(2) We require the following *lemma:* Let x be fixed, and let A be the event that for at least one k with $k \leq n$

(5.6) $$S_k - kp > x.$$

Then

(5.7) $$Pr\{A\} \leq \frac{1}{c} Pr\{S_n - np > x\}.$$

For a proof of the lemma let A_ν be the event that (5.6) holds for $k = \nu$ but not for $k = 1, 2, \cdots, \nu - 1$ (here $1 \leq \nu \leq n$). The events $A_1, A_2, \cdots, A_n$ are mutually exclusive, and A is their union. Hence

(5.8) $$Pr\{A\} = Pr\{A_1\} + \cdots + Pr\{A_n\}.$$

Next, for $\nu < n$ let U_ν be the event that the total number of successes in the trials number $\nu + 1, \nu + 2, \cdots, n$ exceeds $(n - \nu)p$. If both A_ν and U_ν occur, then $S_n > S_\nu + (n - \nu)p > np + x$, and since the

$A_\nu U_\nu$ are mutually exclusive, this implies

$$Pr\{S_n - np > x\}$$
(5.9) $\geq Pr\{A_1 U_1\} + Pr\{A_2 U_2\} + \cdots + Pr\{A_{n-1} U_{n-1}\} + Pr\{A_n\}.$

Now A_ν depends only on the first ν trials, and U_ν only on the following $n - \nu$ trials. Hence A_ν and U_ν are independent, and $Pr\{A_\nu U_\nu\} = Pr\{A_\nu\} Pr\{U_\nu\}$. From the preliminary remark (5.5) we know that $Pr\{U_\nu\} > c$, and since $c < 1$, we get from (5.9) and (5.8)

(5.10) $\qquad Pr\{S_n - np > x\} \geq c \, \Sigma Pr\{A_\nu\} = c \, Pr\{A\}.$

This proves (5.7).

(3) We now prove the part of the theorem relating to (5.4) with $\lambda > 1$. Let γ be a number such that

(5.11) $\qquad\qquad 1 < \gamma < \lambda^{\frac{1}{2}},$

and let n_r be the integer nearest to $\gamma^r (r = 1, 2, \ldots)$. Let B_r be the event that the inequality

(5.12) $\qquad S_n - np > \lambda (2 n_r p q \log \log n_r)^{\frac{1}{2}}$

holds for at least one n with $n_r \leq n < n_{r+1}$. Obviously (5.4) can hold for infinitely many n only if infinitely many B_r occur. Using the first Borel-Cantelli lemma, we see therefore that it suffices to prove that

(5.13) $\qquad\qquad \Sigma Pr\{B_r\}$ converges.

By the lemma

(5.14) $Pr\{B_r\} \leq c^{-1} Pr\{S_{n_{r+1}} - n_{r+1} p > \lambda (2 n_r p q \log \log n_r)^{\frac{1}{2}}\}$

$$= c^{-1} Pr\left\{ S^*_{n_{r+1}} > \lambda \left(2 \frac{n_r}{n_{r+1}} \log \log n_r\right)^{\frac{1}{2}} \right\}.$$

Now $n_r / n_{r+1} \sim \gamma^{-1} > \lambda^{-\frac{1}{2}}$, and hence for sufficiently large r

(5.15) $\qquad Pr\{B_r\} \leq c^{-1} Pr\{S^*_{n_{r+1}} > (2\lambda \log \log n_r)^{\frac{1}{2}}\}.$

From the DeMoivre-Laplace limit theorem [formula (5.2) of Chapter 7] we get, therefore, for large r

(5.16) $\qquad Pr\{B_r\} \leq c^{-1} e^{-\lambda \log \log n_r} = \dfrac{1}{c(\log n_r)^\lambda} \sim \dfrac{1}{c(r \log \gamma)^\lambda}.$

Since $\lambda > 1$, the assertion (5.13) is proved.

(4) Finally, we prove the assertion concerning (5.4) with $\lambda < 1$. This time we choose for γ an integer so large that

(5.17) $$\frac{\gamma - 1}{\gamma} > \eta > \lambda$$

where η is a constant to be determined later. Put

(5.18) $$n_r = \gamma^r.$$

The second Borel-Cantelli lemma applies only to independent events. For this reason we introduce

(5.19) $$\boldsymbol{D}_r = \boldsymbol{S}_{n_r} - \boldsymbol{S}_{n_{r-1}};$$

D_r is the total number of successes following trial number n_{r-1} and up to and including trial n_r; for it we have the binomial distribution $b(k; n, p)$ with $n = n_r - n_{r-1}$. Let A_r be the event

(5.20) $$\boldsymbol{D}_r - (n_r - n_{r-1})p > \eta(2pqn_r \log\log n_r)^{1/2}.$$

We claim that *with probability one infinitely many A_r occur*. Since the various A_r depend on non-overlapping blocks of trials (namely, $n_{r-1} < n \leq n_r$), they are mutually independent, and, according to the second Borel-Cantelli lemma, it suffices to prove that

(5.21) $$\Sigma Pr\{A_r\} \text{ diverges.}$$

Now

(5.22) $Pr\{A_r\}$

$$= Pr\left\{\frac{D_r - (n_r - n_{r-1})p}{\{(n_r - n_{r-1})pq\}^{1/2}} > \eta\left(2\frac{n_r}{n_r - n_{r-1}}\log\log n_r\right)^{1/2}\right\}.$$

Here $n_r/(n_r - n_{r-1}) = \gamma/(\gamma - 1) < \eta^{-1}$, by (5.17). Hence

(5.23) $$Pr\{A_r\} \geq Pr\left\{\frac{D_r - (n_r - n_{r-1})p}{\{(n_r - n_{r-1})pq\}^{1/2}} > (2\eta \log\log n_r)^{1/2}\right\}.$$

Using again the estimate (5.2) of Chapter 7, we find for large r

(5.24) $$Pr\{A_r\} > \frac{1}{2\eta \log\log n_r} e^{-\eta \log\log n_r} = \frac{1}{2\eta (\log\log n_r)(\log n_r)^\eta}.$$

Since $n_r = \gamma^r$ and $\eta < 1$, we find that for large r we have $Pr\{A_r\} > 1/r$, which proves (5.21).

The last step of the proof consists in showing that $\boldsymbol{S}_{n_{r-1}}$ in (5.19) can be neglected. From the first part of the theorem, which has already been proved, we know that to every $\epsilon > 0$ we can find an N so that, with probability $1 - \epsilon$ or better, for all $r > N$,

(5.25) $\qquad |S_{n_{r-1}} - n_{r-1}\, p| < 2(2pq n_{r-1} \log \log n_{r-1})^{1/2}.$

Now suppose that η is chosen so close to 1 that

(5.26) $\qquad\qquad 1 - \eta < \left(\dfrac{\eta - \lambda}{2}\right)^2.$

Then from (5.17)

(5.27) $\qquad\qquad 4n_{r-1} = 4\dfrac{n_r}{\gamma} < n_r(\eta - \lambda)^2$

and hence (5.25) implies

(5.28) $\qquad S_{n_{r-1}} - n_{r-1}p > -(\eta - \lambda)(2pq n_r \log \log n_r)^{1/2}.$

Adding (5.28) to (5.20), we obtain (5.4) with $n = n_r$. It follows that, with probability $1 - \epsilon$ or better, this inequality holds for infinitely many r, and this accomplishes the proof.

The law of the iterated logarithm for Bernoulli trials is a special case of a more general theorem first formulated by Kolmogorov.[4] It is now possible to formulate stronger theorems (cf. problems 5 and 6).

6. Interpretation in Number Theory Language

Let x be a real number in the interval $0 \leq x < 1$, and let

(6.1) $\qquad\qquad x = .a_1 a_2 a_3 \ldots$

be its decimal expansion (so that each a_j stands for one of the digits $0, 1, \cdots, 9$). This expansion is unique except for numbers of the form $a/10^n$, which can be written either by means of a terminating expansion (containing infinitely many zeros) or by means of an expansion containing infinitely many nines. To avoid ambiguities we now agree not to use the latter form.

The decimal expansions are connected with Bernoulli trials with $p = 1/10$, the digit 0 representing success and all other digits failure. If we replace in (6.1) all zeros by the letter S and all other digits by F, then (6.1) represents a possible outcome of an infinite sequence of Bernoulli trials with $p = 1/10$. Conversely, an arbitrary sequence of letters S and F can be obtained in the described manner from the expansion of certain numbers x. In this way every event in the sample space of Bernoulli trials is represented by a certain aggregate of numbers x.

---

[4] A. Kolmogoroff, *Das Gesetz des iterierten Logarithmus*, *Mathematische Annalen*, vol. 101 (1929), pp. 126–135.

For example, the event "success at the nth trial" is represented by all those x whose nth decimal is zero. This is an aggregate of 10^{n-1} intervals each of length 10^{-n}, and the total length of these intervals equals $1/10$, which is the probability of our event. Every particular finite sample sequence of length n corresponds to an aggregate of certain intervals; for example, the sequence SFS is represented by the nine intervals $0.01 \leq x < 0.011$, $0.02 \leq x < 0.021$, $\cdots$, $0.09 \leq x < 0.091$. The probability of each such sample sequence equals the total length of the corresponding intervals on the x-axis. Probabilities of more complicated events are always expressed in terms of probabilities of finite sample sequences, and the calculation proceeds according to the same addition rule which is valid for the usual Lebesgue measure on the x-axis. Accordingly, our probabilities will always coincide with the measure of the corresponding aggregate of points on the x-axis. We have thus a means of translating all limit theorems for Bernoulli trials with $p = 1/10$ into theorems concerning decimal expansions. The phrase "with probability one" is equivalent to "for almost all x" or "almost everywhere."

We have considered the random variable S_n which gives the number of successes in n trials. Here it is more convenient to emphasize the fact that S_n is a function of the sample point, and we write $S_n(x)$ *for the number of zeros among the n first decimals of x*. Obviously $S_n(x)$ is a function of x whose graph is a step polygon whose discontinuities are necessarily points of the form $a/10^n$, where a is an integer. The ratio $S_n(x)/n$ is called the *frequency of zeros* among the first n decimals of x.

In the language of ordinary measure theory the weak law of large numbers asserts that $S_n(x)/n \to 1/10$ in measure, while the strong law states that $S_n(x)/n \to 1/10$ almost everywhere. Khintchine's law of the iterated logarithm shows that

$$(6.2) \qquad \limsup \frac{S_n(x) - n/10}{(n \log \log n)^{1/2}} = (0.3)2^{1/2}$$

for almost all x. It gives an answer to a problem treated in a series of papers initiated by Hausdorff [5] (1913) and Hardy and Littlewood [6] (1914). For a further improvement of this result see problems 5 and 6.

Instead of the digit zero we may consider any other digit and can formulate the strong law of large numbers to the effect that the frequency of each of the ten digits tends to $1/10$ for almost all x. A similar

---

[5] F. Hausdorff, *Grundzüge der Mengenlehre*, Leipzig, 1913.

[6] Hardy and Littlewood, Some Problems of Diophantine Approximation, *Acta Mathematica*, vol. 37 (1914), pp. 155–239.

theorem holds if the base 10 of the decimal system is replaced by any other base. This fact was discovered by Borel (1909) and is usually expressed by saying that almost all numbers are "normal."

7. Problems for Solution

1. Find an integer β such that in rolling dice there are about even chances that a run of three consecutive aces appears before a non-ace run of length β.

2. Consider repeated independent trials with three possible outcomes A, B, C and corresponding probabilities p, q, r ($p + q + r = 1$). Find the probability that a run of α consecutive A's will occur before a B-run of length β.

3. *Continuation.* Find the probability that an A-run of length α will occur before either a B-run of length β or a C-run of length γ.

4. In a sequence of Bernoulli trials let A_n be the event that a run of n consecutive successes occurs between the 2^nth and the 2^{n+1}th trial. If $p \geq 1/2$, there is probability one that infinitely many A_n occur; if $p < 1/2$, then with probability one only finitely many A_n occur.

5. Let $\phi(t)$ be a positive monotonically increasing function, and let n_r be the nearest integer to $e^{r/\log r}$. If

(7.1) $$\sum \frac{1}{\phi(n_r)} e^{-\frac{1}{2}\phi^2(n_r)}$$

converges, then with probability one, the inequality

(7.2) $$S_n > np + (npq)^{\frac{1}{2}}\phi(n)$$

takes place only for finitely many n. [Note that without loss of generality we may suppose that $\phi(n) < 10(\log \log n)^{\frac{1}{2}}$; the law of the iterated logarithm takes care of the larger $\phi(n)$.]

6. Prove [7] that the series (7.1) converges if, and only if,

(7.3) $$\sum \frac{\phi(n)}{n} e^{-\frac{1}{2}\phi^2(n)}$$

converges. [*Hint*: Collect the terms $n_{r-1} < n < n_r$ and note that $n_r - n_{r-1} \sim n_r(1 - 1/\log r)$; furthermore, (7.3) can converge only if $\phi^2(n) > 2 \log \log n$.]

---

[7] Problems 5 and 6 together show that in case of convergence of (7.3) the inequality (7.2) holds with probability one only for finitely many n. Conversely, if (7.3) diverges, the inequality (7.2) holds with probability one for infinitely many n. This converse is much more difficult to prove; cf. W. Feller, The General Form of the So-called Law of the Iterated Logarithm, *Transactions of the American Mathematical Society*, vol. 54 (1943), pp. 373–402, where more general theorems are proved for arbitrary random variables. For the special case of Bernoulli trials with $p = 1/2$ cf. P. Erdös, On the Law of the Iterated Logarithm, *Annals of Mathematics* (2), vol. 43 (1942), pp. 419–436. The law of the iterated logarithm follows for the particular case $\phi(t) = \lambda(2 \log \log t)$.

CHAPTER 9

RANDOM VARIABLES; EXPECTATION

1. Random Variables

According to the definition given in calculus textbooks, the quantity y is called a *function* of the real number x if to every x there corresponds a value y. This definition can be extended to cases where the independent variable is not a real number. Thus we call the distance a function of a pair of points; the perimeter of a triangle is a function defined on the set of triangles; a sequence a_n is a function defined for all positive integers; the binomial coefficient $\binom{x}{k}$ is a function defined for pairs of numbers (x, k) of which the second is a non-negative integer. In the same sense we can say that the number S_n of successes in n Bernoulli trials is a function defined on the sample space; to each of the 2^n points in this space there corresponds a number S_n.

A function defined on a sample space is called a random variable. Throughout the preceding chapters we have been concerned with random variables without using this term. Typical random variables are the number of aces in a hand at bridge, of multiple birthdays in a company of n people, of success runs in n Bernoulli trials. In each case there is a unique rule which associates a number X with any sample point. The classical theory of probability was devoted mainly to a study of the gambler's gain, which is again a random variable; in fact, every random variable can be interpreted as the gain of a real or imaginary gambler in a suitable game. The position of a particle under diffusion, the energy, temperature, etc., of physical systems are random variables; but they are defined in non-discrete sample spaces, and their study is therefore deferred. In the case of a discrete sample space we can actually tabulate any random variable X by enumerating in some order all points of the space and associating with each the corresponding value of X.

The term random variable is somewhat confusing, and random function would be more appropriate (the independent variable being a point in sample space, i.e., outcome of an experiment). The conceptual confusion is increased by the tradition of denoting random variables by single letters without referring to sample points; the same custom now prevails in many branches of mathematics, however.

Let X be a random variable and let $x_1, x_2, x_3, \ldots$ be the values which it assumes (in most of what follows, the x_j will be integers). In general the same value x_j may correspond to several sample points. Their aggregate forms the event that $X = x_j$; its probability is denoted by $Pr\{X = x_j\}$. *The system of relations*

(1.1) $$Pr\{X = x_j\} = f(x_j) \qquad (j = 1, 2, \ldots)$$

defines the (probability) distribution [1] *of the random variable* X. Clearly

(1.2) $$f(x_j) \geq 0, \qquad \Sigma f(x_j) = 1.$$

If a value x is never assumed, we agree to write $Pr\{X = x\} = 0$.

Examples. The number S_n of successes in n Bernoulli trials is a random variable whose distribution is the binomial distribution $b(k; n, p)$. Another example of a random variable is the number X of aces in a random sample of r bridge cards. The possible values of X are 0, 1, 2, 3, 4, and the corresponding probabilities are given by formula (5.3) and Table 3 of Chapter 2. Each line of this table represents a probability distribution.

Consider now two random variables X and Y defined on the same sample space, and denote the values which they assume, respectively, by $x_1, x_2, \ldots$, and $y_1, y_2, \ldots$; let the corresponding probability distribution be $\{f(x_j)\}$ and $\{g(y_k)\}$. The aggregate of points in which the two conditions $X = x_j$ and $Y = y_k$ are satisfied forms an event whose probability will be denoted by $Pr\{X = x_j, Y = y_k\}$. *The system of equations*

(1.3) $$Pr\{X = x_j, Y = y_k\} = p(x_j, y_k) \qquad (j, k = 1, 2, \cdots)$$

[1] For a discrete variable X the probability distribution is the function $f(x_j)$ defined on the aggregate of possible values x_j of X. This term must be distinguished from the term "distribution function," which applies to any non-decreasing function $F(x)$ which tends to 0 as $x \to -\infty$ and to 1 as $x \to \infty$. The distribution function $F(x)$ of X is defined by

$$F(x) = Pr\{X \leq x\} = \sum_{x_j \leq x} f(x_j),$$

the last sum extending over all those x_j which do not exceed x. Thus the distribution function of a variable can be calculated from its probability distribution and vice versa. In this volume we shall not be concerned with distribution functions in general.

is called the joint probability distribution *of X and Y.* It is best exhibited in the form of a double-entry table as exemplified in Table 1. (Note, however, that the numbers of rows and columns need not be equal.) Clearly

(1.4) $$p(x_j, y_k) \geq 0, \quad \sum_{j,k} p(x_j, y_k) = 1.$$

Moreover, for every fixed j

(1.5) $\quad p(x_j, y_1) + p(x_j, y_2) + p(x_j, y_3) + \cdots = Pr\{X = x_j\} = f(x_j)$

and for every fixed k

(1.6) $\quad p(x_1, y_k) + p(x_2, y_k) + p(x_3, y_k) + \cdots = Pr\{Y = y_k\} = g(y_k).$

In other words, by adding the probabilities in individual rows and columns, we obtain the probability distributions of X and Y. They may be exhibited as shown in Table 1 and are then called *marginal distributions.* The adjective "marginal" refers to the outer appearance in the double-entry table and is also used for stylistic clarity when the

TABLE 1

JOINT DISTRIBUTION OF THE NUMBERS OF ACES IN TWO SPECIFIED HANDS AT BRIDGE

Y	X					Marginal Distribution for Y
	0	1	2	3	4	
0	0.055 222	0.124 850	0.093 637	0.027 467	0.002 641	0.303 817
1	.124 850	.202 881	.097 383	.013 734	0	.438 848
2	.093 637	.097 383	.022 473	0	0	.213 493
3	.027 467	.013 734	0	0	0	.041 201
4	.002 641	0	0	0	0	.002 641
Marginal Distribution for X	.303 817	.438 848	.213 493	.041 201	.002 641	

$E(X) = E(Y) = 1; \quad \mathrm{Var}(X) = \mathrm{Var}(Y) = 0.705\,885\ldots,$
$\mathrm{Cov}(X, Y) = -0.235\,295\ldots, \quad \rho(X, Y) = -1/3.$

joint distribution of two variables and also their individual (marginal) distributions appear in the same context. Strictly speaking, the adjective "marginal" is always redundant.

The notion of joint distribution carries over to systems of more than two random variables.

Examples. [For numerical examples of joint distributions cf. Table 1 and the answer to problem 1.]

(a) Equation (5.8) of Chapter 2 gives the probability that a sample of size r contains $X = k_1$ elements of the first class and $Y = k_2$ elements of the second class, that is, the joint distribution of X and Y. We get the marginal distribution $Pr\{X = k_1\}$ if we keep k_1 fixed and add over $k_2 = 0, 1, \cdots, r$. In view of (5.7) of Chapter 2 this operation leads to the ordinary hypergeometric distribution.

(b) In n throws of a die let X, Y, Z, U, V, W be the numbers of ones, twos, $\cdots$, sixes. The joint distribution of these six variables follows from the multinomial distribution (7.2) of Chapter 6. It is given by

$$(1.7) \qquad p(k_1, k_2, k_3, k_4, k_5, k_6) = \frac{n! 6^{-n}}{k_1! k_2! k_3! k_4! k_5! k_6!}$$

provided that $k_1 + k_2 + \cdots + k_6 = n$; for every other combination the probability is zero. For given $X, \cdots, V$ there is only one possible value for W. Keeping k_1, k_2, k_3, k_4 fixed and adding over $k_5 = 0, 1, 2, \cdots, (n - k_1 - k_2 - k_3 - k_4)$, we get the joint distribution of the four variables X, Y, Z, U, or

$$(1.8) \qquad p(k_1, k_2, k_3, k_4) = \frac{n! 2^{n-k_1-k_2-k_3-k_4}}{k_1! k_2! k_3! k_4! (n - k_1 - k_2 - k_3 - k_4)!} 6^{-n}.$$

Adding over $k_4 = 0, 1, \cdots, n - k_1 - k_2 - k_3$, we get the joint distribution of X, Y, Z,

$$(1.9) \qquad p(k_1, k_2, k_3) = \frac{n! 3^{n-k_1-k_2-k_3}}{k_1! k_2! k_3! (n - k_1 - k_2 - k_3)!} 6^{-n}.$$

Proceeding in this way, we arrive at the binomial distribution $b(k_1; n, \frac{1}{6})$ for X alone. In this case all six variables have a common (marginal) distribution.

(c) *Sampling Inspection.* As in Chapter 6, example (7.c), suppose that items are subjected to sampling inspection. We have a double classification: an item is acceptable or defective, and it is or is not inspected. The corresponding probabilities are p, q, and p', q', respectively ($p + q = 1$, $p' + q' = 1$). We are concerned with double

Bernoulli trials in which the four classes ("acceptable and inspected," etc.) have probabilities pp', pq', $p'q$, $p'q'$. Suppose now (as in Dodge's sampling plan) that items are sampled until the first defective item is discovered, and consider the following two random variables: the number N of items passing the inspection desk before this discovery, and the number K of defectives among them (which have passed undiscovered). The event $N = n$, $K = k$ occurs if k out of n items are defective but not inspected, and the $(n + 1)$st item is both defective and inspected. Therefore the joint distribution of N and K is given by

$$(1.10) \quad p(n,k) = Pr\{N = n, K = k\} = \binom{n}{k} p^{n-k}(qq')^k \cdot qp',$$

where $n \geq 0$, $0 \leq k \leq n$ (if $k > n$ the binomial coefficient vanishes). Summing over all k, we get from the binomial formula

$$(1.11) \quad f(n) = (p + qq')^n qp' = (1 - qp')^n qp';$$

this is the probability of the event $N = n$, that n items pass control before the first defective item is discovered. Summing over all $n \geq k$, we get [using formula (9.1) of Chapter 2 and the binomial formula]

$$(1.12) \quad g(k) = (qq')^k qp' \sum_{\nu=0}^{\infty} \binom{-k-1}{\nu} (-p)^\nu = \frac{(qq')^k qp'}{(1-p)^{k+1}} = p'q'^k.$$

Of course, the marginal distributions (1.11) and (1.12) could be derived directly. (To be continued in problem 15.)

With the notation (1.3) the conditional probability of the event $Y = y_k$, given that $X = x_j$, becomes

$$(1.13) \quad Pr\{Y = y_k \mid X = x_j\} = \frac{p(x_j, y_k)}{f(x_j)}.$$

As a glance at Table 1 shows, this conditional probability is in general different from $g(y_k)$. This indicates that inference can be drawn from the values of X to those of Y and vice versa; the two variables are (statistically) *dependent*. The strongest degree of dependence exists when Y is a function of X, that is, when the value of X uniquely determines Y. For example, if a coin is tossed n times and X and Y are the numbers of heads and tails, then $Y = n - X$. Similarly, when $Y = X^2$, we can compute Y from X. In the joint distribution this means that in each row all entries but one are zero. If, on the other hand, $p(x_j, y_k) = f(x_j)g(y_k)$ for all combinations of x_j, y_k, then the events $X = x_j$ and $Y = y_k$ are independent; the joint distribution assumes the form of a multiplication table. In this case we speak of *independent* random variables. They occur in particular in connection with independent trials, for example, if X and Y are the numbers scored in two throws of a die. Note that the joint distribution of X and Y determines the distributions of X and Y, but that we cannot calculate the joint distribution of X and Y from their marginal distributions. If two variables X and Y have the same distribution, they may or may not be independent. For example, the two variables X and Y in Table 1 have the same distribution and are dependent. On the other hand, if X and Y had the same meaning but referred to two independent bridge games, the marginal distributions would be the same but X and Y would be independent, and the joint probability distribution would assume the form of a multiplication table.

All our notions apply also to the case of more than two variables. We recapitulate in the formal

Definition. A random variable X is a function defined on a given sample space, that is, an assignment of a real number to each sample point. The system of equations (1.1) defines the (probability) distribution of X. If a particular number x does not occur among the values assumed by X, then we write $Pr\{X = x\} = 0$. If two or more random variables $X_1, X_2, \cdots, X_n$ are defined on the same sample space, their joint distribution is given by the system of equations which assigns probabilities to all combinations $X_1 = x_{j_1}, X_2 = x_{j_2}$, etc. The variables $X_1, \cdots, X_n$ are called *mutually independent if for any combination of values* $x_{j_1}, x_{j_2}, \cdots, x_{j_n}$

$$(1.14) \quad Pr\{X_1 = x_{j_1}, X_2 = x_{j_2}, \cdots, X_n = x_{j_n}\}$$
$$= Pr\{X = x_{j_1}\}Pr\{X_2 = x_{j_2}\} \cdots Pr\{X_n = x_{j_n}\}.$$

In Chapter 5, section 4, we have defined the sample space corresponding to n mutually independent trials. Comparing this definition to (1.14), we see that *if X_k depends only on the outcome of the kth trial, then the variables $X_1, \cdots, X_n$ are mutually independent.* More generally, if a random variable U depends only on the outcomes of the first k trials, and another variable V depends only on the outcomes of the last $n - k$ trials, then U and V are independent. (Cf. problem 25.)

We may conceive of a random variable as a labelling of the points of the sample space. This procedure is familiar from dice, where the faces are numbered and we speak of numbers as the possible outcomes of individual trials. In conventional mathematical terminology we could say that a random variable X is a mapping of the original sample space onto a new space whose points are $x_1, x_2, \ldots$. *It is therefore legitimate to talk of a random variable* X, *assuming the values* x_1, x_2, ... *with probabilities* $f(x_1)$, $f(x_2)$, ... *without further reference to the old sample space; a new one is formed by the sample points* x_1, x_2, ... *Specifying a probability distribution is equivalent to specifying a sample space whose points are real numbers. Speaking of two independent random variables* X *and* Y *with distributions* $\{f(x_j)\}$ *and* $\{g(y_k)\}$ *is equivalent to referring to a sample space whose points are pairs of numbers* (x_j, y_k) *to which probabilities are assigned by the rule* $Pr\{(x_j, y_k)\}$ $= f(x_j)g(y_k)$. *Similarly, for the sample space corresponding to a set of* n *random variables* $(X_1, \cdots, X_n)$ *we can take an aggregate of points* $(x_{j_1}, x_{j_2}, \cdots, x_{j_n})$ *in the n-dimensional space to which probabilities are assigned by the joint distribution. The variables are mutually independent if their joint distribution is given by* (1.14).

It is clear that the same distribution can occur in conjunction with different sample spaces. If we say that the random variable X assumes the values 0 and 1 with probabilities 1/2, then we refer tacitly to a sample space consisting of the two points 0 and 1. However, the variable X might have been defined by stipulating that it equals 0 or 1 according as the tenth tossing of a coin produces heads or tails; in this case X is defined in a sample space of sequences ($HHT...$), and this sample space has 2^{10} points.

In principle, it is possible to restrict the theory of probability to sample spaces defined in terms of probability distributions of random variables. This procedure avoids references to abstract sample spaces, and also to terms like "trials" and "outcomes of experiments." The reduction of probability theory to random variables is a short-cut to the use of analysis and simplifies the theory in many ways. However, it also has the drawback of obscuring the probability background. The notion of random variable easily remains vague as "something that takes on different values with different probabilities." But random variables are ordinary functions, and this notion is by no means peculiar to probability theory.

Example. (d) Let X be a random variable with possible values $x_1, x_2, \ldots$ and corresponding probabilities $f(x_1), f(x_2), \ldots$. If it helps the reader's imagination, he may always construct a conceptual experiment leading to X. For example, subdivide a roulette wheel into arcs $l_1, l_2, \ldots$ whose lengths are as $f(x_1):f(x_2):\ldots$. Imagine a gambler receiving the (positive or negative) amount x_j if the roulette comes to rest at a point of l_j. Then X is the gambler's gain. In n trials, the gains are assumed to be n independent variables with the common distribution $\{f(x_j)\}$. To obtain two variables with a given joint distribution

$\{p(x_j, y_k)\}$ let an arc correspond to each combination (x_j, y_k) and think of two gamblers receiving the amounts x_j and y_k, respectively.

If $X, Y, Z, \ldots$ are random variables defined on the same sample space, then any function $F(X, Y, Z, \ldots)$ is again a random variable. Its distribution can be obtained from the joint distribution of $X, Y, Z, \ldots$ simply by collecting the terms which correspond to combinations of $(X, Y, Z, \ldots)$ giving the same value of $F(X, Y, Z, \ldots)$.

Example. (e) In the example illustrated by Table 1, the sum $S = X + Y$ represents the total number of aces in two specified bridge hands. It can assume the values 0, 1, 2, 3, 4, and the corresponding probabilities can be found in Table 1. Thus the event $S = 0$ is the same as $(X = 0, Y = 0)$, and its probability is 0.055222. For reasons of symmetry this is also the probability of the event $S = 4$; however, the latter event can occur in five different ways, and using only Table 1 we get for its probability $0.022473 + 2(0.013734 + 0.002641) = 0.055223$. Similarly, $Pr\{S = 1\} = 2(0.124850) = 0.249700$, and this happens to be the same as $Pr\{S = 3\}$. Finally $Pr\{S = 2\} = 0.202881 + 2(0.093637) = 0.390155$. The product XY is another random variable assuming the values 0, 1, 2, 3, 4. The event $XY = 0$ occurs if either X or Y vanishes, and its probability is therefore the sum of all entries in the first row and the first column. The distribution of XY is given by $f(0) = 0.552412$, $f(1) = 0.202881$, $f(2) = 0.194766$, $f(3) = 0.027468$, $f(4) = 0.022473$.

2. Expectations

To achieve reasonable simplicity it is often necessary to describe probability distributions rather summarily by a few "typical values." Among these the expectation or mean is by far the most important. It lends itself best to analytical manipulations, and it is preferred by statisticians because of a property known as sampling stability. Its definition follows the customary notion of an average. If in a certain population n_k families have exactly k children, the total number of families is $n = n_0 + n_1 + n_2 + \ldots$ and the total number of children $m = n_1 + 2n_2 + 3n_3 + \ldots$. The average number of children per family is m/n. The analogy between probabilities and frequencies suggests the following

Definition. *Let X be a random variable assuming the values x_1, $x_2, \ldots$ with corresponding probabilities $f(x_1), f(x_2), \ldots$. The mean or expected value of X is defined by*

$$(2.1) \qquad E(X) = \Sigma x_k f(x_k)$$

provided that the series converges absolutely. In this case we say that X has a finite expectation. If $\Sigma \,|\, x_k\,|\, f(x_k)$ diverges, then we say that X has no finite expectation.

It goes without saying that the most common random variables have finite expectations; otherwise the concept would be impractical. However, variables without finite expectations occur in connection with important recurrence problems in physics. The terms *mean, average,* and *mathematical expectation are synonymous*. We also speak of the *mean of a distribution* instead of referring to the corresponding random variable. The notation $E(X)$ is generally accepted in mathematics and statistics. In physics $\bar{X}$, $<X>$, $<X>_{\text{Av}}$ are common substitutes for $E(X)$.

It should be noted that the definition (2.1) applies also if some among the numbers x_k are equal; for we can collect the corresponding terms in the series without changing the sum. This observation becomes useful in connection with functions such as X^2. This function is a new random variable assuming the values $x_k{}^2$; in general, the probability of $X^2 = x_k{}^2$ is not $f(x_k)$ but $f(x_k) + f(-x_k)$. Nevertheless

$$(2.2) \qquad E(X^2) = \Sigma x_k{}^2 f(x_k)$$

provided that the series converges. More generally we get in the same way the

Theorem. For any function $\phi(x)$ we have a new random variable $\phi(X)$ with

$$(2.3) \qquad E(\phi(X)) = \Sigma \phi(x_k) f(x_k),$$

where the series converges absolutely if, and only if, $E(\phi(X))$ exists. For any constant a we have $E(aX) = aE(X)$.

If several random variables $X_1, \cdots, X_n$ are defined on the same sample space, then their sum $X_1 + \cdots + X_n$ is a new random variable. Its possible values and the corresponding probabilities can be readily found from the joint distribution of the X_ν and thus $E(X_1 + \cdots + X_n)$ can be calculated. A simpler procedure is furnished by the following important

Theorem. If $X_1, X_2, \cdots, X_n$ are random variables with expectations, then the expectation of their sum exists and is the sum of their expectations:

$$(2.4) \qquad E(X_1 + \cdots + X_n) = E(X_1) + \cdots + E(X_n).$$

Proof. It suffices to prove (2.4) for two variables X and Y. Using the notation (1.3), we can write

$$(2.5) \qquad E(X) + E(Y) = \sum_{j,k} x_j p(x_j, y_k) + \sum_{j,k} y_k p(x_j, y_k),$$

the summation extending over all possible values x_j, y_k (which need not be all different). The two series converge absolutely, and their sum can therefore be rearranged to give $\Sigma_{jk} (x_j + y_k) p(x_j, y_k)$. However, this is by definition the expectation of $X + Y$. This accomplishes the proof.

Clearly, no corresponding general theorem holds for products; for example, $E(X^2)$ is generally different from $(E(X))^2$. Thus, if X is the number scored with a balanced die, $E(X) = 7/2$, but $E(X^2) = (1 + 4 + 9 + 16 + 25 + 36)/6 = 91/6$. However, the simple multiplication rule holds for mutually independent variables.

Theorem. If X and Y are mutually independent random variables with finite expectations, then their product is a random variable with finite expectation and

(2.6) $$E(XY) = E(X)E(Y).$$

Proof. To calculate $E(XY)$ we should multiply each possible value $x_j y_k$ with the corresponding probability. We have already remarked that the values x_k in the definition (2.1) need not be different. Hence

(2.7) $$E(XY) = \sum_{j,k} x_j y_k f(x_j) g(y_k),$$
$$= \left\{ \sum_j x_j f(x_j) \right\} \left\{ \sum_k y_k g(y_k) \right\},$$

the rearrangement being justified since the series converge absolutely. This proves the theorem. By induction the same multiplication rule holds for any number of *mutually independent random variables*.

3. Examples and Applications

(a) *Binomial Distribution.* Let S_n be the number of successes in n Bernoulli trials with probability p for success. We know that S_n has the binomial distribution $\{b(k; n, p)\}$ [cf. Chapter 6, (2.1)]. Hence $E(S_n) = \Sigma k b(k; n, p) = np \Sigma b(k - 1; n - 1, p)$. The last sum includes all terms of the binomial distribution for $n - 1$ and hence equals 1. Therefore *the mean of the binomial distribution* is

(3.1) $$E(S_n) = np.$$

The same result could have been obtained without calculation by a method which is often expedient. Let X_k be the number of successes scored at the kth trial. This random variable assumes only the values 0 and 1 with corresponding probabilities q and p. Hence $E(X_k) = 0 \cdot q + 1 \cdot p = p$, and since

(3.2) $$S_n = X_1 + X_2 + \cdots + X_n,$$

we get (3.1) directly from theorem (2.4). [Continuation in examples (4.d) and (5.a).]

(b) *Poisson Distribution.* If X has the Poisson distribution $p(k;\lambda) = e^{-\lambda}\lambda^k/k!$ [cf. Chapter 6, (5.1)], then $E(X) = \Sigma k p(k;\lambda) = \lambda\Sigma p(k-1;\lambda)$. The last series contains all terms of the distribution and therefore adds to unity. Accordingly, λ *is the mean of the Poisson distribution*. [Continuation in example (4.c).]

(c) In a sequence of Bernoulli trials let X be the number of trials up to and including the *first* success. Then X is a random variable whose probability distribution $\gamma_\nu = Pr\{X = \nu\}$ is given by

$$\gamma_\nu = q^{\nu-1}p, \qquad \nu \geq 1. \tag{3.3}$$

This is the *geometric distribution.* For the mean we get

$$\Sigma\nu\gamma_\nu = p(1 + 2q + 3q^2 + \ldots). \tag{3.4}$$

On the right we have the derivative of a geometric series so that the sum is $p(1-q)^{-2} = p^{-1}$. Hence *the mean of the number of trials up to and including the first success is* $E(X) = 1/p$. The number X_k of trials following the $(k-1)$st success and up to and including the kth success is a random variable which obviously has the same distribution as X. The sum $X_1 + \cdots + X_n$ is *the number of trials up to and including the nth success;* its mean is n/p, by the addition theorem (2.4). Note that we have here calculated the mean of a random variable without knowing its distribution. The latter is the *Pascal distribution* to be discussed in Chapter 11, section 3. [Continuation in example (5.b).]

(d) *A Sampling Problem.* A population of N distinct elements is sampled with replacement. Because of repetitions a random sample of size r will in general contain fewer than r distinct elements. As the sample size increases, new elements will enter the sample more and more rarely. We are interested in the sample size S_r necessary for the acquisition of r distinct elements. (As a special case, consider the population of $N = 365$ possible birthdays; here S_r represents the number of people sampled up to the moment where the sample contains r different birthdays. A similar interpretation is possible with random placements of balls into cells. Our problem is of particular interest to collectors of coupons or other items where the acquisition can be compared to random sampling.[2]

---

[2] G. Polya, Eine Wahrscheinlichkeitsaufgabe zur Kundenwerbung, *Zeitschrift für Angewandte Mathematik und Mechanik*, vol. 10 (1930), pp. 96–97. Polya treats a slightly more general problem with different methods.

The first element enters the sample at the first drawing. The number of drawings from the second up to and including the drawing at which a new element enters the sample is a random variable X_1; generally, let X_r be the number of drawings following the selection of the rth element up to and including the selection of the next new element. Then $S_r = 1 + X_1 + \cdots + X_{r-1}$ is the sample size at the moment that the rth element enters the sample. Once the sample contains k different elements the probability of drawing a new one is at each drawing $p = (N-k)/N$. The distribution of X_k is therefore given by (3.3) and $E(X_k) = 1/p = N/(N-k)$. Hence, from the addition theorem (2.4),

$$(3.5) \quad E(S_r) = N \left\{ \frac{1}{N} + \frac{1}{N-1} + \frac{1}{N-2} + \cdots + \frac{1}{N-r+1} \right\}.$$

For $r = N$ we get the expected number of drawings necessary to exhaust the entire population. For $N = 10$ we have $E(S_{10}) = 29.29\ldots$, and $E(S_5) = 6.46\ldots$. This means that we can expect to have covered half the population in about 6 to 7 drawings, whereas the second half requires some 23 more drawings. A reasonable approximation [3] to (3.5) for large N is

$$(3.6) \quad E(S_r) \approx N \log \frac{N}{N-r+1}.$$

In particular, for any fraction $\alpha < 1$ *the expected number of drawings required to obtain a sample containing about the fraction α of the entire population is, for large N, approximately* $N \log \dfrac{1}{1-\alpha}$; the expected number of drawings necessary to have all N elements included in the sample is, approximately, $N \log N$. Note that our results are again obtained without use of the distribution. For sampling without replacement the same method is applicable [cf. problem 14; the present example is continued in section 5, (c); it is treated with different methods in Chapter 11, problems 5 and 6].

(e) *A bowl contains balls numbered 1 to N. Let X be the largest number drawn* in n drawings when random sampling with replacement is used. The event $X \leq k$ means that each of n numbers drawn is less

---

[3] If the tangent and rectangle rules for numerical integration are applied to (3.5), one finds that

$$N \log \frac{N + \frac{1}{2}}{N - r + \frac{1}{2}} > E(S_r) > N \log \frac{N+1}{N-r+1}$$

than or equal to k and therefore $Pr\{X \leq k\} = (k/N)^n$. Hence the probability distribution of X is given by

$$p_k = Pr\{X = k\} = Pr\{X \leq k\} - Pr\{X \leq k-1\}$$
$$= \{k^n - (k-1)^n\}N^{-n}.$$

It follows that

$$(3.7) \quad E(X) = \sum_{k=1}^{N} kp_k = N^{-n} \sum_{k=1}^{N} \{k^{n+1} - (k-1)^{n+1} - (k-1)^n\}$$
$$= N^{-n} \{N^{n+1} - \sum_{k=1}^{N} (k-1)^n\}.$$

For large N the last sum is approximately the area under the curve $y = x^n$ from $x = 0$ to $x = N$, that is, $N^{n+1}/(n+1)$. It follows that for large [4] N

$$(3.8) \qquad E(X) \approx \frac{n}{n+1} N.$$

If a town has $N = 1000$ cars and a sample of $n = 10$ is observed, the expected number of the highest observed license plate (assuming randomness) is about 910. The converse problem of estimating the unknown true number N from the observed maximum in a sample occurs in routine statistical analysis. (Continuation in problems 6–8.)

(f) *Banach's Match-box Problem.* In example (2.h) of Chapter 6, we found the distribution

$$(3.9) \qquad u_r = \binom{2N-r}{N} \frac{1}{2^{2N-r}}$$

for the number X of matches left at the moment when the first box is found empty. We are unable to calculate the expectation $E(X) = \mu$ in a direct way, but the following indirect way is applicable in many similar cases. Using the fact that the u_r add to unity (which is not easily verified), we find

$$(3.10) \quad N - \mu = \sum_{r=0}^{N-1} (N-r)u_r = \sum_{r=0}^{N-1} (N-r) \binom{2N-r}{N-r} \frac{1}{2^{2N-r}}.$$

---

[4] A more precise estimate follows from the tangent and trapezoid rules for numerical integration:

$$N\left\{\frac{n}{n+1} + \frac{1}{2N}\right\} > E(X) > N\left\{1 - \frac{\left(1 - \frac{1}{2N}\right)^{n+1} - \left(\frac{1}{2N}\right)^{n+1}}{n+1}\right\}$$

By a simple operation on the binomial coefficients the last sum is transformed into

$$(3.11) \quad \sum_{r=0}^{N-1} (2N - r) \binom{2N - r - 1}{N - r - 1} \frac{1}{2^{2N-r}}$$

$$= \frac{2N+1}{2} \sum_{r=0}^{N-1} u_{r+1} - \frac{1}{2} \sum_{r=0}^{N-1} (r+1) u_{r+1}.$$

The last sum is identical with the sum defining $\mu = E(X)$. In the first sum all u_r except u_0 occur, and hence the terms add to $1 - u_0$. Thus from (3.10) and (3.11)

$$(3.12) \quad N - \mu = \frac{2N+1}{2}(1 - u_0) - \frac{\mu}{2}$$

or

$$(3.13) \quad \mu = (2N+1)u_0 - 1 = \frac{2N+1}{2^{2N}} \binom{2N}{N} - 1.$$

Using Stirling's formula, we find

$$(3.14) \quad \mu \approx 2\left(\frac{N}{\pi}\right)^{\frac{1}{2}} - 1.$$

In particular, in the distribution of Chapter 6, Table 2, we had $N = 50$. For it $\mu = 7.04\ldots$.

4. The Variance

Let X be a random variable with distribution $\{f(x_j)\}$, and let $r \geq 0$ be an integer. *If the expectation of the random variable X^r, that is,*

$$(4.1) \quad E(X^r) = \Sigma x_j^r f(x_j),$$

exists, then it is called the rth moment of X about the origin. If the series does not converge absolutely, we say that the rth moment does not exist. Since $|X|^{r-1} \leq |X|^r + 1$, it follows that *whenever the rth moment exists so does the $(r-1)$st, and hence all preceding moments.*

Moments play an important role in the general theory, but in the present volume we shall use only the second moment. If it exists, so does the mean

$$(4.2) \quad \mu = E(X).$$

It is then natural to introduce instead of the random variable its deviation from the mean, $X - \mu$. Since $(x - \mu)^2 \leq 2(x^2 + \mu^2)$ we see that the second moment of $X - \mu$ exists whenever $E(X^2)$ exists. We find

(4.3) $$E((X - \mu)^2) = \sum_j (x_j{}^2 - 2\mu x_j + \mu^2) f(x_j).$$

Splitting the right side into three individual sums, we find it equal to $E(X^2) - 2\mu E(X) + \mu^2 = E(X^2) - \mu^2$.

Definition. Let X be a random variable with second moment $E(X^2)$ and let $\mu = E(X)$ be its mean. We define a number called the variance of X by

(4.4) $$\mathrm{Var}(X) = E((X - \mu)^2) = E(X^2) - \mu^2.$$

Its positive square root (or zero) is called the standard deviation of X.

For simplicity we often speak of the variance of a distribution without mentioning the random variable. "Dispersion" is a synonym for the now generally accepted term "variance."

Examples. (a) If X assumes the values $\pm c$, each with probability $1/2$, then $\mathrm{Var}(X) = c^2$.

(b) If X is the number of points scored with a symmetric die, then $\mathrm{Var}(X) = \frac{1}{6}(1^2 + 2^2 + \cdots + 6^2) - (7/2)^2 = 35/12$.

(c) For the *Poisson distribution* $p(k; \lambda)$ the mean is λ [cf. 3, (b)] and hence the variance $\Sigma k^2 p(k; \lambda) - \lambda^2 = \lambda \Sigma k p(k - 1; \lambda) - \lambda^2 = \lambda \Sigma (k - 1) p(k - 1; \lambda) + \lambda \Sigma p(k - 1; \lambda) - \lambda^2 = \lambda^2 + \lambda - \lambda^2 = \lambda$. In this case mean and variance are equal.

(d) For the *binomial distribution* [cf. 3, (a)] a similar computation shows that the variance is

$$\Sigma k^2 b(k; n, p) - (np)^2 = np \Sigma k b(k - 1; n - 1, p) - (np)^2$$
$$= np\{(n - 1)p + 1\} - (np)^2 = npq.$$

The usefulness of the notion of variance will appear only gradually, in particular, in connection with limit theorems (Chapter 10). Here we observe that the variance is a rough *measure of spread*. In fact, if $\mathrm{Var}(X) = \Sigma (x_j - \mu)^2 f(x_j)$ is small, then each term in the sum is small. A value x_j for which $|x_j - \mu|$ is large must therefore have a small probability $f(x_j)$. In other words, in case of small variance large deviations of X from the mean μ are improbable. Conversely, a large variance indicates that not all possible values of X lie near the mean.

Some readers may be helped by the following interpretation in mechanics. Suppose that a unit mass is distributed on the x-axis so that the mass $f(x_j)$ is concentrated at the point x_j. Then the mean μ is the abscissa of the *center of gravity*, and the variance is the *moment of inertia*. Clearly different mass distributions may have the same center of gravity and the same moment of inertia, but it is well known that the most important mechanical properties can be described in terms of these two quantities.

If X represents a measurable quantity like length or temperature, then its numerical values depend on the origin and the unit of measurement. A change of the latter means passing from X to a new variable $aX + b$, where a and b are constants. Clearly $\mathrm{Var}(X + b) = \mathrm{Var}(X)$, and hence

(4.5) $$\mathrm{Var}(aX + b) = a^2 \mathrm{Var}(X).$$

The choice of the origin and unit of measurement is to a large degree arbitrary, and often it is most convenient to take the mean as origin and the standard deviation as unit. We have done so in Chapter 7, when we introduced the normalized number of successes $S_n^* = (S_n - np)/(npq)^{1/2}$. In general, if X has mean μ and variance σ^2 ($\sigma > 0$), then $X - \mu$ has mean zero and variance σ^2, and hence *the variable*

(4.6) $$X^* = \frac{X - \mu}{\sigma}$$

has mean 0 and variance 1. It is called the normalized variable corresponding to X. In the physicist's language, the passage from X to X^* would be interpreted as the introduction of dimensionless quantities.

5. Covariance; Variance of a Sum

Let X and Y be two random variables on the same sample space. Then $X + Y$ and XY are again random variables, and their distributions can be obtained by a simple rearrangement of the joint distribution of X and Y. Our aim now is to calculate $\mathrm{Var}(X + Y)$. For that purpose we introduce the notion of covariance, which will be analyzed in greater detail in section 8. If the joint distribution of X and Y is $\{p(x_j, y_k)\}$, then the expectation of XY is given by

(5.1) $$E(XY) = \Sigma x_j y_k p(x_j, y_k),$$

provided, of course, that the series converges absolutely. Now $|x_j y_k| \leq (x_j^2 + y_k^2)/2$ and therefore $E(XY)$ certainly exists if $E(X^2)$ and $E(Y^2)$ exist. In this case there exist also the expectations

(5.2) $$\mu_x = E(X), \quad \mu_y = E(Y),$$

and the variables $X - \mu_x$ and $Y - \mu_y$ have zero means. For their product we have from the addition rule of section 2

$$(5.3) \quad E((X - \mu_x)(Y - \mu_y)) = E(XY) - \mu_x E(Y) - \mu_y E(X) + \mu_x \mu_y$$
$$= E(XY) - \mu_x \mu_y.$$

Definition. The covariance of X and Y is defined by

$$(5.4) \quad \mathrm{Cov}(X, Y) = E((X - \mu_x)(Y - \mu_y)) = E(XY) - \mu_x \mu_y.$$

This definition is meaningful whenever X and Y have finite variances.

We know from section 2 that for independent variables $E(XY) = E(X)E(Y)$. Hence from (5.4) we have

Theorem 1. If X and Y are independent, then $\mathrm{Cov}(X, Y) = 0$. (The converse is *not* true; cf. section 8.)

We now proceed to prove the fundamental

Theorem 2. If $X_1, \cdots, X_n$ are random variables with finite variances $\sigma_1^2, \cdots, \sigma_n^2$, and $S_n = X_1 + \cdots + X_n$, then

$$(5.5) \quad \mathrm{Var}(S_n) = \sum_{k=1}^{n} \sigma_k^2 + 2 \sum_{j,k} \mathrm{Cov}(X_j, X_k)$$

where the last sum contains each of the $\binom{n}{2}$ pairs (X_j, X_k) with $j < k$ once and only once.

In particular, if the X_j are mutually independent, then the addition rule

$$(5.6) \quad \mathrm{Var}(S_n) = \sigma_1^2 + \sigma_2^2 + \cdots + \sigma_n^2$$

holds.

Proof. Put $\mu_k = E(X_k)$ and $m_n = \mu_1 + \cdots + \mu_n = E(S_n)$. Then $S_n - m_n = \Sigma(X_k - \mu_k)$ and

$$(5.7) \quad (S_n - m_n)^2 = \Sigma(X_k - \mu_k)^2 + 2\Sigma(X_j - \mu_j)(X_k - \mu_k).$$

Taking expectations and applying the addition rule, we get (5.5). Equation (5.6) follows from the preceding theorem.

Examples. (*a*) *Binomial Distribution* $\{b(k; n, p)\}$. In example (3.*a*), the variables X_k are mutually independent. We have $E(X_k^2) = 0 \cdot^2 q + 1 \cdot^2 p = p$, and $E(X_k) = p$. Hence $\sigma_k^2 = p - p^2 = pq$, and from (5.6) we see that *the variance of the binomial distribution is npq*. The same result was derived by direct computation in example (4.*d*).

(b) In example (3.c) the variables X_k are again independent and have the common distribution $\gamma_\nu = q^{\nu-1}p$ ($\nu = 1, 2, \cdots$). This is a *geometric distribution*. For its second moment we find

(5.8) $\quad \Sigma \nu^2 q^{\nu-1} p = qp\Sigma\nu(\nu-1)q^{\nu-2} + p\Sigma\nu q^{\nu-1}$

$$= qp\frac{d^2}{dq^2}\Sigma q^\nu + p\frac{d}{dq}\Sigma q^\nu$$

$$= \frac{2qp}{(1-q)^3} + \frac{p}{(1-q)^2} = \frac{2q}{p^2} + \frac{1}{p}.$$

Since the mean was found to be $1/p$, we have $\sigma_k^2 = \frac{2q}{p^2} + \frac{1}{p} - \frac{1}{p^2}$

$= \frac{q}{p^2}$, and $\mathrm{Var}(S_n) = \frac{nq}{p^2}$. This is *the variance of the number of trials up to and including the nth success* (or the variance of the Pascal distribution; cf. Chapter 11, section 3). Note that once more we have calculated a variance without knowing the distribution.

(c) In the collector's problem (3.d) the variables X_k are still independent, but they no longer have a common distribution. We know that X_k is the number of trials up to and including the first success in a sequence of Bernoulli trials with $p = (N-k)/N$. Hence from the preceding example $E(X_k) = N/(N-k)$, and $\mathrm{Var}(X_k) = kN/(N-k)^2$. Thus

(5.9) $\quad \mathrm{Var}(S_n) = N\left\{\dfrac{1}{(N-1)^2} + \dfrac{2}{(N-2)^2} + \cdots + \dfrac{r-1}{(N-r+1)^2}\right\}.$

(d) *Card Matching*. A deck of n numbered cards is put into random order so that all $n!$ arrangements have equal probabilities. The number of matches (cards in their natural place) is a random variable S_n which assumes the values $0, 1, \cdots, n$. Its probability distribution was derived in Chapter 4. From it the mean and variance could be obtained, but the following way is simpler and more instructive.

Define a random variable X_k which is either 1 or 0, according as card number k is or is not at the kth place. Then $S_n = X_1 + \cdots + X_n$. Now each card has probability $1/n$ to appear at the kth place. Hence $\Pr\{X_k = 1\} = 1/n$ and $\Pr\{X_k = 0\} = (n-1)/n$. Therefore $E(X_k) = 1/n$, and it follows that $E(S_n) = 1$: the average is one match per deck. To calculate the variance we first calculate the variance σ_k^2 of X_k:

(5.10) $\quad\quad\quad\quad \sigma_k^2 = \dfrac{1}{n} - \left(\dfrac{1}{n}\right)^2 = \dfrac{n-1}{n^2}.$

Next we calculate $E(X_jX_k)$. The product X_jX_k is 0 or 1; the latter is true if both card number j and card number k are at their proper places, and the probability for that is $1/n(n-1)$. Hence

(5.11) $$E(X_jX_k) = \frac{1}{n(n-1)},$$

$$\text{Cov}(X_j, X_k) = \frac{1}{n(n-1)} - \frac{1}{n^2} = \frac{1}{n^2(n-1)}.$$

Thus finally

(5.12) $$\text{Var}(S_n) = n\frac{n-1}{n^2} + 2\binom{n}{2}\frac{1}{n^2(n-1)} = 1.$$

We see that for the number of matches both mean and variance are equal to one. This result may be applied to the problem of *card guessing* discussed in Chapter 4, section 4. There we considered three methods of guessing, one of which corresponds to card matching. The second can be described as a sequence of n Bernoulli trials with probability $p = 1/n$, in which case the expected number of correct guesses is $np = 1$ and the variance $npq = (n-1)/n$. The expected numbers are the same in both cases, but the larger variance with the first method indicates greater chance fluctuations about the mean and thus promises a slightly more exciting game. (With more complicated decks of cards the difference between the two variances is somewhat larger but never really big.) With the last mode of guessing the subject keeps calling the same card; the number of correct guesses is necessarily one, and chance fluctuations are completely eliminated (variance 0). We see that the strategy of calling cannot influence the expected number of correct guesses, but has some influence on the magnitude of chance fluctuations.

(e) *Sampling without Replacement.* Suppose that a population consists of b black and g green elements, and that a random sample of size r is taken (without possible repetitions). The number S_k of black elements in the sample is a random variable with the *hypergeometric distribution* (Chapter 2, section 5)

(5.13) $$q_k = \frac{\binom{b}{k}\binom{g}{r-k}}{\binom{b+g}{r}}.$$

The mean and variance can be obtained by direct computation, but the following method is preferable. Define the random variable X_k to assume the values 1 or 0 according as the kth element in the sample is or is not black ($k \leq r$). For reasons of symmetry the probability that $X_k = 1$ is $b/(b+g)$, and hence

(5.14) $$E(X_k) = \frac{b}{b+g}, \quad \operatorname{Var}(X_k) = \frac{bg}{(b+g)^2}.$$

Next, if $j \neq k$, then $X_j X_k = 1$ if the jth and kth elements of the sample are black, and otherwise $X_j X_k = 0$. The probability of $X_j X_k = 1$ is $b(b-1)/(b+g)(b+g-1)$, and therefore

(5.15) $$E(X_j X_k) = \frac{b(b-1)}{(b+g)(b+g-1)},$$

$$\operatorname{Cov}(X_j X_k) = \frac{-bg}{(b+g)^2(b+g-1)}.$$

Thus,

(5.16) $$E(S_r) = \frac{rb}{b+g}, \quad \operatorname{Var}(S_r) = \frac{rbg}{(b+g)^2}\left\{1 - \frac{r-1}{b+g-1}\right\}.$$

In sampling with replacement we would have the same mean, but the variance would be slightly larger, namely, $rbg/(b+g)^2$ [cf. example (a)].

6. Chebyshev's Inequality [5]

It has been pointed out that a small variance indicates that large deviations from the mean are improbable. This statement is made more precise by Chebyshev's inequality, which is an exceedingly useful and handy tool.

Theorem. Let X be a random variable with mean $\mu = E(X)$ and variance $\sigma^2 = \operatorname{Var}(X)$. Then for any $t > 0$

(6.1) $$\Pr\{|X - \mu| \geq t\} \leq \frac{\sigma^2}{t^2}.$$

Proof. The variance is defined in (4.3) by a series with positive terms. Deleting all terms for which $|x_j - \mu| \leq t$ cannot increase the value of the series, and hence

(6.2) $$\sigma^2 \geq \Sigma^*(x_j - \mu)^2 f(x_j)$$

[5] P. L. Chebyshev (1821–1894).

where the star indicates that the summation extends only over those j for which $|x_j - \mu| \geq t$. It is then clear that

(6.3) $\quad \Sigma^*(x_j - \mu)^2 f(x_j) \geq t^2 \Sigma^* f(x_j) = t^2 Pr\{|X - \mu| \geq t\}$

which proves the theorem.

Chebyshev's inequality must be regarded as a theoretical tool rather than a practical method of estimation. Its importance is due to its universality, but no statement of great generality can be expected to yield sharp results in individual cases.

Examples. (a) If X is the number scored in a throw of a true die, then [cf. example (4.b)], $\mu = 7/2$, $\sigma^2 = 35/12$. The maximum deviation of X from μ is $2.5 \approx 3\sigma/2$. The probability of greater deviations is zero, while Chebyshev's inequality only asserts that this probability is smaller than 0.47.

(b) For the geometric distribution $q_k = 2^{-k}$ $(k = 1, 2, \ldots)$ we have [cf. example (5.b)] $\mu = 2$, $\sigma^2 = 2$. Here $Pr\{|X - 2| > 2\} = 2^{-5} + 2^{-6} + 2^{-7} + \ldots = 2^{-4}$ while Chebyshev's inequality gives an upper bound of $1/2$.

(c) For the binomial distribution $\{b(k; n, p)\}$ we have [cf. example (5.a)] $\mu = np$, $\sigma^2 = npq$. For large n we know that

(6.4) $\quad Pr\{|S_n - np| > x(npq)^{1/2}\} \approx 1 - \Phi(x) + \Phi(-x)$.

Chebyshev's inequality states only that the left side is less than $1/x^2$; this is obviously a much poorer estimate than (6.4).

* 7. Kolmogorov's Inequality [6]

As an example of more refined methods we prove:

Let $X_1, \cdots, X_n$ be mutually independent variables with expectations $\mu_k = E(X_k)$ and variances σ_k^2. Put

(7.1) $\quad\quad\quad S_k = X_1 + \cdots + X_k$

and

(7.2) $\quad\quad\quad m_k = E(S_k) = \mu_1 + \cdots + \mu_k,$

$\quad\quad\quad\quad s_k^2 = \text{Var}(S_k) = \sigma_1^2 + \cdots + \sigma_k^2.$

For every $t > 0$ the probability of the simultaneous realization of the n inequalities

---

* Starred sections treat special topics and may be omitted at first reading.

[6] Über die Summen zufälliger Grössen, *Mathematische Annalen*, vol. 99 (1928) pp. 309–319, and vol. 102 (1929), pp. 484–488.

(7.3) $\qquad |S_k - m_k| < ts_n, \quad k = 1, 2, \cdots, n$

is at least $1 - t^{-2}$.

For $n = 1$ this theorem reduces to Chebyshev's inequality. For $n > 1$ Chebyshev's inequality gives the same bound for the probability of the single relation $|S_n - m_n| < ts_n$, so that Kolmogorov's inequality is considerably stronger.

Proof. We want to estimate the probability x that one of the inequalities (7.3) does *not* hold. The theorem asserts that $x \leq t^{-2}$.

Define n random variables Y_k as follows: $Y_\nu = 1$ if

(7.4) $\qquad |S_\nu - m_\nu| \geq ts_n$

but

(7.5) $\qquad |S_k - m_k| < ts_n \quad \text{for} \quad k = 1, 2, \cdots, \nu - 1;$

$Y_\nu = 0$ for all other sample points. In words, Y_ν equals 1 at those points in which the νth of the inequalities (7.3) is the *first* to be violated. Then at any particular sample point at most one among the Y_k is 1, and the sum $Y_1 + Y_2 + \cdots + Y_n$ can assume only the values 0 or 1; it is 1 if and only if one of the inequalities (7.3) is violated, and therefore

(7.6) $\qquad x = Pr\{Y_1 + \cdots + Y_n = 1\}.$

Since $Y_1 + \cdots + Y_n$ is 0 or 1, we have $\Sigma Y_k \leq 1$. Multiplying by $(S_n - m_n)^2$ and taking expectations, we get

(7.7) $\qquad \sum_{k=1}^{n} E(Y_k(S_n - m_n)^2) \leq s_n^2.$

For an evaluation of the terms on the left we put

(7.8) $\qquad U_k = (S_n - m_n) - (S_k - m_k) = \sum_{\nu=k+1}^{n} (X_\nu - \mu_\nu).$

Then

(7.9) $\qquad \begin{aligned} & E(Y_k(S_n - m_n)^2) \\ & = E(Y_k(S_k - m_k)^2) + 2E(Y_k U_k(S_k - m_k)) + E(Y_k U_k^2). \end{aligned}$

However, U_k depends only on $X_{k+1}, \cdots, X_n$, while Y_k and S_k depend only on $X_1, \cdots, X_k$. Hence U_k is independent of $Y_k(S_k - m_k)$ and therefore $E(Y_k U_k(S_k - m_k)) = E(Y_k(S_k - m_k))E(U_k) = 0$, since $E(U_k) = 0$. Thus from (7.9)

(7.10) $\qquad E(Y_k(S_n - m_n)^2) \geq E(Y_k(S_k - m_k)^2).$

But $Y_k \neq 0$ only if $|S_k - m_k| \geq ts_n$, so that $Y_k(S_k - m_k)^2 \geq t^2 s_n^2 Y_k$.

Hence, combining (7.7) and (7.10), we get

(7.11) $$s_n^2 \geq t^2 s_n^2 E(Y_1 + \cdots + Y_n).$$

Since $Y_1 + \cdots + Y_n$ equals either 0 or 1, the expectation to the right equals the probability x defined in (7.6). Thus $xt^2 \leq 1$ as asserted.

*8. The Correlation Coefficient

Let X and Y be any two random variables with means μ_x and μ_y and positive variances σ_x^2 and σ_y^2. We introduce the corresponding normalized variables X^* and Y^* defined by (4.6). Their covariance is called *the correlation coefficient of X, Y and is denoted by $\rho(X, Y)$*. Thus, using (5.4),

(8.1) $$\rho(X,Y) = \text{Cov}(X^*,Y^*) = \frac{\text{Cov}(X,Y)}{\sigma_X \sigma_Y}$$

Clearly this correlation coefficient is independent of the origins and units of measurements, that is, for any constants a_1, a_2, b_1, b_2, with $a_1 > 0$, $a_2 > 0$, we have $\rho(a_1 X + b_1, a_2 Y + b_2) = \rho(X, Y)$.

The use of the correlation coefficient amounts to a fancy way of writing the covariance.[7] Unfortunately, the term correlation is suggestive of implications which are not inherent in it. We know from section 5 that $\rho(X, Y) = 0$ whenever X and Y are independent. It is important to realize that the converse is not true. In fact, *the correlation coefficient $\rho(X, Y)$ can vanish even if Y is a function of X*.

Examples. (a) Let X assume the values ± 1, ± 2 each with probability 1/4. Let $Y = X^2$. The joint distribution is given by $p(-1, 1) = p(1, 1) = p(2, 4) = p(-2, 4) = 1/4$. For reasons of symmetry $\rho(X, Y) = 0$ even though we have a direct functional dependence of Y on X.

(b) Let U and V be *independent* variables with the same distribution, and let $X = U + V$, $Y = U - V$. Then $E(XY) = E(U^2) - E(V^2) = 0$ and $E(Y) = 0$. Hence $\text{Cov}(X, Y) = 0$ and therefore also $\rho(X, Y) = 0$. For example, X and Y may be the sum and difference of points on two dice. Then X and Y are either both odd or both even, and therefore dependent.

It follows that the correlation coefficient is by no means a general measure of dependence between X and Y. However, $\rho(X, Y)$ is connected with the *linear* dependence of X and Y.

---

* Starred sections treat special topics and may be omitted at first reading.

[7] The physicist would define the correlation coefficient as "dimensionless covariance."

Theorem. We have always $|\rho(X, Y)| \leq 1$, and $\rho(X, Y) = \pm 1$ only if $Y = aX + b$, where a and b are constants.

Proof. Let X^* and Y^* be the normalized variables. Then

$$(8.2) \quad \text{Var}(X^* \pm Y^*) = \text{Var}(X^*) \pm 2 \text{Cov}(X^*, Y^*) + \text{Var}(Y^*)$$

$$= 2(1 \pm \rho(X, Y)).$$

The left side cannot be negative; hence $|\rho(X, Y)| \leq 1$. For $\rho(X, Y) = 1$ it is necessary that $\text{Var}(X^* - Y^*) = 0$ which means that the variable $X^* - Y^*$ assumes only one value. In this case $X^* - Y^* = \text{const.}$, and hence $Y = aX + \text{const.}$ with $a = \sigma_Y/\sigma_X$. A similar argument applies to the case $\rho(X, Y) = -1$.

9. Problems for Solution

1. In 5 tosses of a coin let X, Y, Z be, respectively, the number of heads, the number of head runs, the length of the largest head run. Tabulate the 32 sample points together with the corresponding values of X, Y, and Z. By simple counting derive the joint distributions of the pairs (X, Y), (X, Z), (Y, Z) and the distributions of $X + Y$ and XY. Find the means, variances, covariances of the variables.

2. *Birthdays.* For a group of n people find the expected number of days of the year which are birthdays of exactly k people. (Assume 365 days and that all arrangements are equally probable.)

3. *Continuation.* Find the expected number of multiple birthdays. How large should n be to make this expectation exceed 1?

4. A man wants to open his door and has n keys. For reasons which can be only surmised he tries them independently and at random. Find the mean and variance of the number of trials (a) if unsuccessful keys are not eliminated from further selections; (b) if they are. (Assume that only one key fits the door.)

5. Find the covariance of the number of ones and sixes in n throws of a die.

6. In example (3.e) find $E(X^2)$ and hence an asymptotic expression for the variance as $N \to \infty$ (with n fixed).

7. *Continuation.* Find the joint distribution of the largest and smallest observation.

8. *Continuation.* Find the conditional probability that the first two observations are j and k, given that $X = r$.

9. In a sequence of Bernoulli trials let X be the length of the run (of either successes or failures) started by the first trial. Find the distribution of X, $E(X)$, $\text{Var}(X)$.

10. *Continuation.* Let Y be the length of the *second* run. Find the distribution of Y, $E(Y)$, $\text{Var}(Y)$, and the joint distribution of X, Y.

11. *Double hypergeometric distribution.* A population of n elements contains among others, n_1 red and n_2 black elements $(n_1 + n_2 \leq n)$. A random sample of r elements is selected. Let X and Y be the numbers of red and black elements in it. Find the joint distribution, means, variances, and covariance. (Specialize to spades and clubs in bridge.)

12. Let X be the number of runs of red things in a random arrangement of r_1 red and r_2 black things. The probability distribution $\{\pi_k\}$ of X is given in Chapter 3, problem 15. Find $E(X)$, $\mathrm{Var}(X)$.

13. In the Polya urn scheme of Chapter 5, example (2.c), let S_n be the total number of black balls extracted in the first n drawings. Find $E(S_n)$ and $\mathrm{Var}(S_n)$. (Use the results of Chapter 5, problems 16–17, and verify by means of the recursion formula, Chapter 5, problem 19.)

14. In the collector's problem (3.d) let Y_r be the number of drawings required to include r preassigned elements (instead of any r different elements as in the text). Find $E(Y_r)$ and $\mathrm{Var}(Y_r)$.

15. Find $E\left(\dfrac{K}{N+1}\right)$ and $\mathrm{Cov}(K, N)$ in the example (1.c). [In industrial practice the discovered defective item is replaced by an acceptable one so that $K/(N+1)$ is the fraction of defectives and measures the quality of the lot. Note that $E\left(\dfrac{K}{N+1}\right)$ is not $E(K)/E(N+1)$.]

16. *Stratified sampling.* A city has n blocks of which n_j have x_j inhabitants ($n_1 + n_2 + \cdots = n$). Let $m = \Sigma n_j x_j / n$ be the mean number of inhabitants per block and put $a^2 = \Sigma n_j x_j^2 / n - m^2$. In sampling without replacement r blocks are selected at random, and in each the inhabitants are counted. Let $X_1, \cdots, X_r$ be the respective number of inhabitants. Show that

$$E(X_1 + \cdots + X_r) = mr$$

$$\mathrm{Var}(X_1 + \cdots + X_r) = \frac{a^2 r(n-r)}{n-1}.$$

(Note that with sampling with replacement the variance would be larger, namely, $a^2 r$.)

17. *Continuation.*[8] The number x of inhabitants of a city is estimated by the following double sampling procedure. The city is divided into n strata. The (known) number of blocks in stratum j is n_j, so that $n = \Sigma n_j$ is the total number of blocks in the city. The (unknown) number of inhabitants of the kth block in the jth stratum is x_{jk} (so that $x_j = \sum_k x_{jk}$ is the number of inhabitants of the stratum j, and $x = \sum_j x_j$ is the number of inhabitants of the city). Out of the stratum j a random sample of r_j blocks is drawn, and the number of people in each is counted. Let X_{jk} be the number of people in the kth block of the sample taken from the jth stratum. Then $X_j = \sum_k X_{jk}$ is the total number of inhabitants in the blocks sampled in stratum number j. Put $X = \sum \dfrac{n_j}{r_j} X_j$. Show (using the preceding result) that $E(X) = x$ and $\mathrm{Var}(X) = \Sigma \sigma_j^2 \dfrac{n_j^2(n_j - r_j)}{r_j(n_j - 1)}$, where

$$\sigma_j^2 = n_j^{-1} \sum_k n_{jk} \left(\frac{x_{jk} - x_j}{n_j}\right)^2.$$

---

[8] Stratified sampling is usual in many applications, since greater accuracy can be achieved at smaller costs. In population sampling the strata are often further subdivided. There exists an elaborate theory of sampling under various conditions. Cf. *A Chapter in Population Sampling*, by the Sampling Staff, Bureau of the Census, U. S. Government, Washington, 1947.

18.[9] A large number, N, of people are subject to a blood test. This can be administered in two ways. (i) Each person can be tested separately. In this case N tests are required. (ii) The blood samples of k people can be pooled and analyzed together. If the test is *negative*, this *one* test suffices for the k people. If the test is *positive*, each of the k persons must be tested separately, and in all $k+1$ tests are required for the k people.

Assume the probability p that the test be positive is the same for all people and that people are statistically independent.

(a) What is the probability that the test for a pooled sample of k people will be positive?

(b) What is the expected value of the number, X, of tests necessary under plan (ii)?

(c) What should k be to minimize the expected number of tests under plan (ii)? Do not try numerical evaluations, since the problem leads to a rather cumbersome equation for k.

19. Let S_n be the number of successes in n Bernoulli trials. Prove
$$E(|S_n - np|) = 2\nu q b(\nu; n, p)$$
where ν is the integer such that $np < \nu \leq np + 1$.

20. Let S_n be the number of successes in n independent trials with probabilities $p_1, p_2, \cdots, p_n$ of success. Let $a_n = (p_1 + p_2 + \cdots + p_n)/n$ be the average probability of success. Show that for given a_n the maximum of $\mathrm{Var}(S_n)$ is attained when all p_k are equal.[10]

21. If two random variables X and Y assume only two values each, and if $\mathrm{Cov}(X, Y) = 0$, then X and Y are independent.

22. *Generalized Chebyshev inequality.* Let $\phi(x)$ be monotonically increasing, positive, and even, and suppose that $E(\phi(X)) = M$ exists. Prove
$$Pr\{|X| > t\} \leq \frac{M}{\phi(t)}.$$

23. Let $\{X_k\}$ be a sequence of mutually independent random variables with a common distribution. We assume that the X_k assume only positive values and that $E(X_k) = a$ and $E(X_k^{-1}) = b$ exist. Let $S_n = X_1 + \cdots + X_n$. Prove that $E(S_n^{-1})$ is finite and that $E(X_k/S_n) = 1/n$ for $k = 1, 2, \cdots, n$.

24. *Continuation.*[11] Prove that
$$E\left(\frac{S_m}{S_n}\right) = \frac{m}{n}, \quad \text{if } m \leq n$$
$$E\left(\frac{S_m}{S_n}\right) = 1 + (m-n)aE(S_n^{-1}), \quad \text{if } m \geq n.$$

[9] This problem is based on a new technique developed during World War II. See R. Dorfman, The Detection of Defective Members of Large Populations, *Annals of Mathematical Statistics*, vol. 14 (1943), pp. 436–440. In Army practice plan (ii) introduced up to 80 per cent savings.

[10] The variability of p_n may be interpreted as disorder and in this sense *disorder decreases chance fluctuations.* (For example, the number of annual fires in a community may be treated as a random variable; for a given average number a_n the variability is maximal if all households have the same probability of fire.) Trials of the type described in problem 20 are called *Poisson trials* [cf. Chapter 11, example (8.b)].

[11] The observation that 24 can be proved by introducing 23 is due to K. L. Chung.

25. Let $X_1, \ldots, X_n$ be mutually independent random variables. Let U be a function of $X_1, \ldots, X_k$ and V a function of $X_{k+1}, \ldots, X_n$ ($k < n$). Prove that U and V are mutually independent random variables.

Hint: Consider the two sample spaces defined, respectively, by the sets $(X_1, \ldots, X_k)$ and $(X_{k+1}, \ldots, X_n)$.

26. A sequence of Bernoulli trials is continued as long as necessary to obtain r successes, where r is a fixed integer. Let X be the number of trials required. Find [12] $E(r/X)$. (The definition leads to infinite series for which a finite expression can be obtained. The distribution of X is called after Pascal; cf. Chapter 11, section 3.)

27. *Length of random chains.*[13] A chain in the x, y-plane consists of n links, each of unit length. The angle between two consecutive links is $\pm \alpha$ where α is a positive constant; each possibility has probability $1/2$, and the successive angles are mutually independent. The length L_n of the chain is a random variable, and we wish to prove that

$$(9.1) \qquad E(L_n^2) = n \frac{1 + \cos \alpha}{1 - \cos \alpha} - 2 \cos \alpha \frac{1 - \cos^n \alpha}{(1 - \cos \alpha)^2}.$$

Without loss of generality the first link may be assumed to lie in the direction of the positive x-axis. The angle between the kth link and the positive x-axis is a random variable S_{k-1} where $S_0 = 0$, $S_k = S_{k-1} + X_k \alpha$ and the X_k are mutually independent variables, assuming the values ± 1 with probability $1/2$. The projections on the two axes of the kth link are $\cos S_{k-1}$ and $\sin S_{k-1}$. Hence for $n \geq 1$

$$(9.2) \qquad L_n^2 = \left(\sum_{k=0}^{n-1} \cos S_k \right)^2 + \left(\sum_{k=0}^{n-1} \sin S_k \right)^2.$$

Prove by induction successively for $m < n$

$$(9.3) \qquad E(\cos S_n) = \cos^n \alpha, \quad E(\sin S_n) = 0;$$

$$(9.4) \qquad E((\cos S_m) \cdot (\cos S_n)) = \cos^{n-m} \alpha \cdot E(\cos^2 S_m)$$

$$(9.5) \qquad E((\sin S_m) \cdot (\sin S_n)) = \cos^{n-m} \alpha \cdot E(\sin^2 S_m)$$

$$(9.6) \qquad E(L_n^2) - E(L_{n-1}^2) = 1 + 2 \cos \alpha \cdot \frac{1 - \cos^{n-1} \alpha}{1 - \cos \alpha}$$

(with $L_0 = 0$) and hence finally (9.1).

---

[12] This example illustrates the effect of *optional stopping*. If the number n of trials is fixed, the ratio of the number N of successes to the number n of trials is a random variable whose expectation is p. It is often erroneously assumed that the same is true in our example where the number r of successes is fixed and the number of trials depends on chance. If $p = 1/2$ and $r = 2$, then $E(2/X) = 0.614$ instead of 0.5; for $r = 3$ we find $E(3/X) = 0.579$.

[13] This is the two-dimensional analogue to the problem of length of *long polymer molecules* in chemistry. The problem is to illustrate applications to random variables which are not expressible as sums of simple variables.

CHAPTER 10

LAWS OF LARGE NUMBERS

1. Identically Distributed Variables

The limit theorems for Bernoulli trials derived in Chapters 7 and 8 are special cases of general limit theorems which cannot be treated in this volume. However, we shall here discuss at least some cases of the law of large numbers in order to reveal a new aspect of the notion of the expectation of a random variable.

The connection between Bernoulli trials and the theory of random variables becomes clearer when we consider the dependence of the number S_n of successes on the number n of trials. With each trial S_n increases by 1 or 0. To translate this statement into a formula, we write

$$(1.1) \qquad S_n = X_1 + \cdots + X_n,$$

where the random variable X_k equals 1 or 0 according to whether the kth trial results in success or failure. Thus S_n is a sum of n mutually independent random variables, each of which assumes the values 1 and 0 with probabilities p and q. From this it is only one step to consider sums of the form (1.1) where the X_k are mutually independent variables with an arbitrary distribution. The (weak) law of large numbers of Chapter 7, section 3, states that for large n the average proportion of successes S_n/n is likely to lie near p. This is a special case of the following

Law of Large Numbers. *Let $\{X_k\}$ be a sequence of mutually independent random variables with a common distribution. If the expectation* $\mu = E(X_k)$ *exists, then for every $\epsilon > 0$ as $n \to \infty$*

$$(1.2) \qquad Pr\left\{\left|\frac{X_1 + \cdots + X_n}{n} - \mu\right| > \epsilon\right\} \to 0;$$

in words, the probability that the average S_n/n will differ from the expectation by less than an arbitrarily prescribed ϵ tends to one.

In this generality the theorem was first proved by Khintchine.[1]

[1] A. Khintchine, Sur la loi des grands nombres, *Comptes rendus de l'Académie des Sciences*, vol. 189 (1929), pp. 477–479.

Older proofs had to introduce the unnecessary restriction that also the variance $\mathrm{Var}(X_k)$ be finite.[2] For this case, however, there exists a much more precise result which generalizes the DeMoivre-Laplace limit theorem for Bernoulli trials.

Central Limit Theorem. Let $\{X_k\}$ *be a sequence of mutually independent random variables with a common distribution. Suppose that* $\mu = E(X_k)$ *and* $\sigma^2 = \mathrm{Var}(X_k)$ *exist and let* $S_n = X_1 + \cdots + X_n$. *Then for every fixed* α, β $(\alpha < \beta)$

(1.3) $$Pr\left\{\alpha < \frac{S_n - n\mu}{\sigma n^{1/2}} < \beta\right\} \to \Phi(\beta) - \Phi(\alpha);$$

here $\Phi(x)$ is the normal distribution introduced in Chapter 7, section 1. This theorem is due to Lindeberg;[3] Ljapunov and other authors had previously proved it under more restrictive conditions. It must be understood that the above theorem is only a very special case of a much more general theorem which is closely connected with many other limit theorems. We shall defer its general formulation and proof to the second volume. Here we note that (1.3) is much stronger than (1.2), since it gives an estimate for the probability that the discrepancy $\left|\frac{1}{n}S_n - \mu\right|$ be larger than $\sigma/n^{1/2}$. On the other hand, the law of large numbers (1.2) holds even when the random variables X_k have no finite variance so that it is more general than the central limit theorem. For this reason we shall give an independent proof of the law of large numbers, but first we illustrate the two limit theorems.

Examples. (a) Consider a sequence of independent throws of a symmetric die and let X_k be the number scored at the kth throw. Then $E(X_k) = (1 + 2 + 3 + 4 + 5 + 6)/6 = 3.5$, and $\mathrm{Var}(X_k) = (1^2 + 2^2 + 3^2 + 4^2 + 5^2 + 6^2)/6 - (3.5)^2 = 35/12$, and S_n/n is the average score in n throws. The law of large numbers states that for large n this average is likely to be near 3.5. The central limit theorem states that the probability of $|S_n - 3.5n| < \alpha \cdot (35n/12)^{1/2}$ is about $\Phi(\alpha) - \Phi(-\alpha)$. For $n = 1000$ and $\alpha = 1$ we find that there is roughly probability 0.68 that $3450 < S_n < 3550$. Choosing for α the median value $\alpha = 0.6744$, we find that there are roughly equal chances that S_n lies within or without the interval 3500 ± 36.

(b) *Sampling.* Suppose that in a population of N families there are N_k families with exactly k children $(k = 0, 1, \ldots; \Sigma N_k = N)$. If a

---

[2] A. Markov showed that the existence of $E(|X_k|^{1+a})$ for some $a > 0$ suffices.

[3] J. W. Lindeberg, Eine neue Herleitung des Exponentialgesetzes in der Wahrscheinlichkeitsrechnung, *Mathematische Zeitschrift*, vol. 15 (1922), pp. 211–225.

family is chosen at random, the number of children in it is a random variable which assumes the value ν with probability $p_\nu = N_\nu/N$. In sampling with replacement a sample of size n contains n independent random variables or "observations" $X_1, \cdots, X_n$, each with the same distribution; S_n/n is the *sample average*. The law of large numbers tells us that for sufficiently large random samples the sample average is likely to be near $\mu = \Sigma \nu p_\nu = \Sigma \nu N_\nu / N$, which is the population average. The central limit theorem permits us to estimate the probable magnitude of the discrepancy and to determine the sample size necessary for reliable estimates. In practice both μ and σ^2 are unknown. However, it is usually easy to obtain a preliminary estimate of σ^2, and it is always possible to keep to the safe side. If it is desired that there be probability 0.99 or better that the sample average S_n/n differs from the unknown population mean μ by less than $1/10$, then the sample size should be such that

$$(1.4) \qquad Pr\left\{\left|\frac{S_n - n\mu}{n}\right| < \frac{1}{10}\right\} \geq 0.99.$$

Now the root of $\Phi(x) - \Phi(-x) = 0.99$ is $x = 2.57\ldots$, and hence n should be such that $n^{1/2}/10\sigma \geq 2.57$ or $n \geq 660\sigma^2$. A cautious preliminary estimate of σ^2 gives us an idea of the required sample size. Similar situations occur frequently. Thus when the experimenter takes the mean of n measurements he, too, relies on the law of large numbers and uses a sample mean as an estimate for an unknown theoretical expectation. The reliability of this estimate can be judged only in terms of σ^2, and usually we are compelled to use rather crude estimates for σ^2.

(c) *The Poisson Distribution.* In Chapter 7, section 4, we found that for large λ the Poisson distribution $\{p(k; \lambda)\}$ can be approximated by the normal distribution. This is really a direct consequence of the central limit theorem. Suppose that the variables X_k have a Poisson distribution $\{p(k; \gamma)\}$. Then S_n has a Poisson distribution $\{p(k; n\gamma)\}$ with mean and variance equal to $n\gamma$. Writing λ for $n\gamma$, we conclude that *for every $\alpha < \beta$ as $\lambda \to \infty$*

$$(1.5) \qquad \sum_{\alpha\lambda^{1/2} < k-\lambda < \beta\lambda^{1/2}} \frac{e^{-\lambda}\lambda^k}{k!} \to \Phi(\beta) - \Phi(\alpha);$$

the summation extends over all k in the interval $(\lambda + \alpha\lambda^{1/2}, \lambda + \beta\lambda^{1/2})$. This theorem is used in the theory of summability of divergent series and is of general interest; estimates of the difference of the two sides in (1.5) are available from the general theory.

As an application of (1.5) suppose that λ is an integer and let us compute

$$(1.6) \qquad u_r = \sum_{-r \leq k - \lambda \leq r} \frac{e^{-\lambda}\lambda^k}{k!}.$$

To make the limits of summation in (1.5) and (1.6) coincide we may take for $-\alpha$ and β any number between $r\lambda^{-1/2}$ and $(r+1)\lambda^{-1/2}$. As λ increases, this interval of uncertainty decreases, and in the limit the freedom of choice of α and β disappears. However, for moderate values of λ it is most natural to choose for $-\alpha$ and β the midpoint $(r+\frac{1}{2})\lambda^{-1/2}$. We, therefore, compare u_r with the normal approximation

$$(1.7) \qquad u_r' = \Phi((r+\tfrac{1}{2})\lambda^{-1/2}) - \Phi(-(r+\tfrac{1}{2})\lambda^{-1/2}).$$

Table 1 shows that the degree of approximation is highly satisfactory (further numerical comparisons are found in Chapter 7, section 4).

TABLE 1

Showing the Normal Approximation (1.7) to the Poisson Expression (1.6)

λ	r	u_r	u_r'	Difference
9	3	0.760 08	0.756 66	0.0034
16	4	.741 17	.739 41	.0018
25	5	.729 73	.728 67	.0011
49	7	.716 53	.716 01	.0005
100	10	.706 52	.706 28	.0002

Both the law of large numbers and the central limit theorem become meaningless if the expectation μ does not exist, but it is possible to replace them by more general limit theorems. It will be shown later on that most recurrence times connected with physical processes are random variables without finite expectation. Even in the simple coin-tossing game the number of tosses up to the first equalization of the accumulated numbers of heads and tails is a random variable to which the law of large numbers does not apply. The corresponding limit theorems (in particular the arc sine law) will be discussed in detail in Chapter 12, section 5, where it will be found that the fluctuations of such variables have many surprising features and that they are entirely different from the fluctuations described in this chapter.

*2. Proof of the Law of Large Numbers

We proceed in two steps. First we assume that $\sigma^2 = \mathrm{Var}(X_k)$ exists, and note that in this case $\mathrm{Var}(S_n) = n\sigma^2$, by the addition rule [formula (5.6) of Chapter 9]. According to the Chebyshev inequality (Chapter 9, section 6), we have for every $t > 0$

$$(2.1) \qquad Pr\{|S_n - n\mu| > t\} \leq \frac{n\sigma^2}{t^2}.$$

For $t = \epsilon n$ we find that the left side is less than $\sigma^2/\epsilon^2 n$, which quantity tends to zero. This accomplishes the proof.

We now drop the restrictive condition that $\mathrm{Var}(X_k)$ exists. This case is reduced to the preceding one by the *method of truncation* which is an important standard tool used (with various refinements) in many similar cases. We define two new collections of random variables depending on the X_k. Let, for $k = 1, 2, \cdots, n$, and fixed $\epsilon > 0$,

$$(2.2) \qquad \begin{aligned} U_k &= X_k, & V_k &= 0 & \text{if } |X_k| \leq \epsilon n; \\ U_k &= 0, & V_k &= X_k & \text{if } |X_k| > \epsilon n. \end{aligned}$$

Then

$$(2.3) \qquad X_k = U_k + V_k$$

for all k.

Let $\{f(x_j)\}$ be the common probability distribution of the variables X_k. We have assumed that $\mu = E(X_k)$ exists, which means that

$$(2.4) \qquad \Sigma |x_j| f(x_j) = A$$

is finite. Then

$$(2.5) \qquad \mu' = E(U_k) = \sum_{|x_j| \leq \epsilon n} x_j f(x_j),$$

the summation extending over those j for which $|x_j| \leq \epsilon n$. Note that μ' depends on n but is common to $U_1, U_2, \cdots, U_n$. Moreover, $\mu' \to \mu$ as $n \to \infty$, and hence for all n sufficiently large and arbitrary $\delta > 0$

$$(2.6) \qquad |\mu' - \mu| < \delta.$$

Furthermore, from (2.5) and (2.4),

$$(2.7) \qquad \mathrm{Var}(U_k) \leq E(U_k^2) \leq \epsilon n \sum_{|x_j| \leq \epsilon n} |x_j| f(x_j) \leq \epsilon A n.$$

* Starred sections treat special topics and may be omitted at first reading.

The U_k are mutually independent, and their sum $U_1 + U_2 + \cdots + U_n$ can be treated exactly as the X_k in the case of finite variances; applying the Chebyshev inequality, we get the following analogue to (2.1)

$$(2.8) \qquad Pr\left\{\left|\frac{U_1 + \cdots + U_n}{n} - \mu'\right| > \delta\right\} < \frac{\text{Var}(U_k)}{n\delta^2} < \frac{\epsilon A}{\delta^2}.$$

In view of (2.6) this implies

$$(2.9) \qquad Pr\left\{\left|\frac{U_1 + \cdots + U_n}{n} - \mu\right| > 2\delta\right\} < \frac{\epsilon A}{\delta^2}.$$

Next we note that there is a large probability that $V_k = 0$. In fact

$$(2.10) \quad Pr\{V_k \neq 0\} = \sum_{|x_j| > \epsilon n} f(x_j) \leq \frac{1}{\epsilon n} \sum_{|x_j| > \epsilon n} |x_j| f(x_j).$$

Since the series (2.4) converges, the last sum tends to 0 with increasing n. Therefore for n sufficiently large

$$(2.11) \qquad Pr\{V_k \neq 0\} \leq \frac{\epsilon}{n}$$

and hence [cf. Chapter 1, formula (6.6)],

$$(2.12) \qquad Pr\{V_1 + \cdots + V_n \neq 0\} \leq \epsilon.$$

Now $S_n = (U_1 + \cdots + U_n) + (V_1 + \cdots + V_n)$, and therefore from (2.9) and (2.12)

$$(2.13) \quad Pr\left\{\left|\frac{S_n}{n} - \mu\right| > 2\delta\right\} \leq Pr\left\{\left|\frac{U_1 + \cdots + U_n}{n} - \mu\right| > 2\delta\right\}$$
$$+ Pr\{V_1 + \cdots + V_n \neq 0\} \leq \frac{\epsilon A}{\delta^2} + \epsilon.$$

Since ϵ and δ are arbitrary, the right side can be made arbitrarily small, and this proves the assertion.

3. The Theory of "Fair" Games

For a further analysis of the implications of the law of large numbers we shall use the time-honored terminology of gamblers, but our discussion bears equally on less frivolous applications, and our two basic assumptions are more realistic in statistics and physics than in gambling halls. First, we shall assume that our gambler possesses an *unlimited capital* so that no loss can force a termination of the game. (Dropping this assumption leads to the problem of the gambler's *ruin*, which from the very beginning has intrigued students of probability. It is of importance in Wald's sequential analysis and in the theory of stochastic processes, and will be taken up in Chapter 14.) Second, we shall assume that the gambler *does not have the privilege of optional stopping; the number n of trials must be fixed in advance* independently of the development of the game. In practice a player blessed with an unlimited capital would wait for a run of good luck and quit at an opportune moment. Such a player is not interested in the probable fluctuation at a prescribed moment, but only in the maximal fluctuations in the long run. Light is shed on this problem by the law of the iterated logarithm rather than by the law of large numbers (cf. Chapter 8, section 5).

The random variable X_k will now be interpreted as the (positive or negative) gain at the kth trial of a player who keeps playing the same type of game of chance. The sum $S_n = X_1 + \cdots + X_n$ is the accumulated gain in n independent trials. If the player pays for each trial an entrance fee μ' (not necessarily positive), then $n\mu'$ represents the accumulated entrance fees, and $S_n - n\mu'$ the *accumulated net gain*. The law of large numbers applies if $\mu = E(X_k)$ exists. It says roughly that for sufficiently large n the difference $S_n - n\mu$ is likely to be small in comparison to n. Therefore, if the entrance fee μ' is smaller than μ, then, for large n, the player is likely to have a positive gain of the order of magnitude $n(\mu - \mu')$. For the same reason an entrance fee $\mu' > \mu$ is practically sure to lead to a loss. In short, the case $\mu' < \mu$ is *favorable* to the player, while $\mu' > \mu$ is *unfavorable*.

Note that nothing is said about the case $\mu' = \mu$. The *only* possible conclusion in this case is that, for n sufficiently large, *the accumulated gain or loss $S_n - n\mu$ will with overwhelming probability be small in comparison with n*. It is not stated whether $S_n - n\mu$ is likely to be positive or negative, that is, whether the game is favorable or unfavorable. This was overlooked in the classical theory which called $\mu' = \mu$ a "fair" price and a game with $\mu' = \mu$ "*fair.*" Much harm was done by the misleading suggestive power of this name. It must be understood that a "fair" game may be distinctly favorable or unfavorable to the player.

It is clear that "normally" not only $E(X_k)$ but also $\text{Var}(X_k)$ exists. In this case the law of large numbers is supplemented by the central limit theorem, and the latter tells us that, with a "fair" game, the long-run net gain $S_n - n\mu$ is likely to be of the order of magnitude $n^{1/2}$ and that for large n there are about equal odds for this net gain to be positive or negative. Thus, when the central limit theorem applies, the term "fair" appears justified, but even in this case we deal with a limit theorem with emphasis on the words "long run." A closer analysis shows that the convergence in (1.3) deteriorates with increasing variance. If $\sigma^2 = \text{Var}(X_k)$ is large, then the normal approximation may become effective only for exceedingly large n.

To fix ideas, consider a slot machine where the player has a probability of 10^{-6} to win $10^6 - 1$ dollars, and the alternative of losing the entrance fee $\mu' = 1$. Here we have Bernoulli trials, and the game is "fair." In a million trials the player pays as many dollars in entrance fees. He may hit the jackpot 0, 1, 2, ... times. We know from the Poisson approximation to the binomial distribution that, with an accuracy to several decimal places, the probability of hitting the jackpot exactly k times is $e^{-1}/k!$. Thus the player has probability $0.368\ldots$ to lose a million, and the same probability of barely recovering his expenses; he has probability $0.184\ldots$ to gain exactly one million, etc. Here 10^6 trials are equivalent to one single trial in a game with the gain distributed according to a Poisson distribution (which could be realized by matching two large decks of cards; cf. Chapter 4, section 4). Now all fire, automobile, and similar insurance is of the described type; the risk involves a huge sum, but the corresponding probability is very small. Moreover, one plays ordinarily only one trial per year, so that the number n of trials never grows large. For the insured the game is necessarily "unfair," but it may well be economically advantageous: the law of large numbers is of no relevance to him. As for the company, it plays a large number of games, but because of the large variance the chance fluctuations are pronounced. The premiums must be fixed so as to preclude a huge loss in any specific year, and hence the company is concerned with the ruin problem rather than the law of large numbers.

When the variance is infinite, the term "fair games" becomes an absolute misnomer; there is no reason to believe that the accumulated net gain $S_n - n\mu'$ fluctuates around zero. In fact, there exist examples of "fair" games [4] where the probability tends to one that the player will have sustained a net loss. The law of large numbers asserts that this net loss is likely to be of smaller order of magnitude than n. However, nothing more can be asserted. If a_n is an arbitrary sequence such that $a_n/n \to 0$, it is possible to construct a "fair" game where the probability tends to one that at the nth trial the accumulated net loss exceeds a_n. Problem 12 contains an example where the player has a practical assurance that his loss will exceed $n/\log n$. This game is "fair," and the entrance fee is unity. It is difficult to imagine that a

[4] W. Feller, Note on the Law of Large Numbers and 'Fair" Games, *Annals of Mathematical Statistics*, vol. 16 (1945), pp. 301–304.

player will find it "fair" if after 1,000,000 games he is practically certain to have lost more than 150,000 units, and the loss is likely to keep increasing.

*4. The Petersburg Game

If the variables X_k have no finite expectation, the law of large numbers becomes inapplicable and must be replaced by another limit theorem describing the asymptotic behavior of the sum S_n. In the second volume we shall generalize both the law of large numbers and the central limit theorem to random variables without expectations; this condition is typical for recurrence times in many physical processes, and the generalized limit theorems are therefore of more than theoretical interest.[5] The classical theory had no rigorous mathematical formulations at its disposal and had therefore conceptual difficulties with random variables without expectation. The law of large numbers was intimately connected with the concept of probability and was not susceptible of a mathematical analysis. This led even quite recent writers to conclusions which are difficult to understand from the point of view of a formalized theory. The general theory of limit theorems is deferred to the second volume, but it seems appropriate to describe the modern approach, using the time-honored example of the so-called Petersburg paradox.[6]

A single trial in the Petersburg game consists in tossing a true coin until it falls heads; if this occurs at the rth throw the player receives 2^r dollars. In other words, the gain at each trial is a random variable assuming the values 2^1, 2^2, 2^3, ... with corresponding probabilities 2^{-1}, 2^{-2}, 2^{-3}, ... The expectation is formally defined by $\Sigma x_r f(x_r)$ with $x_r = 2^r$ and $f(x_r) = 2^{-r}$, so that each term of the series equals 1. Thus the gain has no finite expectation, and the law of large numbers is inapplicable. Now the game becomes less favorable to the player when amended by the rule that he receives nothing if a trial takes more than N tosses (that is, if the coin falls tails N times in succession). In this amended game the gain has the finite expectation N, and the law of large numbers applies. Therefore, if the player pays a constant entrance fee $\mu' > 0$ for each trial and plays n games, then for n sufficiently large he is almost sure to have a net profit. This

---

* Starred sections treat special topics and may be omitted at first reading.

[5] An interesting special case, connected with the coin-tossing game, is discussed in Chapter 12, section 5. It will give a fair idea about the nature of fluctuations of sums of random variables without finite expectation.

[6] This paradox was discussed by Daniel Bernoulli (1700–1782), who tried in vain to solve it by the concept of moral expectation. Note that Bernoulli trials are named after James Bernoulli.

is true for every μ', but the larger μ', the larger must n be in order that a positive gain be probable. The classical theory concluded that $\mu' = \infty$ is a "fair" entrance fee, but the modern student will hardly understand the mysterious discussions of this "paradox."

It is perfectly possible to determine entrance fees with which the Petersburg game will have all properties of a "fair" game in the classical sense except that these entrance fees will depend on the number of trials instead of remaining constant. Variable entrance fees are undesirable in gambling halls, but there the Petersburg game would be impossible anyway because of limited resources. In the case of a finite expectation $\mu = E(X_k) > 0$, a game is called "fair" if for large n the ratio of the accumulated gain S_n to the accumulated entrance fees $e_n = n\mu'$ is likely to be near 1 (that is, if the difference $S_n - e_n$ is likely to be of smaller order of magnitude than $e_n = n\mu'$). If $E(X_k)$ does not exist, we cannot put $e_n = n\mu'$ but must determine e_n in another way. We shall say that *"a game is fair"* in the classical sense if it is possible to determine accumulated entrance fees e_n so that for every $\epsilon > 0$

$$(4.1) \qquad Pr\left\{\left|\frac{S_n}{e_n} - 1\right| > \epsilon\right\} \to 0.$$

This is the complete analogue of the law of large numbers where $e_n = n\mu'$. The latter is interpreted by the physicist to the effect that the average of n independent measurements is bound to be near μ. In the present instance the average of n measurements is bound to be near e_n/n. Our limit theorem (4.1), when it applies, has a mathematical and operational meaning which is not different from the law of large numbers.

We shall now show [7] that the *Petersburg game becomes "fair" in the classical sense if we put* $e_n = n \operatorname{Log} n$, where $\operatorname{Log} n$ is the logarithm to the base 2, that is, $2^{\operatorname{Log} n} = n$.

Proof. We use again the method of truncation of section 2. Instead of (2.2) we now define the variables U_k and V_k ($k = 1, 2, \cdots, n$) by

$$(4.2) \qquad \begin{aligned} U_k &= X_k, & V_k &= 0 & \text{if} \quad X_k &\leq n \operatorname{Log} n; \\ U_k &= 0, & V_k &= X_k & \text{if} \quad X_k &> n \operatorname{Log} n. \end{aligned}$$

Then again $X_k = U_k + V_k$, and the U_k are mutually independent. For

---

[7] This is a special case of a generalized law of large numbers from which necessary and sufficient conditions for (4.1) can easily be derived; cf. W. Feller, *Acta Scientiarum Litterarum Univ. Szeged*, vol. 8 (1937), pp. 191–201.

every t we have $Pr\{X_k > t\} \leq 2/t$ and hence $Pr\{V_k \neq 0\} < 2/n \, \text{Log } n$, or

(4.3) $$Pr\{V_1 + V_2 + \cdots + V_n > 0\} < \frac{2}{\text{Log } n} \to 0.$$

To verify (4.1) it suffices therefore to prove that

(4.4) $$Pr\{|\, U_1 + \cdots + U_n - n \, \text{Log } n\,| > \epsilon n \, \text{Log } n\} \to 0.$$

Now put $\mu = E(U_k)$ and $\sigma^2 = \text{Var}(U_k)$. These quantities depend on n, but are common to $U_1, U_2, \cdots, U_n$. If r is the largest integer such that $2^r \leq n \, \text{Log } n$, then $\mu = r$ and hence for sufficiently large n

(4.5) $$\text{Log } n < \mu \leq \text{Log } n + \text{Log Log } n.$$

Similarly

(4.6) $$\sigma^2 < E(U_k^2) = 2 + 2^2 + \cdots + 2^r < 2^{r+1} \leq 2n \, \text{Log } n.$$

Since the sum $U_1 + \cdots + U_n$ has mean $n\mu$ and variance $n\sigma^2$, we have by Chebyshev's inequality

(4.7) $$Pr\{|\, U_1 + \cdots + U_n - n\mu\,| > \epsilon n\mu\} \leq \frac{n\sigma^2}{\epsilon^2 n^2 \mu^2} < \frac{2}{\epsilon^2 \, \text{Log } n} \to 0.$$

Now by (4.5) $\mu \sim \text{Log } n$, and hence (4.7) is equivalent to (4.4) and therefore to (4.1).

5. Variable Distributions

Up to now we have considered only the case where the variables X_k have the same distribution. In gambling this case arises when the player keeps playing the same game of chance; however, it is even more interesting to see what happens if the type of game changes at each step. It is not necessary to think of gambling places; the statistician who applies statistical tests is engaged in a dignified sort of gambling, and in his case the distribution of the random variables changes from occasion to occasion.

To fix ideas we shall imagine that an infinite sequence of probability distributions is given. For every n we may then speak of mutually independent variables $X_1, \cdots, X_n$ with the prescribed distributions. We shall assume that the means and variances exist and put

(5.1) $$\mu_k = E(X_k), \quad \sigma_k^2 = \text{Var}(X_k).$$

The sum $S_n = X_1 + \cdots + X_n$ has also finite mean and variance

(5.2) $$m_n = E(S_n), \quad s_n^2 = \text{Var}(S_n)$$

which are given by

(5.3) $\qquad m_n = \mu_1 + \cdots + \mu_n, \quad s_n^2 = \sigma_1^2 + \cdots + \sigma_n^2$

[cf. Chapter 9, formulas (2.4) and (5.6)]. In the special case of identical distribution we had $m_n = n\mu$, $s_n^2 = n\sigma^2$.

The (weak) law of large numbers is said to hold for the sequence $\{X_k\}$ if for every $\epsilon > 0$

(5.4) $\qquad Pr\left\{\dfrac{|S_n - m_n|}{n} > \epsilon\right\} \to 0.$

The sequence $\{X_k\}$ is said to obey the central limit theorem if for every fixed $\alpha < \beta$

(5.5) $\qquad Pr\left\{\alpha < \dfrac{S_n - m_n}{s_n} < \beta\right\} \to \Phi(\beta) - \Phi(\alpha).$

It is one of the salient features of probability theory that both the law of large numbers and the central limit theorem hold for a surprisingly large class of sequences $\{X_k\}$. In particular, *the law of large numbers holds whenever the X_k are uniformly bounded*, that is, whenever there exists a constant A such that

(5.6) $\qquad |X_k| < A$

for all k. More generally, *a sufficient condition for the law of large numbers to hold is that*

(5.7) $\qquad \dfrac{s_n}{n} \to 0.$

This is a direct consequence of the Chebyshev inequality, and the proof given in the opening passage of section 2 applies. Note, however, that the condition (5.7) is not necessary (cf. problem 14).

Various sufficient conditions for the central limit theorem have been discovered, but all were superseded by the *Lindeberg*[8] *theorem according to which the central limit theorem holds whenever for every $\epsilon > 0$ the truncated variables U_k defined by*

(5.8) $\qquad \begin{aligned} U_k &= X_k \quad \text{if} \quad |X_k| \leq \epsilon s_n, \\ U_k &= 0 \quad \text{if} \quad |X_k| > \epsilon s_n, \end{aligned}$

$(k = 1, 2, \cdots, n)$ *satisfy the condition*

(5.9) $\qquad \operatorname{Var}(U_1 + \cdots + U_n) \sim s_n^2.$

[8] J. W. Lindeberg, *loc. cit.* (footnote 3).

Here the sign $\sim$ indicates that $s_n \to \infty$ and that the ratio of the two sides tends to unity.

If the X_k are uniformly bounded [that is, if (5.6) holds] then $U_k = X_k$ for all n which are so large that $s_n > A\epsilon^{-1}$. Therefore the Lindeberg theorem implies that *every uniformly bounded sequence $\{X_k\}$ of mutually independent random variables obeys the central limit theorem*, provided, of course, that $s_n \to \infty$. (The last condition is violated only for degenerate sequences.) It was found that this condition is also necessary for (5.5) to hold.[9] The proof is deferred to the second volume, where we shall also give estimates for the difference between the two sides in (5.5).

In the case where the variables X_k have a common distribution we found the central limit theorem to be stronger than the law of large numbers. This is not so in general, and we shall see that the central limit theorem may apply to sequences which do not obey the law of large numbers.

Examples. (a) Let $\lambda > 0$ be fixed, and let $X_k = \pm k^\lambda$, each with probability $1/2$ (e.g., a coin is tossed, and at the kth throw the stakes are $\pm k^\lambda$). Here $\mu_k = 0$, $\sigma_k^2 = k^{2\lambda}$, and

$$(5.10) \qquad s_n^2 = 1^{2\lambda} + 2^{2\lambda} + 3^{2\lambda} + \cdots + n^{2\lambda} \sim \frac{n^{2\lambda+1}}{2\lambda+1}.$$

The condition (5.7) is satisfied if $\lambda < 1/2$. Therefore the law of large numbers holds if $\lambda < 1/2$; we shall presently see that it does not hold if $\lambda \geq 1/2$.

For $k = 1, 2, \cdots, n$ we have $|X_k| = k^\lambda \leq n^\lambda$, so that for $n > (2\lambda + 1)\epsilon^{-2}$ the truncated variables U_k are identical with the X_k. Hence the Lindeberg condition applies for $\lambda > 0$, and

$$(5.11) \qquad Pr\left\{\alpha < \left(\frac{2\lambda+1}{n^{2\lambda+1}}\right)^{\frac{1}{2}} S_n < \beta\right\} \to \Phi(\beta) - \Phi(\alpha).$$

It follows that S_n is likely to be of the order of magnitude $n^{\lambda+\frac{1}{2}}$, so that the law of large numbers cannot apply for $\lambda \geq 1/2$. We see that in this example *the central limit theorem applies for all $\lambda > 0$, but the law of large numbers only if $\lambda < 1/2$*.

(b) Consider two independent sequences of 1000 tossings of a coin (or emptying two bags of 1000 coins each). We want to investigate

---

[9] W. Feller, Über den zentralen Grenzwertsatz der Wahrscheinlichkeitsrechnung, *Mathematische Zeitschrift*, vol. 40 (1935), pp. 521–559. There also a generalized central limit theorem is derived which may apply to variables without expectations. Note that we are here considering only independent variables; for dependent variables the Lindeberg condition is neither necessary nor sufficient.

the *difference* D of the number of heads. Let the tossings of the two sequences be numbered from 1 to 1000 and from 1001 to 2000, respectively. We define 2000 random variables X_k as follows: if the kth coin falls tails, then $X_k = 0$. If it falls heads, then we put $X_k = 1$ or $X_k = -1$, according to whether $k \leq 1000$ or $k > 1000$. Then $D = X_1 + X_2 + \cdots + X_{2000}$. Moreover, $\mu_k = \pm 1/2$, depending on the sequence to which the coin belongs, $\sigma_k^2 = 1/4$, $m_{2000} = 0$, $s_{2000}^2 = 500$. Therefore the probability that the difference D will lie within the limits $\pm(500)^{1/2}\alpha$ is $\Phi(\alpha) - \Phi(-\alpha)$, approximately. The random variable D is therefore comparable to the deviation $S_{2000} - 1000$ of the number of heads in 2000 tossings from its expected number 1000.

(c) An application to the *theory of inheritance* will illustrate the great variety of conclusions based on the central limit theorem. In Chapter 5 we have studied traits which depend essentially only on one pair of genes (alleles). We conceive of other characters (like height) as the cumulative effect of a great number of pairs of genes. For simplicity, suppose that with respect to each particular pair of genes the individual belongs to one of the three genotypes AA, Aa, or aa. Let x_1, x_2, and x_3 be the corresponding contributions to the height [if $x_2 > (x_1 + x_3)/2$, the gene A is *partially dominant*]. The genotype of an individual is a random event, and hence the contribution of our particular pair of genes to the height is a random variable X, assuming the three values x_1, x_2, x_3 with certain probabilities. The height is the cumulative effect of many such random variables $X_1, X_2, \cdots, X_n$, and since the contribution of each is small, we may in first approximation assume that the height is the *sum* $X_1 + \cdots + X_n$. It is true that not *all* the X_k are mutually independent. However, the central limit theorem holds also for large classes of dependent variables, and, besides, it is plausible that the great majority of the X_k can be treated as independent. These considerations can be rendered more precise; here they serve only as indication of how the central limit theorem explains why many biometric characters, like height, exhibit an empirical distribution which is close to the normal distribution. This theory permits also the prediction of properties of inheritance, e.g., the dependence of the mean height of children on the height of their parents. Such biometric investigations of F. Galton and Karl Pearson laid the foundations of modern statistical theory.

* **6. Applications to Combinatorial Analysis**

We shall give two examples of applications of the central limit theorem to problems which are not directly connected with probability theory.

* Starred sections treat special topics and may be omitted at first reading.

In the following we consider the space whose points are the $n!$ permutations of the elements $a_1, a_2, \cdots, a_n$ and attribute to each permutation probability $1/n!$.

(a) *Inversions.* The element a_k is said to produce r inversions if it precedes r elements among $a_1, a_2, \cdots, a_{k-1}$ (which precede it in the natural order). Thus in the permutation $(a_3 a_6 a_1 a_5 a_2 a_4)$ the element a_2 produces no inversions, a_3 two, a_4 none, a_5 two, a_6 four. The total number of inversions in this case is eight. In $(a_6 a_5 a_4 a_3 a_2 a_1)$ there are fifteen inversions of which a_k produces $k-1$ ($k = 2, 3, 4, 5, 6$). Let X_k be the number of inversions produced by the element a_k. Then X_k is a random variable and $S_n = X_1 + \cdots + X_n$ is the total number of inversions. Now X_k assumes the values $0, 1, \cdots, k-1$, each with probability $1/k$, and therefore

$$\mu_k = \frac{k-1}{2},$$

(6.1)

$$\sigma_k^2 = \frac{1 + 2^2 + \cdots + (k-1)^2}{k} - \left(\frac{k-1}{2}\right)^2 = \frac{k^2 - 1}{12}.$$

The number of inversions produced by a_k does not depend on the relative order of $a_1, a_2, \cdots, a_{k-1}$, and hence the X_k are mutually independent. From (6.1) we get

(6.2) $$m_n = \frac{1 + 2 + \cdots + (n-1)}{2} = \frac{n(n-1)}{4} \sim \frac{n^2}{4}$$

and

(6.3) $$s_n^2 = \frac{1}{12} \sum_{k=1}^{n} (k^2 - 1) = \frac{2n^3 + 3n^2 - 5n}{72} \sim \frac{n^3}{36}.$$

For large n we have $\epsilon s_n > n \geq U_k$, and hence the variables U_k of the Lindeberg condition are identical with X_k. Therefore the central limit theorem applies, and we conclude that the *number N_n of permutations for which the number of inversions lies between the limits* $\frac{n^2}{4} \pm \frac{\alpha}{6} n^{3/2}$ *is, asymptotically, given by* $n!\{\Phi(\alpha) - \Phi(-\alpha)\}$. In particular, for about one-half of all permutations the number of inversions lies between the limits $(n^2/4) \pm (0.11) n^{3/2}$.

(b) *Cycles.* Every permutation can be broken down into cycles, that is, groups of elements permuted among themselves. Thus in $(a_3 a_6 a_1 a_5 a_2 a_4)$ we find that a_1 and a_3 are interchanged, and that the remaining four elements are permuted among themselves; this permutation contains two cycles. If an element is in its natural place, it forms a cycle so that the identical permutation $(a_1, a_2, \cdots, a_n)$ contains as many cycles as elements. On the other hand, the cyclical permutations $(a_2, a_3, \cdots, a_n, a_1)$, $(a_3, a_4, \cdots, a_n, a_1, a_2)$, etc., contain a single cycle each. For the study of cycles it is convenient to describe a permutation by means of arrows indicating to which places the elements have been moved. For example $1 \to 3 \to 4 \to 1$ indicates that a_1 has been moved to the third place, a_3 to the fourth, and a_4 to the first; this completes the first cycle, and we continue with the next element in the natural order, namely, a_2. Thus $(a_4 a_8 a_1 a_3 a_2 a_5 a_7 a_6)$ would be described by $1 \to 3 \to 4 \to 1$; $2 \to 5 \to 6 \to 8 \to 2$; $7 \to 7$; here we have three cycles of lengths 3, 4, 1, respectively.

To study the number of cycles we let the random variables X_k $(k = 1, 2, \cdots, n)$ equal 1 or 0 according to whether a cycle is or is not completed at the kth step in our build-up. Thus in the last example $X_3 = 1$, $X_7 = 1$, $X_8 = 1$ while all other X_k's equal 0. Clearly $X_1 = 1$ if, and only if, a_1 moves to the first place. If a_1 remains at the first place, then $X_2 = 1$ if, and only if, a_2 remains at the second place. In general, at the kth step we have $n - k + 1$ choices of which one and only one leads to the completion of a cycle. It follows that X_k equals 1 with probability [10] $1/(n - k + 1)$ and 0 with probability $(n - k)/(n - k + 1)$. Moreover, the X_k are mutually independent and uniformly bounded. In our case

$$(6.4) \qquad \mu_k = \frac{1}{n - k + 1}, \quad \sigma_k^2 = \frac{n - k}{(n - k + 1)^2}$$

and hence

$$(6.5) \qquad m_n = 1 + \frac{1}{2} + \frac{1}{3} + \cdots + \frac{1}{n} \sim \log n$$

and

$$(6.6) \qquad s_n^2 = \sum_{k=1}^{n} \frac{n - k}{(n - k + 1)^2} \sim \log n.$$

$S_n = X_1 + \cdots + X_n$ is the total number of cycles. *The average number of cycles is m_n; asymptotically, the number of permutations for which the number of cycles lies between $\log n + \alpha(\log n)^{1/2}$ and $\log n + \beta(\log n)^{1/2}$ is given by $n!\{\Phi(\beta) - \Phi(\alpha)\}$.* The refined forms of the central limit theorem give more precise estimates.[11]

---

[10] Formally, the distribution of X_k depends not only on k but also on n. It suffices to reorder the X_k, starting from $k = n$ down to $k = 1$, to have the distribution depend only on the subscript.

[11] A great variety of asymptotic estimates in combinatorial analysis were derived by other methods by V. Gončarov, Du domaine d'analyse combinatoire, *Bulletin de l'Académie Sciences URSS, Sér. Math.* (in Russian, French summary), vol. 8 (1944), pp. 3–48. The present method is simpler but more restricted in scope; cf. W. Feller, The Fundamental Limit Theorems in Probability, *Bulletin of the American Mathematical Society*, vol. 51 (1945), pp. 800–832.

*7. The Strong Law of Large Numbers

The (weak) law of large numbers (5.4) asserts that for every particular sufficiently large n the deviation $|S_n - m_n|$ is likely to be small in comparison to n. It has been pointed out in connection with Bernoulli trials (Chapter 8) that this does not imply that $|S_n - m_n|/n$ remains small for all large n; it can happen that the law of large numbers applies but that $|S_n - m_n|/n$ continues to fluctuate between finite or infinite limits. The law of large numbers permits only the conclusion that large values of $|S_n - m_n|/n$ occur at infrequent moments.

We say that the sequence X_k obeys the strong law of large numbers if to every pair $\epsilon > 0$, $\delta > 0$, there corresponds an N such that there is probability $1 - \delta$ or better that for every $r > 0$ all r inequalities

$$(7.1) \qquad \frac{|S_n - m_n|}{n} < \epsilon, \quad n = N, N+1, \cdots, N+r$$

will be satisfied.

We can interpret (7.1) roughly by saying that with an overwhelming probability $|S_n - m_n|/n$ remains small [12] for all $n > N$.

The Kolmogorov Criterion. *The convergence of the series*

$$(7.2) \qquad \sum \frac{\sigma_k^2}{k^2}$$

is a sufficient condition for the strong law of large numbers to apply to the sequence of mutually independent random variables X_k.

Proof. Let A_ν be the event that for at least one n with $2^{\nu-1} < n \leq 2^\nu$ the inequality (7.1) does *not* hold. Obviously it suffices to prove that for all ν sufficiently large ($\nu > \log N$) and all r

$$Pr\{A_\nu\} + Pr\{A_{\nu+1}\} + \cdots + Pr\{A_{\nu+r}\} < \delta.$$

In other words, we have to prove the convergence of the series $\Sigma Pr\{A_\nu\}$. Now the event A_ν implies that for some n with $2^{\nu-1} < n \leq 2^\nu$

$$(7.3) \qquad |S_n - m_n| \geq \frac{\epsilon}{2} \cdot 2^\nu$$

---

* Starred sections treat special topics and may be omitted at first reading.
[12] The general theory introduces a sample space corresponding to the infinite sequence $\{X_k\}$. The strong law then states that with probability one $|S_n - m_n|/n$ tends to zero. In real variable terminology the strong law asserts convergence almost everywhere, while the weak law is equivalent to convergence in measure.

and by Kolmogorov's inequality (Chapter 9, section 7), we have therefore

(7.4) $$Pr\{A_\nu\} \leq 4\epsilon^{-2} \cdot s^2_{2^\nu} \cdot 2^{-2\nu}.$$

Hence

(7.5) $$\sum_{\nu=1}^{\infty} Pr\{A_\nu\} \leq 4\epsilon^{-2} \sum_{\nu=1}^{\infty} 2^{-2\nu} \sum_{k=1}^{2^\nu} \sigma_k^2 = 4\epsilon^{-2} \sum_{k=1}^{\infty} \sigma_k^2 \sum_{2^\nu \geq k} 2^{-2\nu}$$
$$\leq 8\epsilon^{-2} \sum_{k=1}^{\infty} \frac{\sigma_k^2}{k^2}$$

which accomplishes the proof.

As an application of Kolmogorov's criterion we shall prove the

Theorem. If the sequence of mutually independent random variables X_k have a common distribution $\{f(x_j)\}$ and if $\mu = E(X_k)$ exists, then the strong law of large numbers applies.

This theorem is, of course, stronger than the weak law of section 1. The two theorems are treated independently because of the methodological interest of the proofs. For a converse cf. problem 11.

Proof. We again use the method of truncation. Two new sequences of random variables are introduced by

(7.6) $$U_k = X_k, \quad V_k = 0 \quad \text{if} \quad |X_k| < k,$$
$$U_k = 0, \quad V_k = X_k \quad \text{if} \quad |X_k| \geq k.$$

Then the U_k are mutually independent, and we shall show that they satisfy Kolmogorov's criterion. We have

(7.7) $$\sigma_k^2 \leq E(U_k^2) = \sum_{|x_j| < k} x_j^2 f(x_j).$$

Now for abbreviation put

(7.8) $$a_\nu = \sum_{\nu - 1 \leq |x_j| < \nu} |x_j| f(x_j).$$

Then the series Σa_ν converges since $E(X_k)$ exists. Moreover, from (7.7)

(7.9) $$\sigma_k^2 \leq a_1 + 2a_2 + 3a_3 + \cdots + ka_k$$

and hence

(7.10) $$\sum_{k=1}^{\infty} \frac{\sigma_k^2}{k^2} \leq \sum_{k=1}^{\infty} \frac{1}{k^2} \sum_{\nu=1}^{k} \nu a_\nu = \sum_{\nu=1}^{\infty} \nu a_\nu \sum_{k=\nu}^{\infty} \frac{1}{k^2} < 2 \sum_{\nu=1}^{\infty} a_\nu < \infty.$$

For the expectation we have

(7.11) $$E(U_k) = \mu_k = \sum_{|x_j|<k} x_j f(x_j)$$

so that $\mu_k \to \mu$ and hence $(\mu_1 + \mu_2 + \cdots + \mu_n)/n \to \mu$. Applying the strong law of large numbers to $\{U_k\}$, we find therefore that with probability $1 - \delta$ or better

(7.12) $$\left| n^{-1} \sum_{k=1}^{n} U_k - \mu \right| < \epsilon$$

for all $n \geq N$. It suffices now to prove that the V_n can be neglected, that is, that the probability of one or more V_n with $n > N$ being different from zero tends to 0 with $N \to \infty$. It is easily seen that the first Borel-Cantelli lemma (Chapter 8, section 3) applies with obvious verbal changes, and that it suffices to prove that $\Sigma Pr\{V_n \neq 0\}$ converges. Now

(7.13) $$Pr\{V_n \neq 0\} = \sum_{|x_j| \geq n} f(x_j) \leq \frac{a_{n+1}}{n} + \frac{a_{n+2}}{n+1} + \frac{a_{n+3}}{n+2} + \cdots$$

and hence

(7.14) $$\Sigma Pr\{V_n \neq 0\} \leq \sum_{n=1}^{\infty} \sum_{\nu=n}^{\infty} \frac{a_{\nu+1}}{\nu} = \sum_{\nu=1}^{\infty} \frac{a_{\nu+1}}{\nu} \sum_{n=1}^{\nu} 1 = \sum_{\nu} a_{\nu+1} < \infty,$$

as asserted.

8. Problems for Solution

1. Prove that the law of large numbers applies in example (5.a) also when $\lambda \leq 0$. The central limit theorem holds if $\lambda \geq -1/2$.

2. Decide whether the law of large numbers and the central limit theorem hold for the sequences of mutually independent variables X_k with distributions defined as follows ($k \geq 1$):

   (a) $Pr\{X_k = \pm 2^k\} = \frac{1}{2}$;

   (b) $Pr\{X_k = \pm 2^k\} = 2^{-(2k+1)}$, $Pr\{X_k = 0\} = 1 - 2^{-2k}$;

   (c) $Pr\{X_k = \pm k\} = \frac{1}{2}k^{-\frac{1}{2}}$, $Pr\{X_k = 0\} = 1 - k^{-\frac{1}{2}}$.

3. *Ljapunov's condition* (1901). Suppose that for some fixed $\delta > 0$ one has $E(|X_k|^{2+\delta}) = \lambda_k$ and that λ_k/σ_k^2 is uniformly bounded. Show that Lindeberg's condition is satisfied.

The following six problems treat the weak law of large numbers for dependent variables.

4. Let the $\{X_k\}$ be mutually independent and have a common distribution with mean μ and finite variance. If $S_n = X_1 + \cdots + X_n$, prove that the law of large numbers does not hold for the sequence $\{S_n\}$, but holds for $a_n S_n$ if $a_n \to 0$.

5. Let $\{X_k\}$ be a sequence of random variables such that X_k may depend on X_{k-1} and X_{k+1} but is independent of all other X_j. Show that the law of large numbers holds, provided the X_k have finite variances.

6. If the joint distribution of $(X_1, \cdots, X_n)$ is defined for every n, if the variances are bounded, and all covariances are negative, the law of large numbers applies.

7. *Continuation.* Replace the condition $\text{Cov}(X_j, X_k) \leq 0$ by the assumption that $\text{Cov}(X_j, X_k) \to 0$ uniformly as $|j - k| \to 0$. Prove that the law of large numbers holds.

8. If $|S_n| < cn$ and $\text{Var}(S_n) > \alpha n^2$, then the law of large numbers does not apply to $\{X_k\}$.

9. In the Polya urn scheme [example (2.c) of Chapter 5] let X_k equal 1 or 0 according to whether the kth ball drawn is black or red. Then S_n is the number of black balls in n drawings. Prove that the law of large numbers does not apply to $\{X_k\}$. (*Hint:* Use problem 8 and Chapter 9, problem 13.)

10. Let $\{X_n\}$ be a sequence of mutually independent random variables with a common distribution. Suppose that the X_n have not a finite expectation and let A be a positive constant. The probability is one that infinitely many among the events $|X_n| > An$ occur.

11. *Converse to the strong law of large numbers.* Under the assumption of problem 10 there is probability one that $|S_n| > An$ for infinitely many n.

12. *Example of an unfavorable "fair" game.* Let the possible values of the gain at each trial be $0, 2, 2^2, 2^3, \ldots$; the probability of the gain being 2^k is

$$(8.1) \qquad p_k = \frac{1}{2^k k(k+1)},$$

and the probability of 0 is $p_0 = 1 - (p_1 + p_2 + \ldots)$. The expected gain is

$$(8.2) \qquad \mu = \Sigma 2^k p_k = (1 - \tfrac{1}{2}) + (\tfrac{1}{2} - \tfrac{1}{3}) + (\tfrac{1}{3} - \tfrac{1}{4}) + \ldots = 1.$$

Assume that at each trial the player pays a unit amount as entrance fee, so that after n trials his net gain (or loss) is $S_n - n$, where S_n is the sum of n independent random variables with the above distribution. Show that for every $\epsilon > 0$ *the probability approaches unity that in n trials the player will have sustained a loss greater than $(1 - \epsilon)n/\text{Log}_2 n$, where $\text{Log}_2 n$ denotes the logarithm to the base 2. In symbols, prove that*

$$(8.3) \qquad Pr\left\{S_n - n < -\frac{(1-\epsilon)n}{\text{Log}_2 n}\right\} \to 1.$$

Hint: Use the truncation method of section 4, but replace the bound $n \text{Log } n$ of (4.2) by $n/\text{Log}_2 n$. Show that the probability that $U_k = X_k$ for all $k \leq n$ tends to 1 and prove that

$$(8.4) \qquad Pr\left\{|U_1 + \cdots + U_n - nE(U_1)| < \frac{\epsilon n}{\text{Log}_2 n}\right\} \to 1.$$

$$(8.5) \qquad 1 - \frac{1}{\text{Log}_2 n} \geq E(U_1) \geq 1 - \frac{1+\epsilon}{\text{Log}_2 n}.$$

For details cf. the paper cited in section 2.

13. A converse to Kolmogorov's criterion. If $\Sigma \sigma_k^2/k^2$ diverges, then there exists a sequence $\{X_k\}$ of mutually independent random variables with $\operatorname{Var}\{X_k\} = \sigma_k^2$ for which the strong law of large numbers does not apply. [*Hint:* Prove first that the convergence of $\Sigma Pr\{|X_n| > \epsilon n\}$ is a necessary condition for the strong law to apply].

14. Let $\{X_n\}$ be a sequence of mutually independent random variables such that $X_n = \pm 1$ with probability $(1 - 2^{-n})/2$ and $X_n = \pm 2^n$ with probability 2^{-n-1}. Prove that both the weak and the strong law of large numbers apply to $\{X_k\}$. [*Note.* This shows that the condition (5.7) is not necessary.]

CHAPTER 11

INTEGRAL VALUED VARIABLES
GENERATING FUNCTIONS

1. Generalities

Among discrete random variables those assuming only the integral values $k = 0, 1, 2, \ldots$ are of special importance. In particular, all recurrence and waiting times are of this nature. The study of such variables is facilitated by the powerful method of generating functions which will later be recognized as a special case of the method of characteristic functions on which the theory of probability depends to a large extent. More generally, the subject of generating functions belongs to the domain of operational methods which are widely used in the theory of differential and integral equations. In the theory of probability generating functions have been used since DeMoivre and Laplace, but the power and possibilities of the method are rarely fully utilized.

Definition. *Let $a_0, a_1, a_2, \ldots$ be a sequence of real numbers. If*

(1.1) $$A(s) = a_0 + a_1 s + a_2 s^2 + \ldots$$

converges in some interval $-s_0 < s < s_0$, then the function $A(s)$ is called the generating function of the sequence $\{a_j\}$.

The variable s itself has no significance. If the sequence $\{a_j\}$ is bounded, then a comparison with the geometric series shows that (1.1) converges at least for $|s| < 1$.

Examples. If $a_j = 1$ for all j, then $A(s) = 1/(1 - s)$. The generating function of the sequence $(0, 0, 1, 1, 1, \ldots)$ is $s^2/(1 - s)$. The sequence $a_j = 1/j!$ has the generating function e^s. For fixed n the sequence $a_j = \binom{n}{j}$ has the generating function $(1 + s)^n$. If X is the number scored with a throw of a perfect die, the probability distribution of X has the generating function $(s + s^2 + s^3 + s^4 + s^5 + s^6)/6 = s(1 - s^6)/6(1 - s)$.

11.1] GENERALITIES

Consider now an integral-valued random variable with the probability distribution

(1.2) $$Pr\{X = j\} = p_j, \qquad j = 0, 1, \ldots.$$

It will be convenient to have a special notation for its "tails," and we shall write

(1.3) $$Pr\{X > j\} = q_j,$$

so that

(1.4) $$q_k = p_{k+1} + p_{k+2} + \cdots, \qquad k \geq 0.$$

We introduce the two generating functions

(1.5) $$P(s) = p_0 + p_1 s + p_2 s^2 + p_3 s^3 + \cdots$$

and

(1.6) $$Q(s) = q_0 + q_1 s + q_2 s^2 + q_3 s^3 + \cdots.$$

The two series converge at least for $|s| < 1$ since their coefficients are bounded. Moreover, (1.2) represents a probability distribution so that $P(1) = 1$ and hence $P(s)$ *converges absolutely at least for* $-1 \leq s \leq 1$.

Theorem 1. *For* $-1 < s < 1$ *we have*

(1.7) $$Q(s) = \frac{1 - P(s)}{1 - s}.$$

Proof. From (1.4) we conclude that $q_k = 1 - p_0 - p_1 - \cdots - p_k$. Introduce these expressions into (1.6) and collect like terms. The positive terms add to $1 + s + s^2 + \cdots = 1/(1-s)$. The term $-p_j$ appears as a summand of $q_j, q_{j+1}, q_{j+2}, \ldots$ so that in (1.6) $-p_j$ appears as factor of $s^j + s^{j+1} + s^{j+2} + \cdots = s^j/(1-s)$. Hence $Q(s) = (1-s)^{-1}(1 - p_0 - p_1 s - p_2 s^2 - \cdots) = (1-s)^{-1}\{1 - P(s)\}$, as asserted in (1.7).

Theorem 2. *The expectation $E(X)$ can be calculated either from the probability distribution* (1.2) *or in terms of the "tails"* (1.3). *Thus*

(1.8) $$E(X) = \sum_{j=1}^{\infty} j p_j = \sum_{k=0}^{\infty} q_k.$$

In terms of the generating functions

(1.9) $$E(X) = P'(1) = Q(1).$$

It is understood that the two series in (1.8) may diverge. Since all terms of the series are positive, we may without danger speak of an *infinite expectation*. In this case (1.9) is to be replaced by the statement that $P'(s)$ and $Q(s)$ tend to ∞ as $s \to 1$.

Proof. If $E(X)$ is finite, then the relation $E(X) = P'(1)$ follows from the definition. That $Q(1) = P'(1)$ follows by differentiating the identity $1 - P(s) = (1-s)Q(s)$ of theorem 1. If $E(X)$ is infinite, then $P'(s)$ and $Q(s)$ increase over all bounds as $s \to 1$. The equations (1.8) are, of course, the same as (1.9). The fact that $\Sigma q_k = \Sigma j p_j$ can be proved independently by the argument used to prove theorem 1.

Theorem 3. *If $E(X^2) = \Sigma n^2 p_n$ is finite, then*

(1.10) $\qquad E(X^2) = P''(1) + P'(1) = 2Q'(1) + Q(1)$

and hence

(1.11) $\qquad \operatorname{Var}(X) = P''(1) + P'(1) - P'^2(1)$
$$= 2Q'(1) + Q(1) - Q^2(1).$$

The variance is infinite if and only if $P''(s) \to \infty$ as $s \to 1$.

Proof. The first expression in (1.10) follows from

(1.12) $\qquad E(X^2) = \Sigma k^2 p_k = \Sigma k(k-1) p_k + \Sigma k p_k.$

That the two representations in (1.10) are equivalent is seen upon differentiating the identity $1 - P(s) = (1-s)Q(s)$ twice.

2. Convolutions

Let X and Y be non-negative independent integral-valued random variables with the probability distributions $Pr\{X = j\} = a_j$ and $Pr\{Y = j\} = b_j$. The event $(X = j, Y = k)$ has probability $a_j b_k$. The sum $S = X + Y$ is a new random variable, and the event $S = r$ is the union of the events

$$(X = 0, Y = r), \quad (X = 1, Y = r-1), \quad (X = 2, Y = r-2), \quad \cdots,$$
$$(X = r, Y = 0).$$

These events are mutually exclusive, and therefore the distribution $c_r = Pr\{S = r\}$ is given by

(2.1) $\qquad c_r = a_0 b_r + a_1 b_{r-1} + a_2 b_{r-2} + \cdots + a_{r-1} b_1 + a_r b_0.$

The operation (2.1), leading from the two sequences $\{a_k\}$ and $\{b_k\}$ to a new sequence $\{c_k\}$, occurs so frequently that it is convenient to introduce a special name and notation for it.

Definition. *Let $\{a_k\}$ and $\{b_k\}$ be any two number sequences (not necessarily probability distributions). The new sequence $\{c_r\}$ defined by (2.1) is called the convolution* [1] *of $\{a_k\}$ and $\{b_k\}$ and will be denoted by*

(2.2) $$\{c_k\} = \{a_k\}*\{b_k\}$$

Examples. (a) If $a_k = b_k = 1$ for all $k \geq 0$, then $c_k = k + 1$. If $a_k = k$, $b_k = 1$, then $c_k = 1 + 2 + \cdots + k = k(k+1)/2$. Finally, if $a_0 = a_1 = \frac{1}{2}$, $a_k = 0$ for $k \geq 2$, then $c_k = (b_k + b_{k-1})/2$, etc.

Let now the sequences $\{a_k\}$ and $\{b_k\}$ have generating functions $A(s) = \Sigma a_k s^k$ and $B(s) = \Sigma b_k s^k$. The product $A(s)B(s)$ can be obtained by termwise multiplication of the power series for $A(s)$ and $B(s)$. Collecting terms with equal powers of s, we find that the coefficient c_r of s^r in the expansion of $A(s)B(s)$ is given by (2.1). We have thus the

Theorem. *If $\{a_k\}$ and $\{b_k\}$ are sequences with generating functions $A(s)$ and $B(s)$, and $\{c_k\}$ is their convolution, then the generating function $C(s) = \Sigma c_k s^k$ is the product*

(2.3) $$C(s) = A(s)B(s).$$

If X and Y are non-negative integral-valued mutually independent random variables with generating functions $A(s)$ and $B(s)$, then their sum $X + Y$ has the generating function $A(s)B(s)$.

Let now $\{a_k\}$, $\{b_k\}$, $\{c_k\}$, $\{d_k\}$, ... be any sequences. We can form the convolution $\{a_k\}*\{b_k\}$, and then the convolution of this new sequence with $\{c_k\}$, etc. The generating function of $\{a_k\}*\{b_k\}*\{c_k\}*\{d_k\}$ is $A(s)B(s)C(s)D(s)$, and this fact shows that the order in which the convolutions are performed is immaterial. For example, $\{a_k\}*\{b_k\}*\{c_k\} = \{c_k\}*\{b_k\}*\{a_k\}$, etc. *Thus the convolution is an associative and commutative operation* (exactly as the summation of random variables).

In the study of sums of independent random variables X_n the special case where the X_n have a common distribution is of particular interest. If $\{a_j\}$ is the common probability distribution of the X_n, then the distribution of $S_n = X_1 + \cdots + X_n$ will be denoted by $\{a_j\}^{n*}$. Thus

(2.4) $$\{a_j\}^{2*} = \{a_j\}*\{a_j\}, \quad \{a_j\}^{3*} = \{a_j\}^{2*}*\{a_j\}, \ldots$$

and generally

(2.5) $$\{a_j\}^{n*} = \{a_j\}^{(n-1)*}*\{a_j\}.$$

[1] Some American and British writers prefer the word *faltung*, which is of German origin. The French equivalent is *composition*.

In words, $\{a_j\}^{n*}$ *is the sequence of numbers whose generating function is* $A^n(s)$. In particular, $\{a_j\}^{1*}$ is the same as $\{a_j\}$, and $\{a_j\}^{0*}$ is the sequence whose generating function is $A^0(s) = 1$, that is, the sequence $(1, 0, 0, 0, \ldots)$.

Examples. (b) *Binomial Distribution.* The generating function of the binomial distribution $b(k; n, p) = \binom{n}{k} p^k q^{n-k}$ is

$$(2.6) \qquad \sum_{k=0}^{n} \binom{n}{k} (ps)^k q^{n-k} = (q + ps)^n.$$

The fact that the generating function is the nth power of $q + ps$ shows that $\{b(k; n, p)\}$ is the distribution of a sum $S_n = X_1 + \cdots + X_n$ of n independent random variables with the common generating function $q + ps$; each variable X_j assumes the value 0 with probability q and the value 1 with probability p. Thus

$$(2.7) \qquad \{b(k; n, p)\} = \{b(k; 1, p)\}^{n*}.$$

We have used the representation $S_n = X_1 + \cdots + X_n$ several times [cf. Chapter 9, examples (3.a) and (5.a)]. The above argument can be reversed and used for a new derivation of the binomial distribution. The multiplicative property $(q + ps)^m (q + ps)^n = (q + ps)^{m+n}$ shows also that

$$(2.8) \qquad \{b(k; m, p)\} * \{b(k; n, p)\} = \{b(k; m + n, p)\}.$$

We have thus a new derivation of the result of Chapter 6, problem 22. Also the formulas $E(S_n) = np$ and $\text{Var}(S_n) = npq$ can be derived in a simple way by differentiation of the generating function $(q + ps)^n$ in accordance with (1.9) and (1.11).

(c) *Poisson Distribution.* The generating function of the distribution $p(k; \lambda) = e^{-\lambda} \lambda^k / k!$ is

$$(2.9) \qquad \sum_{k=0}^{\infty} e^{-\lambda} \frac{(\lambda s)^k}{k!} = e^{-\lambda + \lambda s}.$$

It follows that

$$(2.10) \qquad \{p(k; \lambda)\} * \{p(k; \mu)\} = \{p(k; \lambda + \mu)\},$$

which can also be proved directly (cf. Chapter 6, problem 23). By differentiation we find that both mean and variance of the Poisson distribution are λ; this result has been proved previously by direct computation [cf. Chapter 9, example (4.c)].

3. The Geometric and the Pascal Distributions

We say that the *random variable* X has a *geometric distribution if*

(3.1) $$Pr\{X = k\} = q^k p, \qquad k = 0, 1, 2, \ldots$$

where p and q are positive constants with $p + q = 1$. The corresponding generating function is

(3.2) $$p \sum_{k=0}^{\infty} (qs)^k = \frac{p}{1 - qs}.$$

Using theorems 2 and 3 of section 1, it is easily found that

(3.3) $$E(X) = \frac{q}{p}, \quad \text{Var}(X) = \frac{q}{p^2}.$$

To fix ideas consider a sequence of n Bernoulli trials with probability p for success. The probability that at least one success occurs and that the first success is preceded by exactly k failures ($k \leq n - 1$) is $q^k p$. A passage to the limit leads us to interpret X as the number of failures preceding the first success when the trials are continued as long as necessary for a success to occur, but this interpretation refers to an infinite sample space. The advantage of the formal definition (3.1) is that we need not worry about the structure of the original sample space.

We can in a similar way study the number of trials preceding the rth success. Let X_k be the number of failures following the $(k - 1)$th and preceding the kth success. Then $X_1 + X_2 + \cdots + X_r + r$ is the number of trials up to and including the rth success. Strictly speaking, these variables are defined only in the sample space corresponding to unending sequences of Bernoulli trials studied in Chapter 8. However, we can study the distribution directly without any reference to a particular sample space. The notion of Bernoulli trials implies that the variables $X_1, \ldots, X_r$ are mutually independent and that they have the common distribution (3.1). The sum

(3.4) $$S_r = X_1 + \cdots + X_r$$

can be interpreted as the *number of failures preceding the rth success.* Then

(3.5) $$Pr\{S_r = k\} = f(k; r, p)$$

is the probability of the *rth success occurring at the trial number $r + k$.* This means that among the first $r + k - 1$ trials there are exactly k

failures and the following, or $(r+k)$th, trial results in success; the corresponding probabilities are $\binom{r+k-1}{k} p^{r-1} q^k$ and p, so that

(3.6) $$f(k; r, p) = \binom{r+k-1}{k} p^r q^k.$$

The same result follows directly from (3.4). In fact, from (3.2) we find that the generating function of S_r is

(3.7) $$\left(\frac{p}{1-qs}\right)^r.$$

Using the binomial formula (6.6) of Chapter 2, it is seen that the coefficient of s^k is

(3.8) $$f(k; r, p) = \binom{-r}{k} p^r (-q)^k, \qquad k = 0, 1, 2, \ldots,$$

and this formula agrees with (3.6) (cf. Chapter 2, problem 1 of section 9).

The distribution $f(k; r, p)$ defined by either (3.6) or (3.8) *is called the Pascal distribution*. It depends on the two parameters r and p, where r is an integer and $0 < p < 1$ ($q = 1 - p$). Its generating function is (3.7). *The Pascal distribution is the r-fold convolution of the geometric distribution* (3.1), or in symbols

(3.9) $$\{f(k; r, p)\} = \{q^k p\}^{r*}.$$

Conversely, the geometric distribution (3.1) is the special case $r = 1$ of the general Pascal distribution. It follows that the *mean and variance of the Pascal distribution are rq/p and rq/p^2*. Moreover,

(3.10) $$\{f(k; r_1, p)\} * \{f(k; r_2, p)\} = \{f(k; r_1 + r_2, p)\}.$$

The Pascal distribution remains meaningful if r is not an integer provided that $r \geq 0$. It occurs in many connections and is often called the *negative binomial distribution*. [An alternative form of the Pascal distribution with non-integral r is given by the Polya distribution (8.12) of Chapter 6, where $\lambda \rho$ corresponds to r and $1/(1 + \rho)$ to q.]

4. Relation to Holding or Waiting Times

If a sequence of Bernoulli trials is performed at the rate of one per second, then the number X of failures preceding the first success represents a waiting time. Its probability distribution has a curious property which puzzles unprepared minds.

Suppose that in a particular sequence no success occurred in the first ν trials, so that $X > \nu$. The waiting time from this trial to the next success is independent of the number of preceding failures. Hence the probability that the waiting time will be prolonged by an additional k seconds is independent of ν and equals the initial probability of the total length exceeding k seconds. It is obvious that this property is not shared by waiting times encountered in phenomena such as waiting lines before counters and lifetimes of machines. For example, the information that no streetcar has passed for five minutes ordinarily increases our expectation that a car will pass during the next minute, and a reasonable theory of waiting times must reflect this common-sense statement. However, there exist many phenomena where the waiting times share the described property of the waiting time for the first success. A typical example is provided by a type of conversation in telephone booths which has often been cited as an ideal illustration of incoherence. If the booth is equipped with a tolerably comfortable seat eliminating physical fatigue, then the conversation must be regarded as typical of a process which depends entirely on momentary impulses. Whatever happens bears no relation to the past, and the termination is an instantaneous chance effect. The probability that it will occur within the next minute must be independent of the length of the foregoing chatter, so that the length of the conversation is analogous to the waiting time for the first success.

The general theory of stochastic processes (Chapters 15 and 17) will reveal that we are here dealing with an important but particular case of Markov processes. There is only one possible probability distribution for waiting times with the described property; in the case of discrete trials it is the geometric distribution, and in the case of a continuous time it is the exponential distribution. Here we are limited to discrete trials and proceed to show that our property is characteristic of the geometric distribution.

Suppose, then, that a waiting time X can assume the values 0, 1, 2, ... with probabilities $p_0, p_1, p_2, \ldots$. Let the distribution of X have the following property: *the conditional probability that the waiting time terminates at the kth trial, assuming that it has not terminated before, equals p_0 (the probability at the first trial).* We claim that $p_k = (1 - p_0)^k p_0$, so that X has a geometric distribution.

For a proof we introduce again the "tails"

$$q_k = p_{k+1} + p_{k+2} + p_{k+3} + \ldots = Pr\{X > k\}.$$

Our hypothesis is $X > k - 1$, and its probability is q_{k-1}. The conditional probability of $X = k$ is therefore p_k/q_{k-1}, and the assumption is that for all $k \geq 1$

$$\frac{p_k}{q_{k-1}} = p_0. \tag{4.1}$$

Now $p_k = q_{k-1} - q_k$, and hence

$$\frac{q_k}{q_{k-1}} = 1 - p_0. \tag{4.2}$$

Since $q_0 = p_1 + p_2 + \ldots = 1 - p_0$, it follows that $q_k = (1 - p_0)^{k+1}$, and hence $p_k = q_{k-1} - q_k = (1 - p_0)^k p_0$, as asserted.

It will be useful to indicate how a simple passage to the limit leads from the geometric distribution $\{q^k p\}$ to the exponential distribution. Suppose that each trial takes a time Δt, so that k trials take time $k\Delta t$. We shall let $\Delta t \to 0$ and at the same time $k \to \infty$, so that the total time consumed remains constant, $k\Delta t = t$. If p and q were to remain constant, the first success would occur sooner and sooner, and in the limit we would be sure to have waiting time zero. In a sensible limiting process we must therefore let $p \to 0$ and $q \to 1$, exactly as was done in the derivation of the Poisson distribution in Chapter 6. The mean of the geometric distribution was found to be q/p [cf. (3.3)]. If each trial takes time Δt then the mean of the waiting time is $(q/p)\Delta t$, and the process remains physically meaningful if $(q/p)\Delta t$ is a constant. Since $p \to 0$, $q \to 1$, we can put

$$p \sim \lambda \Delta t. \tag{4.3}$$

Then

$$q^k p \sim (1 - \lambda \Delta t)^{t/\Delta t} \lambda \Delta t. \tag{4.4}$$

Taking logarithms, the first factor is seen to tend to $e^{-\lambda t}$, and hence

$$q^k p \sim e^{-\lambda t} \lambda \Delta t. \tag{4.5}$$

For the *probability that the first success occurs after time* $t = k \cdot \Delta t$ we get in the limit

$$q^k \sim (1 - \lambda \Delta t)^{t/\Delta t} \to e^{-\lambda t}. \tag{4.6}$$

This is the *exponential distribution*. We have encountered it as the first term in the *Poisson distribution* (Chapter 6), where $e^{-\lambda t}$ was interpreted as the probability of no event within time t, which is just another way of saying that the waiting time exceeds t. More generally, with our passage to the limit $k\Delta t = t$, $p \sim \lambda \Delta t$, we have from (3.6)

$$(4.7) \qquad f(k; r, p) \sim e^{-\lambda t} \frac{(\lambda t)^{r-1}}{(r-1)!} \cdot \lambda \Delta t.$$

The factor $e^{-\lambda t}(\lambda t)^{r-1}/(r-1)!$ is the Poisson expression for the probability of exactly $r-1$ events within time t; the factor $\lambda \Delta t$ stands for the probability of an additional event within the following interval of length Δt. We have thus a new argument leading to the Poisson distribution and relating it to the theory of waiting times (or of addition of random variables). This theory will be continued in Chapter 17 and, more systematically, in volume 2.

5. Compound Distributions

Let $\{X_k\}$ be a sequence of mutually independent random variables with the common distribution $Pr\{X_k = j\} = f_j$ $(j = 0, 1, 2, \ldots)$. We are often interested in sums $S_N = X_1 + X_2 + \cdots + X_N$, where the number N of terms is a random variable which is independent of the X_j and has a given distribution $Pr\{N = n\} = g_n$. The distribution of S_N follows from the fundamental formula for conditional probabilities [Chapter 5, (1.6)]

$$(5.1) \quad Pr\{S_N = j\} = \sum_{n=0}^{\infty} Pr\{N = n\} \, Pr\{X_1 + \cdots + X_n = j\};$$

here the distribution of $X_1 + \cdots + X_n$ is given by the n-fold convolution of $\{f_j\}$ with itself, so that (5.1) can also be written in the form

$$(5.2) \qquad Pr\{S_N = j\} = \sum_{n=0}^{\infty} g_n \, \{f_j\}^{n*}.$$

Distributions of this type are called *compound distributions*. The most important case is that where N has a Poisson distribution while the variables X_k can assume only the values 1 and 0 with probabilities p and q. In this case $S_n = X_1 + \cdots + X_n$ has the binomial distribution $b(j; n, p) = \binom{n}{j} p^j q^{n-j}$, and $g_n = e^{-\lambda} \cdot \lambda^n / n!$. Then

$$(5.3) \qquad Pr\{S_N = j\} = e^{-\lambda} p^j \sum_{n=0}^{\infty} \frac{\lambda^n}{n!} \binom{n}{j} q^{n-j}$$

$$= \frac{e^{-\lambda} p^j \lambda^j}{j!} \sum_{n=j}^{\infty} \frac{(\lambda q)^{n-j}}{(n-j)!}.$$

The last sum equals $e^{\lambda q} = e^{\lambda(1-p)}$, and hence *the composition of a Poisson distribution for N and a binomial distribution for $X_1 + \cdots + X_n$*

leads to the compound distribution

(5.4) $$Pr\{S_n = j\} = e^{-\lambda p}\frac{(\lambda p)^j}{j!} ;$$

that is, a Poisson distribution with parameter λp.

Examples. (*a*) We saw in Chapter 6, example (6.*c*), that X-rays produce chromosome breakages in cells: for a given dosage and time of exposure the number *N* of breakages in each cell is a random variable with a Poisson distribution. Each breakage has a fixed probability q of healing; with probability $p = 1 - q$ a breakage does not heal, in which case the cell dies. The probability that j out of *N* breakages in a cell will not heal is given by the binomial distribution $b(j; N, p)$, and hence the number S_n of observable (unhealed) breakages has the distribution (5.4).[2]

(*b*) Compound distributions of the form (5.2) occur in particular in connection with composite populations such as encountered in Chapter 5, section 2. For example, if g_n is the probability of a family having exactly n children, and if the sex ratio of boys to girls is $p:q$ ($p + q = 1$), then the probability of a family having exactly j boys is the compound binomial distribution

(5.5) $$\sum_{n=0}^{\infty} g_n \binom{n}{j} p^j q^{n-j}.$$

Again, if an insect has probability g_n of laying exactly n eggs and each egg has a probability p of survival, then (assuming statistical independence) (5.5) is the probability that exactly j eggs will survive.

(*c*) Let X_j represent the damage caused by the *j*th lightning hit. If g_n is the probability of exactly n strikes, then (5.2) is the probability distribution of the accumulated damage. Similar accumulated chance effects occur in insurance, physics, engineering, and elsewhere.

Formula (5.2) is not pleasing, but it can be simplified by the use of generating functions. Let $f(s) = \Sigma f_j s^j$ and $g(s) = \Sigma g_n s^n$ be the generating functions of the X_k and *N*, respectively. Then for a fixed n the generating function of the n-fold convolution $\{f_j\}^{n*}$ is $f^n(s)$ (cf. section 2), and hence the generating function of (5.2) is $\Sigma g_n f^n(s)$. This is the function $g(s)$ with s replaced by $f(s)$. Hence we have the

[2] Cf. D. G. Catcheside, Genetic Effects of Radiations, *Advances in Genetics*, edited by M. Demerec, vol. 2, Academic Press, New York, 1948, pp. 271–358, in particular p. 339.

Theorem. *The generating function of the compound distribution* (5.2) *is the compound function* $g(f(s))$.

Examples. (d) In the special case (5.4) the generating function of the X_j was $q + ps$, and $g(s) = e^{-\lambda + \lambda s}$. Hence $g(f(s)) = e^{-\lambda + \lambda q + \lambda ps}$ $= e^{-\lambda p + \lambda ps}$, in agreement with (5.4).[3]

(e) The generating function of (5.5) is $g(q + ps)$. Suppose, in particular, that $\{g_n\}$ is the geometric distribution $\{(1-\gamma)\gamma^n\}$. Then $g(s) = (1-\gamma)/(1-\gamma s)$ and $g(q+ps) = (1-\alpha)/(1-\alpha s)$ with $\alpha = \gamma p/(1-\gamma q)$. This implies that the combination of a geometric and a binomial distribution leads to a new geometric distribution (see, for an example, problem 9 of Chapter 5).

As an application let us calculate the expectation of the distribution (5.2). By the chain rule of differentiation this expectation is $g'(f(1))f'(1)$. However, $f'(1) = E(X_j)$ and $g'(f(1)) = g'(1) = E(N)$. Hence *the expectation of the distribution* (5.2) *is the product* $E(N)E(X_j)$.

*6. Chain Reactions

We shall now analyze a chance process which serves as a simplified model of many empirical processes and also provides an excellent illustration of one way in which generating functions are useful. In words the process may be described as follows.

We consider particles which are able to produce new particles of like kind. A single particle forms the original, or zero, generation. Any particle has probability p_k $(k = 0, 1, 2, \ldots)$ of creating exactly k new particles; the direct descendants of the nth generation form the $(n+1)$th generation. The particles of each generation act independently of each other.

Three typical illustrations may precede a rigorous formulation in terms of random variables.

Examples. (a) *Nuclear Chain Reactions.* This application became familiar in connection with the atomic bomb.[4] The particles are

---

* Starred sections treat special topics and may be omitted at first reading.

[3] For limit theorems concerning sums $X_1 + \ldots + X_N$ where N is a random variable cf. H. Robbins, The Asymptotic Distribution of the Sum of a Random Number of Variables, *Bulletin of the American Mathematical Society*, vol. 54 (1948), pp. 1151–1161. For other general classes and properties of compound distributions cf. W. Feller, On a General Class of "Contagious" Distributions, *Annals of Mathematical Statistics*, vol. 14 (1943), pp. 389–400.

[4] The following description follows E. Schroedinger, Probability Problems in Nuclear Chemistry, *Proceedings of the Royal Irish Academy*, vol. 51, sect. A, No. 1 (December, 1945). There the assumption of spatial homogeneity is removed.

neutrons, which are subject to chance hits by other particles. On fission, each particle gives birth to a fixed number, m, of direct descendants. Let p be the probability that a particle sooner or later scores a hit; then $q = 1 - p$ is the probability that the particle remains inactive (is removed or absorbed in a different way). In this case the only possible numbers of descendants are 0 and m, and the corresponding probabilities are q and p (i.e., $p_0 = q$, $p_m = p$, $p_j = 0$ for all other j). At worst, the first particle remains inactive and the process never starts. At best, there will be m particles of the first generation, m^2 of the second, and so on. If p is near one, the number of particles is likely to increase very rapidly. Mathematically, this number may increase indefinitely. Physically speaking, for very large numbers of particles the probabilities of fission cannot remain constant, and also statistical independence no longer holds. However, for ordinary chain reactions, the mathematical description "indefinitely increasing number of particles" may be translated by "explosion."

(b) *Survival of Family Names.* Here (as often in life), only male descendants count; they play the role of particles, and p_k is the probability of a newborn boy's producing exactly k boys. Our scheme introduces two artificial simplifications. Fertility is subject to secular trends, and therefore the distribution $\{p_k\}$ in reality changes from generation to generation. Moreover, common inheritance and common environment are bound to produce similarities among brothers, a fact which is contrary to our assumption of statistical independence. Our model can be refined to take care of these objections, but the essential features remain unaffected. We shall derive the probability of finding k carriers of the family name in the nth generation and, in particular, the probability of an extinction of the line. Survival of family names appears to have been the first chain reaction studied by probability methods. The problem was first treated by F. Galton (1889); for a detailed account the reader is referred to A. Lotka's book.[5] Lotka shows that American experience is reasonably well described by the distribution $p_0 = 0.4825$, $p_k = (0.2126)(0.5893)^{k-1} (k \geq 1)$, which, except for the first term, is a geometric distribution.

(c) *Genes and Mutations.* Every gene of a given organism (Chapter 5, section 5) has a chance to reappear in 1, 2, 3, ... direct descendants of that organism, and our scheme describes the process, neglecting, of course, variations within the population and with time. This scheme is of particular use in the study of mutations,[6] or changes of form in a

---

[5] Théorie analytique des associations biologiques, Vol. 2, *Actualités scientifiques et industrielles*, No. 780 (1939), pp. 123–136, Hermann et Cie, Paris.

[6] R. A. Fisher, *The Genetical Theory of Natural Selection*, Oxford, 1930, pp. 73ff.

gene. A spontaneous mutation produces a single gene of the new kind, which plays the role of a zero-generation particle. Our theory leads to estimates of the chances of survival and of the spread of the mutant gene. To fix ideas concerning the forms which the distribution $\{p_k\}$ may assume, consider (following R. A. Fisher) a corn plant which is father to some 100 seeds and mother to an equal number. If the population size remains constant, an average of 2 among these 200 seeds will develop to a plant. Each seed has probability 1/2 to receive a particular gene. The probability of a particular mutant gene to be represented in exactly k new plants is therefore comparable to the probability of exactly k successes in 200 Bernoulli trials with probability $p = 1/200$. Here the Poisson approximation applies, and it appears reasonable to assume that $\{p_k\}$ is, approximately, a Poisson distribution with mean 1. If the gene carries a biological advantage, we are led to assume a Poisson distribution with mean $\lambda > 1$.

For a mathematical description of the chain reaction we introduce the random variable X_n representing the number of particles in the nth generation. Then $X_0 = 1$, while X_1 has the given probability distribution $\{p_k\}$. The number of descendants of each of the X_1 particles of the first generation is a random variable with the same distribution. Therefore $X_2 = U_1 + U_2 + \cdots + U_{X_1}$ where each U_j has the distribution $\{p_k\}$. By assumption these U_k are mutually independent. Now

$$(6.1) \qquad P(s) = \sum_{k=0}^{\infty} p_k s^k$$

is the common generating function of the U_j and X_1. Therefore the generating function of X_2 is $P_2(s) = P(P(s))$ by the theorem of the preceding section. In like manner, the X_3-particles of the third generation are second-order descendants of the X_1-particles of the first generation, so that X_3 is the sum of X_1 mutually independent variables each having the same distribution as X_2. This means that the generating function of X_3 is $P_3(s) = P(P_2(s))$. The same argument shows that in general *the generating function $P_{n+1}(s)$ of the number X_{n+1} of particles in the $(n+1)$st generation is defined recursively by*

$$(6.2) \qquad P_1(s) = P(s), \quad P_{n+1}(s) = P(P_n(s)).$$

In example (a) $P(s) = q + ps^m$; and hence $P_2(s) = q + p(q + ps^m)^m$, $P_3(s) = q + p\{q + p(q + ps^m)^m\}^m$, etc. For a Poisson distribution $P(s) = e^{-\lambda(1-s)}$, $P_2(s) = e^{-\lambda + \lambda e^{-\lambda + \lambda s}}$, etc. These formulas are not very pleasing, but (6.2) permits us to draw important general conclusions.

We first inquire into the probability x_n that the process terminates at or before the nth generation. This means that $X_n = 0$, and hence $x_n = P_n(0)$. It is clear from the definition that x_n can only increase. As a matter of fact, we have $x_1 = P(0) = p_0$, and $x_{n+1} = P(x_n)$. Now $P(s)$ is an increasing function, and therefore $x_2 = P(x_1) > P(0) = x_1$; by induction $x_{n+1} = P(x_n) > P(x_{n-1}) = x_n$. It follows that $x_n \to \zeta$ where ζ satisfies

$$(6.3) \qquad \zeta = P(\zeta).$$

We claim that ζ is the *smallest* root of (6.3). In fact, if η is any other root, then, since $P(s)$ is increasing, $x_1 = P(0) < P(\eta) = \eta$; by induction if $x_n < \eta$, then also $x_{n+1} = P(x_n) < P(\eta) = \eta$ so that $\zeta \leq \eta$, which proves that no smaller root than ζ can exist.

We have now to investigate the roots of the equation $s = P(s)$. Clearly $s = 1$ is always a root. If there exist two roots s_1 and s_2, then the difference ratio $\{P(s_2) - P(s_1)\}/(s_2 - s_1)$ equals one, and by the mean value theorem there exists a point σ between s_1 and s_2 for which $P'(\sigma) = 1$. However, for $0 < s < 1$ the function $P'(s)$ increases steadily, and hence there exists at most one value σ between 0 and 1 for which $P'(\sigma) = 1$. In other words, there can exist at most one pair of roots of $s = P(s)$ lying within the interval $0 \leq s \leq 1$. The end point $s = 1$ is always a root. For the existence of a second root s with $0 \leq s < 1$ it is necessary that $P'(\sigma) = 1$ for some value $\sigma < 1$ and hence $P'(1) > 1$. On the other hand, if no root $s < 1$ exists, then $p_0 = P(0) > 0$ and hence $P(s) > s$ for $0 \leq s < 1$. Since $P(1) = 1$, it follows that $P'(1) \leq 1$. Thus a root $s < 1$ of $s = P(s)$ exists if and only if $P'(1) > 1$; this root is unique. Now $P'(1) = \Sigma k p_k$ is the expected number of direct descendants of a particle. We can, therefore, formulate the fundamental result:

Let $\mu = \Sigma k p_k$ *be the mean of the number of direct descendants of a single particle. If* $\mu \leq 1$, *then the probability tends to one that the process will terminate before the nth generation* (that is, $X_n = 0$). *If* $\mu > 1$, *then there exists a unique root* $\zeta < 1$ *of* (6.3), *and* ζ *is the limit of the probability that the process terminates after finitely many generations.*

The difference $1 - \zeta$ can be called the probability of an infinitely prolonged process. Usually x_n converges to ζ rapidly, so that if the process terminates it is likely to proceed for only very few generations. In practice, therefore, ζ is the probability of a rapid extinction. In the example (c) we may call $1 - \zeta$ the probability that a mutant gene establishes itself. If we start with r particles instead of a single one, the probability that all r descendant lines die out is ζ^r, and the probability of at least one being successful is $1 - \zeta^r$. Even if ζ is relatively large, $1 - \zeta^r$ is near 1 if the initial number r is large. In the nuclear chain reaction of example (a) this is always the case, and hence we can say: if $\mu > 1$, the probability of an explosion is near 1, while for $\mu \leq 1$ the probability is 1 that the process stops after a finite number of generations.

We can also find the expected size of the nth generation $E(X_n) = P'_n(1)$. Since $P_n(s) = P(P_{n-1}(s))$, we find

$$P'_n(1) = P'(P_{n-1}(1))P'_{n-1}(1) = P'(1)P'_{n-1}(1) = \mu\, E(X_{n-1}),$$

and generally by induction

(6.4) $$E(X_n) = \mu^n.$$

Hence, if $\mu > 1$, we should expect an exponential growth. This argument can be amplified. It is easily seen that not only $P_n(0) \to \zeta$ but also $P_n(s) \to \zeta$ for all $s < 1$. This means that the coefficients of s, s^2, s^3, ... tend to zero. *After a large number of generations the probability that no descendants exist is near ζ, and the probability that the number of descendants exceeds any preassigned bound is near $1 - \zeta$;* it is exceedingly improbable to find a moderate number of descendants.

We have found only the distribution of the individual variables X_n. For a general theory we require also the joint distributions of all combinations $(X_1, X_2, \cdots, X_n)$, and these can be easily derived (cf. problem 21). The sequence $\{X_n\}$ of (mutually dependent) random variables describes our stochastic process. The whole infinite sequence and probability relations in it must be considered if we desire properties of other random variables connected with the process, such as the lifetime (number of generations before extinction) in the case $\mu \leq 1$, etc. Further results concerning the behavior of X_n are found in a paper by Harris, which, however, uses deep complex variable methods.[7]

7. Partial Fraction Expansions

It is frequently practically impossible to find explicit expressions for a required probability distribution $\{p_k\}$ but simple to find the corresponding generating function $P(s)$. The chain reactions treated in the preceding section illustrate the fact that much useful information can be obtained directly from the generating function. Moreover, if the generating function is known, it is usually possible to derive from it simple approximations to the required probabilities. The exact expressions for $\{p_k\}$ are in many cases so complicated that approximations

[7] T. E. Harris, Branching Processes, *Annals of Mathematical Statistics*, vol. 19 (1948), pp. 474–494.

are preferable. Perhaps the most useful method is that of partial fractions which we proceed to describe in the simplest case.

Suppose that the generating function is rational, that is,

(7.1) $$P(s) = \frac{U(s)}{V(s)}$$

where $U(s)$ and $V(s)$ are polynomials without common roots. For simplicity let us first assume that the degree of $U(s)$ is lower than the degree of $V(s)$. Assume that $V(s)$ is of degree m and that its m (real or imaginary) roots $s_1, \cdots, s_m$ are all distinct. Then

(7.2) $$V(s) = (s - s_1)(s - s_2) \cdots (s - s_m),$$

and it is known from algebra that $P(s)$ can be decomposed into *partial fractions*

(7.3) $$P(s) = \frac{\rho_1}{s_1 - s} + \frac{\rho_2}{s_2 - s} + \cdots + \frac{\rho_m}{s_m - s}$$

where $\rho_1, \rho_2, \cdots, \rho_m$ are constants. The roots $s_1, s_2, \cdots, s_m$ are found by solving the equation $V(s) = 0$. To find the constant ρ_1 we multiply (7.3) by $s_1 - s$; we see that as $s \to s_1$ the product $(s_1 - s)P(s)$ tends to ρ_1. On the other hand, from (7.1) and (7.2) we get

(7.4) $$(s_1 - s)P(s) = \frac{-U(s)}{(s - s_2)(s - s_3) \cdots (s - s_m)}.$$

As $s \to s_1$ the numerator tends to $-U(s_1)$ and the denominator to $(s_1 - s_2)(s_1 - s_3) \cdots (s_1 - s_m)$, which is the same as $V'(s_1)$. Thus $\rho_1 = -U(s_1)/V'(s_1)$. The same argument applies to all roots, so that for $k \leq m$

(7.5) $$\rho_k = \frac{-U(s_k)}{V'(s_k)}.$$

Unfortunately, extensive numerical calculation is usually required to put (7.1) into the form (7.3). However, once the expansion (7.3) is obtained, we can easily derive an exact expression for the coefficient *of s^n in $P(s)$.* Note that we can write

(7.6) $$\frac{1}{s_k - s} = \frac{1}{s_k} \cdot \frac{1}{1 - s/s_k}.$$

For $|s| < |s_k|$ we can expand the last fraction into a geometric series

(7.7) $$\frac{1}{1 - s/s_k} = 1 + \frac{s}{s_k} + \left(\frac{s}{s_k}\right)^2 + \left(\frac{s}{s_k}\right)^3 + \cdots.$$

Introducing these expressions into (7.3), we find for the *coefficient* p_n *of* s^n

(7.8) $$p_n = \frac{\rho_1}{s_1^{n+1}} + \frac{\rho_2}{s_2^{n+1}} + \cdots + \frac{\rho_m}{s_m^{n+1}}.$$

Thus, to get p_n we have first to find the roots $s_1, \ldots, s_m$ of the denominator and then to determine the coefficients $\rho_1, \ldots, \rho_m$ from (7.5).

In (7.8) we have an *exact* expression for the probability p_n. The labor involved in calculating all m roots is usually prohibitive, and therefore formula (7.8) is primarily of theoretical interest. Fortunately a single term in (7.8) almost always provides a satisfactory approximation. In fact, suppose that s_1 is a root which is *smaller* in absolute value than all other roots. Then the first denominator in (7.8) is smallest. Clearly, as n increases, the proportionate contributions of the other terms decrease and the first term preponderates. In other words, *if s_1 is a root of $V(s) = 0$ which is smaller in absolute value than all other roots, then, as $n \to \infty$,*

(7.9) $$p_n \sim \frac{\rho_1}{s_1^{n+1}}$$

(where the sign $\sim$ indicates that the ratio of the two sides tends to 1). Usually this formula provides surprisingly good approximations even for relatively small values of n. The main advantage of (7.9) lies in the fact that it requires the computation of only one root of an algebraic equation.

Examples. (*a*) Let q_n be the probability that in n tosses of an ideal coin no run of three consecutive heads appears [note that $\{q_n\}$ is not a probability distribution; if p_n is the probability that the first run of three consecutive heads ends at the nth trial, then $\{p_n\}$ is a probability distribution, and q_n represents its "tails," $q_n = p_{n+1} + p_{n+2} + \ldots$].

We can easily show that q_n satisfies the recurrence formula

(7.10) $$q_n = \tfrac{1}{2}q_{n-1} + \tfrac{1}{4}q_{n-2} + \tfrac{1}{8}q_{n-3}.$$

In fact, the event that n trials produce no sequence HHH can occur only when the trials begin with T, HT, or HHT. The probabilities that the following trials lead to no run HHH are q_{n-1}, q_{n-2}, and q_{n-3}, respectively, and the right side of (7.10) therefore contains the probabilities of the three mutually exclusive ways in which the event "no run HHH" can occur.

Evidently $q_0 = q_1 = q_2 = 1$, and hence the q_n can be calculated successively from (7.10). To obtain the generating function $Q(s)$

$= \Sigma q_n s^n$ we multiply both sides by s^n and add for $n = 3, 4, 5, \ldots$. We get

$$Q(s) - 1 - s - s^2$$

(7.11) $$= \frac{s}{2}\{Q(s) - 1 - s\} + \frac{s^2}{4}\{Q(s) - 1\} + \frac{s^3}{8}Q(s)$$

or

(7.12) $$Q(s) = \frac{2s^2 + 4s + 8}{8 - 4s - 2s^2 - s^3}.$$

The denominator has the real root $s_1 = 1.0873778\ldots$ and two complex roots. For $|s| < s_1$ we have

$$|4s + 2s^2 + s^3| < 4s_1 + 2s_1^2 + s_1^3 = 8,$$

and the same inequality holds also when $|s| = s_1$ unless $s = s_1$. Hence the other two roots exceed s_1 in absolute value. Thus, from (7.9)

(7.13) $$q_n \sim \frac{1.236840}{(1.0873778)^{n+1}},$$

where the numerator equals $(2s_1^2 + 4s_1 + 8)/(4 + 4s_1 + 3s_1^2)$. Table 1 shows that (7.13) gives excellent approximations for all n in question. (The general theory of runs is developed in Chapter 13, section 1.)

TABLE 1

Illustration to Partial Fraction Expansions

	q_n		r_n	
n	True	Approximate	True	Approximate
3	0.875	0.884 69	0.875	0.856 07
4	.8125	.813 60	.750	.751 15
5	.75	.748 22	.656 25	.659 09
6	.6875	.688 10	.578 12	.578 31
7	.632 81	.632 80	.507 81	.507 43
8	.582 03	.581 95	.445 31	.445 24
9	.535 16	.535 19	.390 63	.390 67
10	.492 19	.492 18	.342 77	.342 79
11	.452 64	.452 63	.300 78	.300 78
12	.416 26	.416 26	.263 91	.263 91

q_n is the probability that in n tosses of a coin the sequence HHH does not appear, and r_n the corresponding probability for the sequence HTH. The true values are calculated from (7.10) and (7.14), the approximations from (7.13) and (7.16).

(b) Let now r_n be the probability that in n tossings of a coin the sequence HTH does not appear. This example is similar to the preceding one, but note that r_n is not the same as the probability q_n that no run HHH appears. In this case it is no longer easy to obtain a recursion formula analogous to (7.10): the general theory of Chapter 13 will show that

(7.14) $\quad r_n = r_{n-1} - \tfrac{1}{4}r_{n-2} + \tfrac{1}{8}r_{n-3}, \qquad r_0 = r_1 = r_2 = 1.$

(The reader may try to verify this formula: cf. problem 2 of Chapter 13.) Proceeding as before, we find for the generating function

(7.15) $$R(s) = \frac{8 + 2s^2}{8 - 8s + 2s^2 - s^3}$$

and hence

(7.16) $\quad r_n \sim 1.444248(1.139680)^{-n-1},$

approximately. (Cf. Table 1.)

(c) Another numerical example will be found in Chapter 13, section 5.

It is easy to remove the restrictions under which we have derived the asymptotic formula (7.9). To begin with, the degree of the numerator in (7.1) may exceed the degree m of the denominator. Let $U(s)$ be of degree $m + r$ ($r \geq 0$); a division reduces $P(s)$ to a polynomial of degree r plus a fraction $U_1(s)/V(s)$ in which $U_1(s)$ is a polynomial of a degree lower than m. The polynomial affects only the first $r + 1$ terms of the distribution $\{p_n\}$, and $U_1(s)/V(s)$ can be expanded into partial fractions as explained above. Thus (7.9) remains true. Secondly, the restriction that $V(s)$ should have only simple roots is unnecessary. It is known from algebra that every rational function admits of an expansion into partial fractions. If s_k is a double root of $V(s)$, then the partial fraction expansion (7.3) will contain an additional term of the form $a/(s - s_k)^2$, and this will contribute a term of the form $a(n + 1)s_k^{-(n+2)}$ to the exact expression (7.8) for p_n. However, this does not affect the asymptotic expansion (7.9), provided only that s_1 is a simple root. We note this result for future reference as a

Theorem. *If $P(s)$ is a rational function with a root s_1 of the denominator which is smallest in absolute value and is a simple root, then the coefficient p_n of s^n is given asymptotically by $p_n \sim \rho_1 s_1^{-(n+1)}$, where ρ_1 is defined in (7.5).*

A similar asymptotic expansion exists also in the case where s_1 is a multiple root. Finally, the restriction to rational functions was introduced only for convenience. It is known from the theory of complex variables that a much more general class of functions admits of partial fraction expansions, and this is one of the sources of usefulness of generating functions and characteristic functions in general. (Cf. also problem 22.)

* 8. The Continuity Theorem

We know from Chapter 6 that the Poisson distribution $\{e^{-\lambda}\lambda^k/k!\}$ is the limiting form of the binomial distribution with the probability p depending on n in such a way that $np \to \lambda$ as $n \to \infty$. Then $b(k; n, p) \to e^{-\lambda}\lambda^k/k!$ The generating function of $\{b(k; n, p)\}$, is $(q + ps)^n = \{1 - \lambda(1 - s)/n\}^n$. Taking logarithms, we see directly that this generating function tends to $e^{-\lambda(1-s)}$, which is the generating function of the Poisson distribution. We now show that this situation prevails in general; a sequence of probability distributions converges to a limiting distribution if and only if the corresponding generating functions converge. Unfortunately, this theorem is of limited applicability, since the most interesting limiting forms of discrete distributions are continuous distributions (for example, the normal distribution appears as a limiting form of the binomial distribution).

Continuity Theorem. *Suppose that for every fixed n the sequence $a_{0,n}, a_{1,n}, a_{2,n}, \ldots$ is a probability distribution, that is,*

$$(8.1) \qquad a_{k,n} \geq 0, \quad \sum_{k=0}^{\infty} a_{k,n} = 1.$$

In order that for every fixed k

$$(8.2) \qquad a_{k,n} \to a_k$$

as $n \to \infty$ it is necessary and sufficient that for every s with $0 \leq s < 1$

$$(8.3) \qquad A_n(s) \to A(s)$$

where

$$(8.4) \qquad A_n(s) = \sum_{k=0}^{\infty} a_{k,n} s^k, \quad A(s) = \sum_{k=0}^{\infty} a_k s^k$$

are the corresponding generating functions.

Note. If (8.2) holds, then automatically $0 \leq a_k < 1$ and $\Sigma a_k \leq 1$. The generating function $A(s)$ exists therefore at least for $|s| \leq 1$. However, the limiting sequence $\{a_k\}$ is not necessarily a probability distribution; for example, if the first n terms of the distribution $\{a_{k,n}\}$ vanish, then the limiting sequence vanishes identically. For $\{a_k\}$ to

---
* Starred sections treat special topics and may be omitted at first reading.

be a probability distribution it is necessary and sufficient that $\Sigma a_k = 1$ or $A(1) = 1$.

Proof.[8] First, suppose that (8.2) holds. For fixed s ($0 < s < 1$) and fixed ϵ we can choose r so that $s^r/(1-s) < \epsilon$. Then

$$(8.5) \qquad |A_n(s) - A(s)| \leq \sum_{k=0}^{r} |a_{k,n} - a_k| s^k + 2\epsilon.$$

The sum on the right contains only finitely many terms each of which tends to zero. Hence $|A_n(s) - A(s)|$ is arbitrarily small for n sufficiently large. Next, assume that (8.3) holds. We use the well-known fact [9] that it is always possible to find a subsequence $\{a_{k,n}\}$ of the given sequence of distributions which converges. If (8.2) were not true, then it would be possible to extract two subsequences converging to two different limiting sequences $\{a_k^*\}$ and $\{a_k^{**}\}$, and the corresponding subsequences of $\{A_n(s)\}$ would converge to $A^*(s) = \Sigma a_k^* s^k$ and $A^{**}(s) = \Sigma a_k^{**} s^k$, respectively. However, this is impossible in view of the assumption (8.3). Therefore (8.3) implies (8.2).

Examples. (a) *The Pascal Distribution.* We saw in section 3 that the generating function of the Pascal distribution $\{f(k; r, p)\}$ is given by $(p/1 - qs)^r$. Now let λ be fixed, and let $p \to 1$, $q \to 0$, so that $q = \lambda/r$ ($r \to \infty$). Then

$$(8.6) \qquad \left(\frac{p}{1-qs}\right)^r = \left(\frac{1-\lambda/r}{1-\lambda s/r}\right)^r.$$

Passing to logarithms, we see that the right side tends to $e^{-\lambda+\lambda s}$, which is the generating function of the Poisson distribution $\{e^{-\lambda}\lambda^k/k!\}$. Hence *if $r \to \infty$ and $rq \to \lambda$, then*

$$(8.7) \qquad f(k; r, p) \to e^{-\lambda}\frac{\lambda^k}{k!}.$$

Note that this is a new limit theorem: in section 4 we found another limit for the case where $p \to 0$ and $q \to 1$, but r remained constant.

(b) *Poisson Trials.* This name is customary for the following generalization of the scheme of Bernoulli trials. *Each of n mutually*

---

[8] The theorem is a special case of the continuity theorem for Laplace-Stieltjes transforms, and the proof follows the general pattern. In the literature the continuity theorem for generating functions is usually stated and proved under unnecessary restrictions.

[9] This is easily established by the "method of diagonals" due to G. Cantor and found in all books on set theory. The statement is, incidentally, a special case of a well-known theorem of Helly.

independent trials results in success (S) or failure (F); the corresponding probabilities in the kth trial are p_k and q_k $(p_k + q_k = 1)$. The number S_n of successes in n trials is a random variable which can be written as the sum

$$(8.8) \qquad S_n = X_1 + \cdots + X_n$$

of n mutually independent random variables X_k with the distributions

$$(8.9) \qquad Pr\{X_k = 0\} = q_k, \quad Pr\{X_k = 1\} = p_k.$$

The generating function of X_k is $q_k + p_k s$, and hence the generating function of S_n

$$(8.10) \qquad P(s) = (q_1 + p_1 s)(q_2 + p_2 s) \cdots (q_n + p_n s).$$

Clearly

$$(8.11) \qquad E(S_n) = p_1 + p_2 + \cdots + p_n,$$
$$\mathrm{Var}(S_n) = p_1 q_1 + p_2 q_2 + \cdots + p_n q_n.$$

As an application of this scheme let us assume that each house in a city has a small probability p_k of burning on a given day. The sum $p_1 + \cdots + p_n$ is the expected number of fires in the city, n being the number of houses. We have seen in Chapter 6 that if all p_k are equal and if the houses are statistically independent, then the number of fires is a random variable whose distribution is near the Poisson distribution. We show now that this conclusion remains valid also under the more realistic assumption that the probabilities p_k are not equal. This result should increase our confidence in the Poisson distribution as an adequate description of phenomena which are the cumulative effect of many improbable events ("successes"). Examples are accidents and telephone calls.

We use the now-familiar model of an increasing number n of variables where the probabilities p_k depend on n in such a way that the largest p_k tends to zero, but the sum

$$(8.12) \qquad p_1 + p_2 + \cdots + p_n = \lambda$$

remains constant. Then from (8.10)

$$(8.13) \qquad \log P(s) = \sum_{k=1}^{n} \log\{1 - p_k(1 - s)\}.$$

Since $p_k \to 0$, we can use the fact that $\log(1 - x) = -x - \theta x$, where $\theta \to 0$ as $x \to 0$. It follows that

$$(8.14) \quad \log P(s) = -(1-s)\left\{\sum_{k=1}^{n}(p_k + \theta_k p_k)\right\} \to -\lambda(1-s),$$

so that $P(s)$ tends to the generating function of the Poisson distribution. Hence, S_n *has in the limit a Poisson distribution*. We conclude that for large n and moderate values of $\lambda = p_1 + p_2 + \cdots + p_n$ the distribution of S_n can be approximated by a Poisson distribution. Estimates of the error involved can be derived, but we shall not go into such details. (Cf. also problem 20 of Chapter 9.)

9. Problems for Solution

It is understood that the random variables occurring in the following problems assume only non-negative integral values.

1. Let X be a random variable with generating function $P(s)$. Find the generating functions of $X + 1$ and $2X$.

2. *Continuation.* Find the generating functions of (a) $Pr\{X \leq n\}$, (b) $Pr\{X < n\}$, (c) $Pr\{X \geq n\}$, (d) $Pr\{X > n + 1\}$, (e) $Pr\{X = 2n\}$.

3. In a sequence of Bernoulli trials let u_n be the probability that the first combination SF occurs at trials number $n-1$ and n. Find the generating function, mean, and variance.

4. In a sequence of n Bernoulli trials let u_n be the probability of an *even* number of successes. Prove the recursion formula $u_n = qu_{n-1} + p(1 - u_{n-1})$. From it derive the generating function and hence an explicit formula for u_n. [Note that this formula is considerably simpler than the obvious formula $u_n = b(0; n, p) + b(2; n, p) + \cdots$.]

5. In the sampling problem (3.d) of Chapter 9, find the generating function of the variable S_r (r fixed). Verify formulas (3.5) and (5.9) of Chapter 9 for the mean and variance.

6. *Continuation.* The following is an alternative method for deriving the same result. Let $p_n(r) = Pr\{S_r = n\}$. Prove the recursion formula

$$(9.1) \quad p_{n+1}(r) = \frac{r-1}{N} p_n(r) + \frac{N-r+1}{N} p_n(r-1).$$

Derive the generating function directly from (9.1).

7. Solve the two preceding problems for r preassigned elements (instead of r arbitrary ones).

8.[10] Let the sequence of Bernoulli trials up to the first failure be called a *turn*.

---

[10] Problems 8–10 have a direct bearing on the *game of billiards*. The probability p of success is a measure of the player's skill. The player continues to play until he fails. Hence the number of successes he accumulates is the length of his "turn." The game continues until one player has scored N successes. Problem 8 therefore gives the probability distribution of the number of turns one player needs to score k successes, problem 9 the average duration, and problem 10 the probability of a tie between two players. For further details cf. O. Bottema and S. C. Van Veen, Kansberekningen bij het biljartspel, *Nieuw Archief voor Wiskunde* (in Dutch), vol. 22 (1943), pp. 16–33 and 123–158.

Find the generating function and the probability distribution of the accumulated number S_r of successes in r turns.

9.[10] *Continuation.* Let R be the number of successive turns up to the νth success (that is, the νth success occurs during the Rth turn). Prove that $Pr\{R = r\}$
$= p^\nu q^{r-1} \binom{r+\nu-2}{\nu-1}$. Find $E(R)$ and $Var(R)$.

10.[10] *Continuation.* Consider *two* sequences of Bernoulli trials with probabilities p_1, q_1, and p_2, q_2, respectively. Show that the probability that the same number of turns will lead to the Nth success can be exhibited in either of the forms:

$$(p_1 p_2)^N \sum_{\nu=1}^{\infty} \binom{N+\nu-2}{\nu-1}^2 (q_1 q_2)^{\nu-1} = (p_1 p_2)^N (1 - q_1 q_2)^{1-2N} \sum_{k=0}^{N-1} \binom{N-1}{k}^2 (q_1 q_2)^k.$$

11. Let $\{X_k\}$ be mutually independent variables, each assuming the values $0, 1, 2, \cdots, a-1$ with probabilities $1/a$. Let $S_n = X_1 + \cdots + X_n$. Show that the generating function of S_n is

$$P(s) = \left\{\frac{1 - s^a}{a(1 - s)}\right\}^n$$

and hence

$$Pr\{S_n = j\} = \frac{1}{a^n} \sum_{\nu=0}^{\infty} (-1)^{\nu+j+a\nu} \binom{n}{\nu} \binom{-n}{j - a\nu}.$$

(Only finitely many terms in the sum are different from zero.)

Note: For $a = 6$ we get the probability of scoring the sum $j + n$ in a throw with n dice. The solution goes back to DeMoivre.

12. *Continuation.* The probability $Pr\{S_n \leq j\}$ has the generating function $P(s)/(1 - s)$ and hence

$$Pr\{S_n \leq j\} = \frac{1}{a^n} \sum_{\nu} (-1)^\nu \binom{n}{\nu} \binom{j - a\nu}{n}.$$

13. *Continuation: the limiting form.* If $a \to \infty$ and $j \to \infty$, so that $j/a \to x$, then

$$Pr\{S_n \leq j\} \to \frac{1}{n!} \sum_{\nu} (-1)^\nu \binom{n}{\nu} (x - \nu)^n,$$

the summation extending over all ν with $0 \leq \nu < x$.

Note: This result is due to Laplace. In the theory of geometric probabilities the right-hand side represents the distribution function of the sum of n independent random variables with "uniform" distribution in the interval $(0, 1)$.

14. Let u_n be the probability that the number of successes in n Bernoulli trials is divisible by 3. Find a recurrence relation for u_n and hence the generating function.

15. *Continuation: alternative method.* Let v_n and w_n be the probabilities that S_n is of the form $3v + 1$ and $3v + 2$, respectively (so that $u_n + v_n + w_n = 1$). Find three simultaneous recurrence relations and hence three equations for the generating functions.

16. Let X and Y be independent variables with generating functions $U(s)$ and $V(s)$. Show that $Pr\{X - Y = j\}$ is the coefficient of s^j in $U(s)V(1/s)$, where $j = 0, \pm 1, \pm 2, \ldots$.

17. *Moment generating functions.* Let X be a random variable with generating function $P(s)$, and suppose that $\Sigma p_n s^n$ converges for some $s_0 > 1$. Then all mo-

ments $m_r = E(X^r)$ exist, and the generating function $F(s)$ of the sequence $m_r/r!$ converges at least for $|s| < \log s_0$. Moreover

$$F(s) = \sum_{r=0}^{\infty} \frac{m_r}{r!} s^r = P(e^s).$$

Note: $F(s)$ is usually called the *moment generating function*, although in reality it generates $m_r/r!$.

18. Prove the following formula for the "tails" of the Pascal distribution

$$\sum_{k=n}^{\infty} f(k; r, p) = \sum_{k=0}^{r-1} f(k; n-r, q).$$

Note: No computations are necessary.

19. *Compound Poisson distribution.*[11] Let the random variable X assume the values $\lambda_1, \lambda_2, \ldots$ with probabilities $u_1, u_2, \ldots$ ($\lambda_j > 0$, $\Sigma u_j = 1$). Show that

$$p_k = \frac{1}{k!} \sum_{j=0}^{\infty} u_j e^{-\lambda_j} (\lambda_j)^k$$

is a probability distribution. Find the generating function and prove that its mean equals $E(X)$ and that its variance equals $\mathrm{Var}(X) + E(X)$.

20. In the chain reaction problem calculate $\mathrm{Var}(X_n)$.

21. *Continuation.* If $n > m$, show that $E(X_m X_n) = \mu^{n-m} E(X_m^2)$.

22. Suppose that $A(s) = \Sigma a_n s^n$ is a rational function $U(s)/V(s)$ and that s_1 is a root of $V(s)$, which is smaller in absolute value than all other roots. If s_1 is of multiplicity r, show that

$$a_n \sim \frac{\rho_1}{s_1^{n+r}} \binom{n+r-1}{r-1}$$

where $\rho_1 = -r! U(s_1)/V^{(r)}(s_1)$.

[11] The word "compound" is used here in a slightly more general sense than in section 5. Such distributions are of great importance. A typical interpretation is as follows. The number of raisins in a cake may be assumed to have a Poisson distribution (Chapter 6, end of section 5). Different kinds of cake are characterized by different values of the mean. If these varieties are mixed in the proportions $u_1:u_2:u_3:\ldots$, then p_k is the probability of finding exactly k raisins in a cake if selected at random.

CHAPTER 12

RECURRENT EVENTS:
THEORY

1. Definition

We consider a sequence of repeated trials with possible results E_j ($j = 1, 2, \ldots$). In all examples of this chapter the trials will be independent, but the theorems will be formulated so as to apply also to dependent trials and in particular to Markov chains (Chapter 15). As usual, we suppose that it is in principle possible to continue the trials indefinitely and that the probabilities $Pr\{(E_{j_1}, \cdots, E_{j_n})\}$ of the outcomes of the first n trials are defined consistently for all n. We shall investigate classes of events defined by certain repetitive patterns. Roughly speaking, a pattern $\mathcal{E}$ qualifies for our theory if it has the following two properties. For every possible outcome of n trials $(E_{j_1}, E_{j_2}, \cdots, E_{j_n})$ it can be uniquely determined whether $\mathcal{E}$ has occurred and, if so, at which trial or trials. Moreover, every time $\mathcal{E}$ occurs the trials start from scratch in the sense that the trials following an occurrence of $\mathcal{E}$ are an exact replica of the entire sequence of trials. Before giving a rigorous definition we shall illustrate the notion by a few simple

Examples. (a) *Return to Equilibrium in Coin Tossing.* In a sequence of independent tossings of a symmetric coin let $\mathcal{E}$ stand as an abbreviation for "the accumulated numbers of heads and tails are equal." For the time-honored bettor who at each trial loses or gains a unit amount, the occurrence of $\mathcal{E}$ means that his accumulated gain is zero. Here it is clear that the occurrence of $\mathcal{E}$ means a return to the initial situation or, as we prefer to call it, a *return to equilibrium*. Clearly $\mathcal{E}$ can occur only at an even number of trials. If u_n is the probability that $\mathcal{E}$ occurs at the nth trial, then $u_1 = u_3 = u_5 = u_7 = \ldots = 0$; for $\mathcal{E}$ to occur at the $2n$th trial it is necessary and sufficient that the $2n$ trials produce n heads and n tails, and hence

$$(1.1) \qquad u_{2n} = \binom{2n}{n} \frac{1}{2^{2n}}.$$

12.1] DEFINITION 239

If the coin is tossed until $\mathcal{E}$ occurs for the first time, we get the following sequences (arranged according to length): *HT, TH; HHTT, TTHH; HHHTTT, HHTHTT, TTTHHH, TTHTHH; HHHHTTTT, HHHTHTTT, HHHTTHTT, HHTHHTTT, HHTHTHTT, TTTTHHHH, TTTHTHHH, TTTHHTHH, TTHTTHHH, TTHTHTHH;* etc. If f_n is the probability that $\mathcal{E}$ occurs *for the first time* at the nth trial, then obviously $f_n = 0$ whenever n is odd. An inspection of the above sequence shows that $f_2 = 1/2, f_4 = 1/8, f_6 = 1/16, f_8 = 5/128$. A further enumeration leads to the values $f_{10} = 7/256, f_{12} = 21/1024, f_{14} = 33/2048, \ldots$. The general formula is not discernible, but will be deduced later [cf. (3.19)].

(b) *Success Runs.* We consider Bernoulli trials with probability p for success. Let $\mathcal{E}$ stand for "three consecutive successes." This description is insufficient inasmuch as it does not tell whether in an uninterrupted sequence of five successes $\mathcal{E}$ occurs once, three times, or not at all. For our purposes we must define $\mathcal{E}$ so that the trials start from scratch whenever $\mathcal{E}$ occurs. This means that we must count only non-overlapping runs of three successes each: in n trials $\mathcal{E}$ occurs as often as there are non-overlapping runs of exactly three successes. In other words, $\mathcal{E}$ occurs for the first time when for the first time three successes appear in succession, and then the counting starts anew. Thus, in the sequence $SSSSSSSSF$ the recurrent event $\mathcal{E}$ occurs at the third and the sixth places. If f_n is the probability that $\mathcal{E}$ occurs for the first time at the nth trial, then $f_1 = f_2 = 0, f_3 = p^3, f_4 = f_5 = f_6 = qp^3, f_7 = (1 - p^3)qp^3, f_8 = (1 - p^3 - qp^3)qp^3, \ldots$. The general theory of success runs and similar repetitive patterns will be taken up in Chapter 13.

(c) Consider consecutive tossings of a symmetric die and let $\mathcal{E}$ stand for "ones, twos, $\cdots$, sixes appeared in equal numbers." This is similar to example (a), but, as we shall see, there is one important difference. In example (a) we have $f_1 + f_2 + f_3 + \ldots = 1$, which may be interpreted by saying that $\mathcal{E}$ is certain to occur sooner or later. In the present case this is not so. In fact, it will be found that the probability $f_1 + f_2 + f_3 + \ldots$ that $\mathcal{E}$ ever occurs is only about 0.022. [Cf. example (3.d).]

Our examples show that it would be most natural to consider the sample space of infinite sequences of trials [Chapter 8, section 1]. However, this is not necessary, and we may study recurrent events also in spaces of finitely many trials. The essential property of recurrent events is that after every occurrence of $\mathcal{E}$ the trials start from scratch. This means that all events preceding an occurrence of $\mathcal{E}$ should be statistically independent of all subsequent events. A more precise description is contained in the

Definition. *We speak of a recurrent event* $\mathcal{E}$ *if the following conditions are satisfied.*

(1) *There exists a rule which for every sequence* $(E_{j_1}, E_{j_2}, \cdots, E_{j_n})$ *of possible outcomes uniquely determines whether or not* $\mathcal{E}$ *occurs at the last trial. This rule depends on the sequence and not on the following trials.*

(2) *Let* $(E_{j_1}, E_{j_2}, \cdots, E_{j_m})$ *be a sequence in which* $\mathcal{E}$ *occurs at the last trial and let* $(E_{k_1}, \cdots, E_{k_n})$ *be an arbitrary sequence. Then* $\mathcal{E}$ *occurs at the last [or $(m+n)$th] trial of the combined sequence* $(E_{j_1}, \cdots, E_{j_m}, E_{k_1}, \cdots, E_{k_n})$ *if and only if it occurs at the last [or nth] trial of* $(E_{k_1}, \cdots, E_{k_n})$.

(3) *In this case*

$$(1.2) \quad Pr\{(E_{j_1}, \cdots, E_{j_m}, E_{k_1}, \cdots, E_{k_n})\} = Pr\{(E_{j_1}, \cdots, E_{j_m})\} Pr\{(E_{k_1}, \cdots, E_{k_n})\}.$$

Given any particular sequence $(E_{j_1}, E_{j_2}, \cdots, E_{j_n})$ of possible outcomes, we can consider the n subsequences (E_{j_1}), (E_{j_1}, E_{j_2}), $(E_{j_1}, E_{j_2}, E_{j_3})$, $\cdots$, $(E_{j_1}, \cdots, E_{j_n})$ and mark those in which $\mathcal{E}$ occurs at the last trial. If this is true of k among these subsequences, then we say that $\mathcal{E}$ occurs in $(E_{j_1}, \cdots, E_{j_n})$ exactly k times. Thus, in example (1.a), $\mathcal{E}$ occurs in the sequence $HHTHTTHTTHT$ three times, namely, at trials number 6, 8, and 10.

Condition (3) implies the following property: if $\mathcal{E}$ occurs at trial number m, then the conditional probability that it occurs again at the trial number $m+n$ equals the probability that $\mathcal{E}$ occurs at the nth trial.

Example. (d) Consider a sequence of Bernoulli trials and suppose that in the first trials successes alternate with failures. It is obviously legitimate to say that in the sequence $SFSFSFSFSFSF$ the pattern SFS occurred at trials number 3, 5, 7, 9, and 11. However, if we wish to define a recurrent event $\mathcal{E}$ characterized by the pattern SFS, then the counting is regulated by condition (2). We are required to consider all subsequences, and since the third trial completes the first pattern SFS, there is no doubt that $\mathcal{E}$ occurs at the third trial. Now condition (2) starts to operate. According to it, $\mathcal{E}$ occurs at trial number $n+3$ if and only if it occurs at the nth trial of the reduced sequence $FSFSFSFSF$ (in which the first three trials are omitted). Since here $\mathcal{E}$ occurs for the first time at the fourth trial it occurs in the original sequence for the second time only at the seventh trial. This implies that the third occurrence of ε takes place at the eleventh trial, etc.

Consider now the sample space of n trials, that is, the aggregate of all sequences $(E_{j_1}, E_{j_2}, \cdots, E_{j_n})$. In it we may consider an event such as "ε occurs exactly three times"; it consists of those sample points $(E_{j_1}, \cdots, E_{j_n})$ in which ε occurs exactly three times. Similarly, events such as "ε occurs for the first time at the third trial," "ε occurs an even number of times," etc., are well defined.

In particular, we may talk of the probability f_k that ε occurs for the first time at the kth trial ($k \leq n$). Then $f_1 + f_2 + \cdots + f_k$ is the probability that ε occurs in the first k trials. If $\Sigma f_k = 1$, this probability tends to one, and we call ε *certain;* if $\Sigma f_k < 1$, ε will be called *uncertain.* [In examples (1.a) and (1.b) ε is certain, in (1.c) uncertain.] Finally we shall say that ε *has period* $t > 1$ if it can occur only at trials number $t, 2t, 3t, \cdots, (t > 1)$. In example (1.a) ε can occur only at even trials, hence $t = 2$; similarly, in example (1.c) $t = 6$. We summarize these ideas in the following definition.

Classification. Let f_k *be the probability that* ε *occurs for the first time at the kth trial. The recurrent event* ε *is called certain or uncertain according as* $\Sigma f_k = 1$ *or* $\Sigma f_k < 1$. *Let t be the greatest integer such that* ε *can occur only at trials number* $t, 2t, 3t, \cdots$ *(then $f_k = 0$ whenever k is not divisible by t). We say that* ε *has the period t if $t > 1$. If $t = 1$ then* ε *is called aperiodic.*

2. Recurrence Times

Suppose that ε is a *certain* recurrent event, that is, let

(2.1) $$\Sigma f_j = 1.$$

In the preceding section we have considered the number of trials up to and including the *first occurrence* of ε. This is obviously a random variable X_1 which is completely defined only in the sample space of infinite sequences of trials. However, it is clear that its probability distribution must be given by

(2.2) $$Pr\{X_1 = j\} = f_j.$$

More generally, consider the random variable X_k *defined as the number of trials following the $(k - 1)$st occurrence of* ε *up to and including the kth occurrence. This is the kth recurrence time of* ε. Again, this random variable is defined completely only in the sample space of infinite sequences of trials. We could effect a simple limiting process, but it is clear from the very definition of recurrent events that all variables

X_k have the common distribution (2.2) and that they are mutually independent. We can avoid any further reference to infinite sample space if we accept the following fact.

With every certain recurrent event $\mathcal{E}$ there is associated an infinite sequence of mutually independent random variables X_k, the recurrence times. They have the common distribution $\{f_j\}$, and

$$(2.3) \qquad S_n = X_1 + \cdots + X_n$$

is to be interpreted as the number of trials up to and including the nth occurrence of $\mathcal{E}$.

We put

$$(2.4) \qquad \mu = \sum_{j=1}^{\infty} jf_j$$

and call μ the *mean recurrence time of $\mathcal{E}$.* Clearly $E(X_k) = \mu$. Note that the mean recurrence time may be infinite; in fact, in the most interesting physical applications μ is infinite.

In the case of an uncertain $\mathcal{E}$ put

$$(2.5) \qquad f = \Sigma f_j$$

so that $f < 1$. Now $\{f_j\}$ is no longer a probability distribution. However, all preceding remarks remain valid if we agree to describe a non-occurrence of $\mathcal{E}$ by saying that the recurrence time is infinite. This means that *the recurrence times X_k now are generalized random variables which (with probability $1 - f$) assume the (improper) value ∞.* Thus $X_k = j$ with probability f_j and $X_k = \infty$ with probability $1 - f$; the latter is an abbreviation for saying that there is probability $1 - f$ that the $(k - 1)$st occurrence of $\mathcal{E}$ (if any) is not followed by a kth occurrence. The probability distribution of S_n is obtained in the customary way, adding the rule "infinity plus any value equals infinity." (This is possible since we are adding positive variables and no subtractions occur.)

The probability that $\mathcal{E}$ does not occur at all is $1 - f$. The probability that it occurs once and only once is (because of the independence of recurrence times) $f(1 - f)$. Proceeding in this way, we find $f^n(1 - f)$ for the probability that $\mathcal{E}$ occurs exactly n times. The probability that $\mathcal{E}$ occurs at most n times is therefore

$$(2.6) \qquad (1 - f)(1 + f + f^2 + \cdots + f^n) = 1 - f^{n+1}.$$

As $n \to \infty$ the right side tends to 1. This fact can be described by saying that *with probability one an uncertain recurrent event occurs only a finite number of times, while a certain $\mathcal{E}$ is bound to occur infinitely often.*

The first half of this statement is the correct interpretation of (2.6) in the sample space of infinitely many trials.

3. Fundamental Theorems

We keep the notation f_j for the probability that $\mathcal{E}$ occurs for the *first time* at trial number j. At the same time we shall study the *probability u_j that $\mathcal{E}$ occurs at the jth trial* (not necessarily for the first time). The quantities u_n and f_n are related by the fundamental equation

$$(3.1) \quad u_n = f_n + f_{n-1}u_1 + f_{n-2}u_2 + \cdots + f_2 u_{n-2} + f_1 u_{n-1}.$$

In fact, if $\mathcal{E}$ occurs at the nth trial, then either it occurs for the first time (probability f_n), or it occurred at some previous trial. The probability that $\mathcal{E}$ occurs for the first time at trial number $n - \nu$ and again at the nth trial is obviously $f_{n-\nu} u_\nu$. These events are mutually exclusive, and therefore (3.1) holds.

Equation (3.1) is a recurrence relation from which the u_n can be calculated successively if the f_j are known and vice versa. It assumes a more symmetric form if we agree to define

$$(3.2) \quad u_0 = 1, \quad f_0 = 0.$$

With this definition (3.1) becomes

$$(3.3) \quad u_n = u_0 f_n + u_1 f_{n-1} + u_2 f_{n-2} + \cdots + u_{n-1} f_1,$$

but this equation is valid only for $n \geq 1$. The right side is the convolution of the two sequences $\{f_j\}$ and $\{u_j\}$, and this suggests introducing the generating functions

$$(3.4) \quad U(s) = \sum_{j=0}^{\infty} u_j s^j, \quad F(s) = \sum_{j=1}^{\infty} f_j s^j.$$

The generating function of the right side in (3.3) is $U(s)F(s)$. (Cf. Chapter 11, section 2.) On the left we have the sequence $\{u_n\}$ with the term $u_0 = 1$ missing, so that the generating function of the left side is $U(s) - 1$. Thus $U(s) - 1 = U(s)F(s)$, and we have the

Theorem 1. *For $k \geq 1$ let f_k be the probability that $\mathcal{E}$ occurs for the first time at the kth trial, and u_k the probability that $\mathcal{E}$ occurs at the kth trial; for $k = 0$ put $f_0 = 0$ and $u_0 = 1$. The generating functions (3.4) are then related by the identity*

$$(3.5) \quad U(s) = \frac{1}{1 - F(s)}.$$

Theorem 2. *The recurrent event* $\mathcal{E}$ *is uncertain if and only if*

$$(3.6) \qquad u = \sum_{j=0}^{\infty} u_j$$

is finite. In this case the probability f that $\mathcal{E}$ *ever occurs is given by*

$$(3.7) \qquad f = \frac{u-1}{u}.$$

Note. We can interpret u_j as the expectation of a random variable which equals 1 or 0 according to whether $\mathcal{E}$ does or does not occur at the jth trial. Hence $u_1 + u_2 + \cdots + u_n$ is the expected number of occurrences of $\mathcal{E}$ in n trials, and $u - 1$ can be interpreted as the expected number of occurrences of $\mathcal{E}$ in infinitely many trials.

Proof. If $\mathcal{E}$ is uncertain, then the series for $F(s)$ converges at $s = 1$ and $F(1) = f < 1$. Now the series for $U(s)$ has only non-negative coefficients, and since $U(s)$ approaches $1/(1-f)$ as $s \to 1$ it follows from Abel's theorem that the series converges for $s = 1$ and $U(1) = 1/(1-f)$. Conversely, if $\mathcal{E}$ is certain, then $F(s) \to 1$ as $s \to 1$, and hence $U(s) \to \infty$ so that Σu_j diverges.

We now come to the main result of the theory from which we shall derive, among others, the ergodic properties of Markov chains. The proof [1] is of an elementary nature, but since it does not contribute to an understanding of the applications of the theorem we defer it to section 7 (cf. also problem 1).

Theorem 3. *If* $\mathcal{E}$ *is a certain recurrent event and not periodic, then*

$$(3.8) \qquad u_n \to \frac{1}{\mu}$$

where $\mu = \Sigma n f_n$ *is the mean recurrence time* ($u_n \to 0$ *if the mean recurrence time is infinite*).

Theorem 4. *If* $\mathcal{E}$ *is certain and has period* $\lambda > 1$, *then as* $n \to \infty$

$$(3.9) \qquad u_{n\lambda} \to \frac{\lambda}{\mu},$$

while $u_k = 0$ *for every k which is not divisible by* λ.

Proof. If $\mathcal{E}$ has period λ, then only powers of s^λ appear in $F(s)$, and $F(s^{1/\lambda}) = F_1(s)$ is again a power series. Since $F_1(1) = 1$, we may

---

[1] P. Erdös, W. Feller, and H. Pollard, A Theorem on Power Series, *Bulletin of the American Mathematical Society*, vol. 55, pp. 201–204 (1949).

consider $F_1(s)$ as the generating function of a recurrent event to which theorem 3 applies. It follows that the coefficients of $U_1(s) = 1/\{1 - F_1(s)\}$ tend to $1/\mu_1$, where

$$(3.10) \qquad \mu_1 = F_1'(1) = \frac{1}{\lambda} F'(1) = \frac{\mu}{\lambda}.$$

(Clearly μ and μ_1 are either both finite or both infinite.) However, the coefficient of s^n in $U_1(s)$ is the coefficient of $s^{n\lambda}$ in $U(s)$, and this proves (3.9).

Examples. (a) Consider a sequence of Bernoulli trials and let $\mathcal{E}$ stand for "success." Then $u_n = p$ when $n \geq 1$. Theorem 3 states that the mean recurrence time for successes is $\mu = 1/p$. This can be verified directly. The probability that the first success occurs at the nth trial is $f_n = q^{n-1}p$, and the mean of this geometric distribution already has been found to be $1/p$. In order to verify the fundamental identity (3.5) note that we have $F(s) = ps(1 + qs + q^2s^2 + \ldots) = ps/(1 - qs)$. Moreover $u_0 = 1$ and $u_n = p$ for $n \geq 1$, so that $U(s) = 1 + p(s + s^2 + s^3 + \cdots) = 1 + ps/(1 - s)$. Hence $U(s) = (1 - qs)/(1 - s) = 1/(1 - F(s))$.

(b) *The Classical Gambling; Return to Equilibrium.* The time-honored gambler of probability textbooks wins or loses a unit amount with probabilities p and q, respectively. Let $\mathcal{E}$ stand for "the accumulated net gain is zero," so that the gambler is back at the initial position. In our terminology we are concerned with Bernoulli trials, and $\mathcal{E}$ stands for "the accumulated numbers of successes and failures are equal." Clearly $\mathcal{E}$ is a recurrent event [cf. example (a) of section 1 where $p = q$]. Since this recurrent event can occur only at an even-numbered trial, it has period 2. For $\mathcal{E}$ to occur at trial number $2n$ it is necessary and sufficient that the $2n$ trials result in n successes and n failures. Hence

$$(3.11) \qquad u_{2n} = \binom{2n}{n} p^n q^n.$$

We know from the normal approximation to the binomial distribution (Chapter 7), and we can also readily verify using Stirling's formula, that

$$(3.12) \qquad \binom{2n}{n} 2^{-2n} \sim \frac{1}{(\pi n)^{1/2}},$$

so that

$$(3.13) \qquad u_{2n} \sim \frac{(4pq)^n}{(\pi n)^{1/2}}.$$

If $p \neq 1/2$, then $4pq < 1$ and the series Σu_{2n} converges faster than the geometric series with ratio $4pq$. Therefore, *if $p \neq 1/2$, the return to equilibrium is uncertain.* If $p = q = 1/2$, then $u_{2n} \sim (\pi n)^{-1/2}$; the series Σu_{2n} diverges, but $u_{2n} \to 0$. Therefore, for $p = q = 1/2$ *the return to equilibrium is a certain recurrent event, but the mean recurrence time is infinite.*

We can get even more information from the generating functions. Using the readily verified formula

$$(3.14) \qquad \binom{2n}{n} = \binom{-1/2}{n} \cdot (-4)^n$$

(cf. Chapter 2, problem 2 of section 9) and the binomial series (Chapter 2, 6.6), we get from (3.11)

$$(3.15) \qquad U(s) = \sum_{n=0}^{\infty} u_{2n} s^{2n} = (1 - 4pqs^2)^{-1/2}.$$

If $p \neq 1/2$, then $u = U(1) = (1 - 4pq)^{-1/2} = |p - q|^{-1}$. From (3.7) we conclude that *the probability f that the accumulated numbers of successes and failures will ever equalize is given by*

$$(3.16) \qquad f = 1 - |p - q|.$$

(*This is the probability of at least one return to equilibrium.*)

From (3.5) we get for *the generating function of the recurrence times*

$$(3.17) \qquad F(s) = 1 - (1 - 4pqs^2)^{1/2}.$$

This formula is most interesting in the case $p = q = 1/2$. Then

$$(3.18) \qquad F(s) = 1 - (1 - s^2)^{1/2}$$

and the binomial expansion shows that

$$(3.19) \qquad f_{2n} = (-1)^{n+1} \binom{1/2}{n} = \frac{1}{n}\binom{2n-2}{n-1} 2^{-2n+1}$$

(f_n vanishes whenever n is odd). Equation (3.19) *gives the distribution of the recurrence times for the return to equilibrium in the classical coin-tossing game.* From (3.18) it follows again that the expected value of this recurrence time is infinite. A few surprising features of these recurrence times will be discussed in section 5. Numerical values for the first seven f_{2n} are given at the end of example (1.*a*).

(*c*) *Ties in Multiple Coin Games.* We consider repeated independent tossings of *two* coins and say that $\mathcal{E}$ has occurred whenever the accumulated number of heads (and therefore of tails) is the same for both coins. Clearly

$$(3.20) \qquad u_n = \frac{1}{2^{2n}} \left\{ \binom{n}{0}^2 + \binom{n}{1}^2 + \binom{n}{2}^2 + \cdots + \binom{n}{n}^2 \right\}.$$

Using formula (9.8) of Chapter 2 and then the normal approximation to the binomial distribution, we find that

$$(3.21) \qquad u_n = \binom{2n}{n} 2^{-2n} \sim \frac{1}{(n\pi)^{1/2}}.$$

Hence Σu_n diverges, but $u_n \to 0$. Therefore $\mathcal{E}$ *is certain but has infinite mean recurrence time.*

Let us now more generally consider the simultaneous tossing of r coins, and let $\mathcal{E}$ stand for the recurrent event that *all r coins are in the same phase* (accumulated numbers of heads are the same for all coins). Then

$$(3.22) \qquad u_n = \frac{1}{2^{rn}} \left\{ \binom{n}{0}^r + \binom{n}{1}^r + \cdots + \binom{n}{n}^r \right\}.$$

To estimate u_n note that the maximal term of the binomial distribution $\binom{n}{k} 2^{-n}$ is of the order of magnitude $(2/\pi n)^{1/2}$ and is smaller than $n^{-1/2}$. Therefore

$$(3.23) \qquad u_n < n^{-\frac{r-1}{2}} 2^{-n} \left\{ \binom{n}{0} + \binom{n}{1} + \cdots + \binom{n}{n} \right\} = n^{-\frac{r-1}{2}}.$$

Accordingly Σu_n converges if $r \geq 4$. For $r = 2$ we saw that Σu_n diverges. A special consideration is necessary for the case $r = 3$. From the normal approximation to the binomial distribution we know that for sufficiently large n and values of k lying between $\frac{1}{2}n - n^{1/2}$ and $\frac{1}{2}n + n^{1/2}$ we have $\binom{n}{k} 2^{-n} > cn^{-1/2}$, where c is a positive constant (say e^{-4}). Therefore, when $r = 3$,

$$(3.24) \qquad u_n > 2n^{1/2}(c^3 n^{-3/2}) = 2c^3/n,$$

and hence Σu_n diverges. In other words, *with two or three coins it is certain that they will sooner or later (and hence arbitrarily often) show the same accumulated number of heads. For four or more coins the same recurrent event is uncertain.*

(d) *Dice.* In example (1.c) we considered the recurrent event $\mathcal{E}$ that the accumulated numbers of aces, twos, threes, etc., are equal.

Obviously $\mathcal{E}$ has period 6 and $u_{6n} = (6n)!(n!)^{-6}6^{-6n}$. Using Stirling's formula, it is readily found that u_{6n} is of the order of magnitude $n^{-5/2}$, so that Σu_n converges. Hence $\mathcal{E}$ is uncertain. From (3.7) it is easy to calculate that the probability of a recurrence is about 0.022.

4. Application of the Central Limit Theorem

In (2.3) we have defined the random variable S_n giving the number of trials up to and including the nth occurrence of $\mathcal{E}$. Often it is more convenient to fix the number of trials and consider the number of occurrences of $\mathcal{E}$ as a random variable. This leads to defining the *variable N_r as the number of occurrences of $\mathcal{E}$ in r trials*. The probability distributions of S_n and N_r are related by the obvious identity:

$$(4.1) \qquad Pr\{N_r \geq k\} = Pr\{S_k \leq r\}.$$

Now let us suppose that $\mathcal{E}$ is certain and that the distribution $\{f_n\}$ has finite mean μ and finite variance σ^2. Then by the central limit theorem (Chapter 10, section 1) for every fixed α

$$(4.2) \qquad Pr\left\{\frac{S_k - k\mu}{\sigma k^{1/2}} < \alpha\right\} \to \Phi(\alpha)$$

where $\Phi(x)$ is the normal distribution function. Using (4.1), we can reformulate (4.2) as a limit theorem for N_r. If we let k and r both tend to infinity so that [2]

$$(4.3) \qquad r - k\mu \sim \alpha\sigma k^{1/2},$$

then

$$(4.4) \qquad Pr\{N_r \geq k\} \to \Phi(\alpha).$$

To write this relation in a more familiar form we introduce the reduced variable N_r^* defined by

$$(4.5) \qquad N_r^* = \frac{N_r - r/\mu}{\sigma r^{1/2}\mu^{-3/2}}.$$

The inequality $N_r \geq k$ is now the same as

$$(4.6) \qquad N_r^* \geq \frac{k - r/\mu}{\sigma r^{1/2}\mu^{-3/2}}.$$

Using (4.3), the right side is seen to be asymptotically equal to $-\alpha$, and hence (4.4) can be written in the form

$$(4.7) \qquad Pr\{N_r^* \geq -\alpha\} \to \Phi(\alpha)$$

[2] In this book the sign $\sim$ indicates that the *ratio* of the two sides tends to unity.

or

(4.8) $$Pr\{N_r^* < -\alpha\} \to 1 - \Phi(\alpha).$$

We have thus proved the

Theorem. *If the recurrent event $\mathcal{E}$ is certain and its recurrence times have finite mean μ and finite variance σ^2, then the number S_k of trials up to the kth occurrence of $\mathcal{E}$ and the number N_r of occurrences of $\mathcal{E}$ in the first r trials are asymptotically normally distributed as indicated by the relations* (4.2) *and* (4.8).

The limit theorem (4.8) makes it plausible that

(4.9) $$E(N_r) \sim \frac{r}{\mu}, \quad \mathrm{Var}(N_r) \sim \frac{r\sigma^2}{\mu^3}$$

but an exact proof requires additional arguments. (Cf. problems 7 and 11.) Note, incidentally, that (4.8) is an example of the central limit theorem applied to a sequence of variables other than sums of independent variables.

The usefulness of the above theorem will be illustrated in Chapter 13, where it will be applied to the theory of runs. In deriving it, we have assumed that the recurrence times have a finite variance σ^2. This restriction at first appears mild, but it turns out that the most interesting recurrence times in various physical processes have infinite mean and variance. In this case formulas (4.2) and (4.8) become meaningless and must be replaced by more general limit theorems.[3] In the following section we discuss a typical example which will reveal many surprising features. It will be seen that the types of fluctuations in the cases of finite and infinite mean recurrence times are not at all similar.

5. Fluctuations in the Coin-tossing Game; the Arc Sine Law

We now discuss two interesting theorems which have important conceptual and statistical implications. They will reveal the fact that widespread beliefs concerning chance fluctuations are based on misconceptions and may lead to dangerous fallacies. We shall here discuss only the coin-tossing game, but our results are typical of a wide class of fluctuation phenomena, and the two limit theorems are valid under much more general conditions.

Suppose that in a coin-tossing game with unit stakes Peter bets on heads and Paul on tails. Denote Peter's accumulated gain at the nth

---

[3] W. Feller, Fluctuation Theory of Recurrent Events, *Transactions of the American Mathematical Society*, vol. 67, pp. 98–119 (1949).

trial by S_n, so that Paul's accumulated gain is $-S_n$. If n is odd, then one of the two gains is positive and the other negative, so that one of the players is in the lead. If n is even, then it is possible that $S_n = 0$, in which case we speak of a tie. The frequency of such ties is so small that they play no role in the following considerations. However, a constant reference to ties renders the exposition clumsy, and it is therefore desirable to eliminate them from our language. Accordingly, in the case of a tie we agree to say that Peter leads if he has led in the preceding trial. In other words, *Peter leads if either* $S_n > 0$ *or* $S_n = 0$ *but* $S_{n-1} > 0$. With this convention, periods of Peter's lead will alternate with periods of Paul's lead, and *the lead passes from one player to the other whenever S_n changes sign.* Note that, if Peter leads at an odd trial, he automatically leads at the following trial.

To fix ideas, let the coin be tossed $n = 20$ times. The number r of trials at which Peter leads may be any even number between 0 and 20.

TABLE 1

DISTRIBUTION OF LEADS IN 20 TRIALS

r	a_r	b_r
0	0.1762	0.3524
2	.0927	.5379
4	.0736	.6851
6	.0655	.8160
8	.0617	.9394
10	.0606	
12	.0617	
14	.0655	
16	.0736	
18	.0927	
20	.1762	

A coin is tossed twenty times; a_r is the probability that Peter leads in exactly r trials; b_r is the probability that the less fortunate player leads in at most r trials.

One feels intuitively that 10 ought to be the most probable number, but this is not so. On the contrary, Table 1 shows that 10 is the least probable number and that the extreme tails $r = 0$ and $r = 20$ are most probable. In more than one-third of all cases we must expect that one of the two players will remain in the lead throughout the game.

This result is very surprising, but a similar statement is true for an arbitrary number of trials. If Peter leads in r out of n trials, we call r/n *the fraction of time during which Peter is in the lead.* Table 2 illustrates the general situation for a large number of trials. It is supposed that a coin is tossed once per second for a total of 1 year, and the fractions of time during which the two players are in the lead are denoted by Z and $1 - Z$. For definiteness let Z be the smaller of the two fractions so that Z lies in the interval from 0 to $\frac{1}{2}$. Table 2 gives the values x such that the relation $Z \leq x$ has probabilities 0.9, 0.8, etc. The following features are noteworthy.

The probability that the less fortunate player leads for more than 5 months is less than 1/10, and the probability that he leads for a total of less than 35 days exceeds 0.4. Turning to the significance levels commonly used by statisticians, we note that in one out of twenty cases the less fortunate player will lead for less than $13\frac{1}{2}$ hours, and in one out of a hundred cases for a total of 32 minutes. It is difficult to imagine a player who would find himself in the red for 364 out of 365 days without complaining, and yet this should occur in about one

TABLE 2

ILLUSTRATING THE ARC SINE LAW

p	t_p
0.9	153.95 days
.8	126.10 days
.7	99.65 days
.6	75.23 days
.5	53.45 days
.4	34.85 days
.3	19.89 days
.2	8.93 days
.1	2.24 days
.05	13.5 hours
.02	2.16 hours
.01	32.4 minutes

A coin is tossed once per second for a total of 365 days; let Z be the fraction of time during which the less fortunate player is in the lead. Then t_p is the interval such that the event $Z < t_p$ has probability p, approximately.

out of sixteen cases. Few people would believe that a true coin will lead to preposterous sequences where S_n does not change sign in millions of trials, and yet this should occur rather frequently. We have here the simplest example of the behavior of chance fluctuations in processes depending on *cumulative* chance effects. Processes of this kind are usual in economical and sociological phenomena, and our findings should serve as a warning to those who easily discover "obvious" secular trends or deviations from average norms.

We now give the formulas from which the two tables are constructed but postpone the proofs to the next section.

Theorem 1.[4] *The probability that Peter leads in $2r$ out of $2n$ trials is*

(5.1) $$p_{2r,2n} = \binom{2r}{r}\binom{2n-2r}{n-r}2^{-2n}.$$

Here $r/n = Z$ is the fraction of time during which Peter leads. The probability that $Z \leq k/n$ is obviously $p_{0,2n} + p_{2,2n} + \cdots + p_{2k,2n}$. From theorem 1 we shall derive

Theorem 2. The Arc Sine Law.[5] *Let n be large, and let Z be the fraction of time during which Peter leads. Then the probability that $Z < t$ is given approximately by*

(5.2) $$f(t) = \frac{2}{\pi}\arcsin t^{1/2},$$

with the error tending to zero as $n \to \infty$.

Here t may be any fixed number in the interval $(0, 1)$. The fraction of time during which the less fortunate player leads lies necessarily in the interval $(0, \tfrac{1}{2})$, and the corresponding probabilities are obtained by doubling $f(t)$. Figure 7 illustrates the arc sine law; Table 2 was computed from it by converting the fraction t to days. For a surprising counterpart to the arc sine law see problem 4.

The following theorem throws light on the same situation from another point of view. The explanation of the arc sine law lies in the fact that an enormous number of trials are frequently required before the numbers of heads and tails equalize. This statement is rendered more precise by the following theorem, in which $\Phi(x)$ is the normal distribution function.

---

[4] K. L. Chung and W. Feller, On Fluctuations in Coin-tossing, *Proceedings of the National Academy of Sciences*, vol. 35 (1949), pp. 605–608. This theorem was suggested by the discovery of E. Sparre Andersen that (5.1) gives the probability of exactly $2r$ changes of sign in the sequence of the first $2n$ partial sums $X_1 + \cdots + X_{2k}$ where the X_k are mutually independent random variables with a common continuous distribution (cf. On the Number of Positive Sums of Random Variables, *Skandinavisk Aktuarietidskrift*, 1950). The formula is *not* true for general discrete variables. Our derivation leads also to a formula for $p_{k,2n+1}$.

[5] Paul Lévy [Sur certains processus stochastiques homogènes, *Compositio Mathematica*, vol. 7 (1939), pp. 283–339] found the arc sine law for certain continuous diffusion processes and referred to the connection with the coin-tossing game. A general arc sine law for the number of positive partial sums in a sequence of mutually independent random variables was proved by P. Erdös and M. Kac, On the Number of Positive Sums of Independent Random Variables, *Bulletin of the American Mathematical Society*, vol. 53 (1947), pp. 1011–1020.

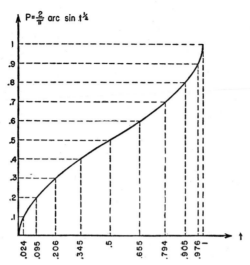

Figure 7. *The Arc Sine Law.*

Theorem 3.[6] *Consider n independent coin-tossing games each of which is continued until for the first time the accumulated numbers of heads and tails are equal. Let $Y_1, Y_2, \cdots, Y_k$ be the durations of these n games and*

(5.3) $$\bar{Y}_k = \frac{Y_1 + \cdots + Y_k}{k}$$

their average. Then for every fixed positive α we have approximately

(5.4) $$Pr\{\bar{Y}_k \leq k\alpha\} \approx 2\{1 - \Phi(\alpha^{-\frac{1}{2}})\},$$

and the difference of the two sides tends to zero as $n \to \infty$.

The surprising feature of this theorem is the following. In the language of the theory of observations $Y_1, \cdots, Y_k$ are k independent measurements on the same physical quantity [namely, the recurrence time of the return to equilibrium as discussed in examples (1.*a*) and (3.*b*)]. The standard theories would lead us to expect that the average $\bar{Y}_k$ will, with increasing k, settle down to a "true" value, which would be called the mean value of the recurrence time. In our case the mean recurrence time is *infinite* [example (3.*b*)], and therefore $\bar{Y}_k$ increases

---

[6] In the form of a limit theorem this theorem is due to P. Lévy (cf. the reference in footnote 5). The present derivation leads to excellent numerical approximations even when $n = 1$. Theorem 3 is a special case of a class of limit theorems which replace the central limit theorem in the case where the recurrence times have no finite variance. (Cf. the reference in footnote 3.)

over all bounds. The surprising feature is that $\bar{Y}_k$ has a probable order of magnitude k, so that the sum $Y_1 + \cdots + Y_k$ increases like k^2. In fact, the value α for which the right side of (5.4) equals 1/2 is $\alpha = 2.198\ldots$. We conclude that each of the two relations

TABLE 3
Probability Distributions of the Averages $\bar{Y}_k$ of Theorem 3

p	a_1	a_{10}	a_{100}
0.5	3	23	221
.4	5	37	365
.3	7	69	675
.2	15	157	1,559
.1	63	635	6,337
.05	255	2,545	25,441
.01	6,359	63,593	635,930

a_k is the number of trials for which $\Pr\{\bar{Y}_k > a_k\}$ comes nearest to the probability p given in the first column.

TABLE 4
Distribution of the First Return to Equilibrium Together with Approximation (5.4)

n	$\Pr\{Y_1 < 2n\} =$ $f_2 + f_4 + \cdots + f_{2n}$	$2\{1 - \Phi((2n)^{-1/2})\}$
1	0.5	0.4795
2	.625	.6170
3	.6875	.6831
4	.7266	.7237
5	.7539	.7518
6	.7744	.7728
7	.7905	.7893
8	.8036	.8026
9	.8145	.8137
10	.8238	.8231

$\bar{Y}_k < 2.198k$ and $\bar{Y}_k > 2.198k$ has a probability near 1/2. In 10 observations the *average* duration of the game is as likely as not to exceed 22; in 100 observations the same average is equally likely to fall short of or exceed 220; and in 1000 observations the average has about even chances to exceed 2200. This phenomenon contrasts sharply with the familiar theory of the "stability" of the means of a large number of "good" measurements. The unorthodox behavior of our averages $\bar{Y}_k$ is due to the fact that in n measurements of our recurrence times one or two are likely to be overwhelmingly large as compared with all others. This fact reveals the danger in the practice of many experimentalists of "throwing away" excessively large observations. With this frequently advocated method the physicist is bound to obtain spurious finite values for recurrence times even though they have infinite means. Note also that the St. Petersburg game discussed in Chapter 10 is closely related to the present considerations. Fortunately we are able to replace the classical paradoxes by limit theorems which can be checked experimentally.

*6. Proof of the Theorems of Section 5

The following proofs are of methodological interest as they depend on double generating functions. Similar ideas are used in the more advanced theory, but an understanding of them is not required for the present volume.

Proof of Theorem 1. By $p_{2r,2n}$ we denoted the probability that Peter leads in exactly $2r$ out of the first $2n$ trials. For $n = 0$ this probability is not defined, but it is convenient to put $p_{0,0} = 1$. Similarly, we put $p_{2r,2n} = 0$ if $r > n$. Peter is betting on heads.

Consider first the case $0 < r < n$. The event "Peter leads in $2r$ and Paul leads in $2n - 2r$ trials" can occur in the following mutually exclusive ways. (1) The first trial results in tails; the first equalization of the numbers of heads and tails occurs at the trial number $2k$ ($k = 1, 2, \cdots, n$). Then Paul leads in the first $2k$ trials, and therefore Peter leads in exactly $2r$ out of the last $2n - 2k$ trials. (2) The first trial results in heads, and the first equalization occurs at the trial number $2k$ ($k = 1, \cdots, r$). Then Peter leads in the first $2k$ trials and in exactly $2r - 2k$ out of the last $2n - 2k$ trials.

We now compute the corresponding probabilities. The probability f_{2k} that the first equalization occurs at the trial number $2k$ is given by formula (3.19); the corresponding generating function is given by (3.18). Whenever the numbers of heads and tails are equal, the game starts anew, and therefore the probability of the event (1) is $\frac{1}{2} \cdot f_{2k} \cdot p_{2r,2n-2k}$. Similarly, the probability of the event (2) is $\frac{1}{2} \cdot f_{2k} \cdot p_{2r-2k,2n-2k}$. Hence

$$(6.1) \qquad p_{2r,2n} = \frac{1}{2} \sum_{k=1}^{n} f_{2k} p_{2r,2n-2k} + \frac{1}{2} \sum_{k=1}^{n} f_{2k} p_{2r-2k,2n-2k}.$$

* Starred sections may be omitted at first reading.

This argument is not valid when $r = 0$ or $r = n$, since then the enumeration of the possibilities mentioned under (1) and (2) is not complete.[7] The events that Peter never leads and that he leads all the time can occur also if no equalization of heads and tails occurs during the first $2n$ trials. Consequently, for $r = 0$ and $r = n$ we have to add to the right side of (6.1) the quantity

$$(6.2) \qquad q_{2n} = \frac{1}{2}(f_{2n+2} + f_{2n+4} + f_{2n+6} + \ldots).$$

We now introduce the generating functions

$$(6.3) \qquad P_{2n}(s) = \sum_{r=0}^{n} p_{2r,2n} s^{2r}.$$

Multiply (6.1) by s^{2r} and add over all r, remembering that for $r = 0$ and $r = n$ the right side should be increased by the quantity (6.2). We get

$$(6.4) \quad P_{2n}(s)$$
$$= \frac{1}{2} \sum_{k=1}^{n} f_{2k} P_{2n-2k}(s) + \frac{1}{2} \sum_{k=1}^{n} f_{2k} s^{2k} P_{2n-2k}(s) + (1 + s^{2n}) q_{2n}.$$

To solve this equation we introduce the generating function of the sequence $P_{2n}(s)$, namely,

$$(6.5) \qquad P(s, t) = \sum_{n=0}^{\infty} P_{2n}(s) t^{2n}.$$

Multiply (6.4) by t^{2n} and add over all n. To the left we get $P(s, t)$. The first sum on the right is the convolution of the sequences $\{f_k\}$ (in which all odd terms vanish) and $P_{2n}(s)$, so that the generating function of this sum is the product of $P(s, t)$ and the generating function of $\{f_k\}$ [the latter is the function $F(t)$ defined in (3.18)]. The same argument applies to the second sum except that the sequence $\{f_k\}$ is now replaced by $\{f_k s^k\}$, the generating function of which is clearly $F(st)$. Finally, the generating function of the sequence q_{2n} is

[7] If $r = 0$ the event (2) cannot occur, but the terms $p_{2r-2k,2n-2k}$ vanish automatically. Similarly, for $r = n$ we have $p_{2r,2n-2k} = 0$, so that in either case there remains only one sum in (6.1).

PROOF OF THEOREMS OF SECTION 5

(6.6)
$$\sum_{n=0}^{\infty} q_{2n}t^{2n} = \frac{1}{2}\sum_{n=0}^{\infty} t^{2n}(f_{2n+2} + f_{2n+4} + \cdots)$$
$$= \frac{1}{2}\sum_{k=1}^{\infty} f_{2k}(1+t^2+\cdots+t^{2k-2})$$
$$= \frac{1}{2}\sum_{k=1}^{\infty} f_{2k}\frac{1-t^{2k}}{1-t^2} = \frac{1}{2}\frac{1-F(t)}{1-t^2}$$
$$= \frac{1}{2}(1-t^2)^{-1/2}.$$

The generating function of the sequence $\{s^{2n}q_{2n}\}$ is obtained from the last formula on replacing t by st. Hence

(6.7) $\quad 2P(s,t) = \{F(t) + F(st)\}P(s,t) + (1-t^2)^{-1/2} + (1-s^2t^2)^{-1/2}.$

Now $1 - F(t) = (1-t^2)^{1/2}$ and therefore

(6.8) $\qquad\qquad P(s,t) = (1-t^2)^{-1/2}(1-s^2t^2)^{-1/2}.$

The required probability $p_{2r,2n}$ is the coefficient of $s^{2r}t^{2n}$ in $P(s,t)$. Using the binomial expansion, we find

(6.9) $\qquad\qquad p_{2r,2n} = \binom{-1/2}{r}\binom{-1/2}{n-r}(-1)^n$

and this formula reduces to (5.1) [cf. formula (9.2) of Chapter 2]. This proves theorem 1.

Proof of the Arc Sine Law. To evaluate the probability (5.1) we reduce the binomial coefficients to factorials and use Stirling's formula. We get (as we did in Chapter 7)

(6.10) $\qquad\qquad p_{2r,2n} \sim \dfrac{1}{\pi r^{1/2}(n-r)^{1/2}},$

where the ratio of the two sides tends to unity as $r \to \infty$ and $n - r \to \infty$.

Now let $0 < \alpha < \beta < 1$ be fixed, and let us evaluate the probability $x(\alpha, \beta)$ that the fraction of time r/n during which Peter leads lies between α and β. For fixed n this probability is obtained by adding $p_{2r,2n}$ for all r in the interval $\alpha n < r < \beta n$. Hence

(6.11) $\qquad x(\alpha, \beta) \sim \pi^{-1} \sum \dfrac{1}{\{(r/n)(1-r/n)\}^{1/2}} \cdot n^{-1}$

the sum extending over all r in the interval $(\alpha n, \beta n)$. This sum is a Riemann sum approximating the integral

$$(6.12) \qquad \pi^{-1} \int_\alpha^\beta \frac{dx}{\{x(1-x)\}^{1/2}} = 2\pi^{-1}\{\arc\sin \beta^{1/2} - \arc\sin \alpha^{1/2}\}.$$

We cannot apply this formula to $\alpha = 0$ since (6.10) is not correct for small r. However, it follows from (6.12) that

$$x(0, t) = 2\pi^{-1} \arc\sin t^{1/2} + C,$$

where C is a constant. For reasons of symmetry we have obviously $x(0, 1/2) = 1/2$, and hence $C = 0$. This proves (5.2).

Proof of Theorem 3. We introduce the probability distribution of the sum $Y_1 + \cdots + Y_n$

$$(6.13) \qquad q_{k,2n} = Pr\{Y_1 + \cdots + Y_k = 2n\}.$$

Since each of the variables Y_ν has the generating function $F(s)$ given by (3.18), the generating function of their sum is

$$(6.14) \qquad F^k(s) = \{1 - (1-s^2)^{1/2}\}^k.$$

To get $q_{k,2n}$ we have to find the coefficient of s^{2n}. Our first aim is to prove that [8]

$$(6.15) \qquad q_{k,2n} = \frac{1}{2^{2n-k}} \binom{2n-k}{n} \frac{k}{2n-k}.$$

From the obvious relation $F^k(s) = 2F^{k-1}(s) - s^2 F^{k-2}(s)$ we get the recursion equations

$$(6.16) \qquad q_{k,2n} = 2q_{k-1,2n} - q_{k-2,2n-2}.$$

This formula enables us to calculate all $q_{k,2n}$ if the $q_{1,2n}$ and $q_{2,2n}$ are known. Now $q_{1,2n}$ is the probability that the first return to equilibrium occurs at the trial number $2n$. Therefore $q_{1,2n} = f_{2n}$ with f_{2n} defined in (3.19). This formula checks with (6.15) for $k = 1$. Next, $q_{2,2n}$ is the coefficient of s^{2n} in $F^2(s) = 2F(s) - s^2$, so that $q_{2,2} = 0$ and $q_{2,2n} = 2q_{1,2n}$ for $n > 1$. Again this formula checks with (6.15). To prove that (6.15) holds for all k it suffices therefore to show that the quantities (6.15) satisfy the recurrence relations (6.16), and this is easily verified. This proves (6.15).

[8] Note that (6.15) gives an explicit formula for the probability that the kth return to the equilibrium occurs at the $2n$th trial.

We can evaluate (6.15), using the normal approximation to the binomial distribution. If $n \to \infty$ and $k(2n-k)^{-1/2}$ remains bounded, then

$$(6.17) \qquad \binom{2n-k}{n} 2^{-2n+k} \sim \left\{\frac{1}{2}\pi(2n-k)\right\}^{-1/2} e^{-k^2/(2(2n-k))},$$

and therefore

$$(6.18) \qquad q_{k,2n} \sim \left(\frac{2}{\pi}\right)^{1/2} \frac{k}{(2n-k)^{3/2}} e^{-k^2/2(2n-k)}.$$

Here the degree of the approximation increases rapidly as $n \to \infty$.

Using (6.18), we can evaluate the probability

$$(6.19) \qquad x_\alpha = Pr\{\overline{Y}_k > k\alpha\} = \sum_{2n > \alpha k^2} q_{k,2n},$$

the summation extending over all n exceeding $\alpha k^2/2$. When k is large and $n > \alpha k^2/2$, we may in the limit replace $2n - k$ by $2n$ and get from (6.18) the approximation

$$(6.20) \qquad x_\alpha \sim \frac{1}{2\pi^{1/2}} \sum_{n > \alpha k^2/2} \frac{1}{(n/k^2)^{3/2}} e^{-k^2/4n} \cdot \frac{1}{k^2}.$$

On the right we have a Riemann sum (with $\Delta x = k^{-2}$) approximating the integral

$$(6.21) \qquad \frac{1}{2\pi^{1/2}} \int_{\alpha/2}^{\infty} t^{-3/2} e^{-1/(4t)} \, dt,$$

which by the change of variables $2t = x^{-2}$ is transformed into $2\Phi(\alpha^{-1/2}) - 1$. This verifies (5.4). We have proved only that the difference of the two sides in (5.4) tends to zero as $k \to \infty$. Actually the approximation is excellent even when k is small. (It can be further improved by keeping the terms $2n - k$ instead of replacing them by $2n$. This changes the limits in the integral only slightly.) Table 4 gives the distribution of Y_1 together with the approximation (5.4) when α is an even integer. It will be noticed that even the first term gives a surprisingly good approximation.

*7. Proof of Theorem 3 of Section 3

In section 3 we have omitted the proof of theorem 3. This theorem can be formulated either as a "Tauberian" theorem on power series or in an elementary way as follows. *Given a sequence* $\{f_n\}$ *such that*

---

* Starred sections may be omitted at first reading.

$f_0 = 0$, $f_n \geq 0$, $\Sigma f_n = 1$, and that the greatest common divisor of those n for which $f_n > 0$ is one. Let $u_0 = 1$ and define u_n for $n \geq 1$ by

(7.1) $$u_n = f_1 u_{n-1} + f_2 u_{n-2} + \cdots + f_n u_0$$

[this is (3.3)]. Then $u_n \to 1/\mu$, where $\mu = \Sigma n f_n$ (and $u_n \to 0$ if $\Sigma n f_n$ diverges).

For the proof put

(7.2) $$r_n = f_{n+1} + f_{n+2} + \cdots,$$

so that [by formula (1.8) of Chapter 11]

(7.3) $$\mu = \Sigma r_n.$$

From (7.2) we get $r_0 = 1$, $f_1 = r_0 - r_1$, $f_2 = r_1 - r_2$, etc. Substituting these values into (7.1) we find that $r_0 u_n + r_1 u_{n-1} + \cdots + r_n u_0 = r_0 u_{n-1} + r_1 u_{n-2} + \cdots + r_{n-1} u_0$. If the left side is called A_n, then the right side is A_{n-1}, and our equation states that all A_n are equal. Now $A_0 = r_0 u_0 = 1$, and hence $A_n = 1$ for all n. Thus we have for every n

(7.4) $$r_0 u_n + r_1 u_{n-1} + \cdots + r_n u_0 = 1.$$

From (7.1) it follows by induction that $u_n \leq 1$. Hence there exists a number $\lambda = \limsup u_n$ such that for any $\epsilon > 0$ and all sufficiently large n we have $u_n < \lambda + \epsilon$, while there exists some sequence $n_1, n_2, n_3, \ldots$ such that $u_{n_\nu} \to \lambda$. Choose an integer $j > 0$ such that $f_j > 0$. We claim that $u_{n_\nu - j} \to \lambda$. If this were not so, we could find arbitrarily large subscripts n such that simultaneously

(7.5) $$u_n > \lambda - \epsilon, \quad u_{n-j} < \lambda' < \lambda.$$

Now let N be so large that $r_N < \epsilon$. Since $u_k \leq 1$, we have then from (7.1) for $n > N$

(7.6) $$u_n \leq f_0 u_n + f_1 u_{n-1} + \cdots + f_N u_{n-N} + \epsilon.$$

For sufficiently large n each u_k on the right side is less than $\lambda + \epsilon$, and $u_{n-j} < \lambda'$. Hence

(7.7) $$\begin{aligned} u_n &< (f_0 + f_1 + \cdots + f_{j-1} + f_{j+1} + \cdots + f_N)(\lambda + \epsilon) \\ &\quad + f_j \lambda' + \epsilon \\ &\leq (1 - f_j)(\lambda + \epsilon) + f_j \lambda' + \epsilon \\ &< \lambda + 2\epsilon - f_j(\lambda - \lambda'). \end{aligned}$$

If we choose ϵ so small that $f_j(\lambda - \lambda') > 3\epsilon$, then the last inequality contradicts the first one in (7.5), so that the assumption $\lambda' < \lambda$ is impossible.

This proves that, whenever $u_{n_\nu} \to \lambda$, also $u_{n_\nu - j} \to \lambda$. Repeating the argument, we see: *if $f_j > 0$ and $u_{n_\nu} \to \lambda = \limsup u_n$, then also*

$$u_{n_\nu - j} \to \lambda, \quad u_{n_\nu - 2j} \to \lambda, \quad u_{n_\nu - 3j} \to \lambda, \quad \text{etc.}$$

For simplicity let us first consider the case where $f_1 > 0$. Then we can take $j = 1$ and conclude that $u_{n_\nu - k} \to \lambda$ for every fixed k. From (7.4) we find for $n = n_\nu$

(7.8) $$1 \geq r_0 u_{n_\nu} + r_1 u_{n_\nu - 1} + \cdots + r_N u_{n_\nu - N}.$$

For fixed N every $u_{n_\nu - k} \to \lambda$, so that $1 \geq \lambda(r_0 + r_1 + \cdots + r_N)$. Since N is arbitrary, we conclude that $1 \geq \lambda\mu$ or $\lambda \leq 1/\mu$.

Next, let $\gamma = \liminf u_n$. The same argument shows that, for every sequence n_ν for which $u_{n_\nu} \to \gamma$, also $u_{n_\nu - k} \to \gamma$. If N is large enough that $r_N < \epsilon$, then from (7.4)

(7.9) $$1 \leq r_0 u_{n_\nu} + \cdots + r_N u_{n_\nu - N} + \epsilon;$$

herein $u_{n_\nu - k} \to \gamma$ so that $1 \leq (r_0 + \cdots + r_N)\gamma + \epsilon$ and hence $\mu\gamma \geq 1$. However, by definition, $\gamma \leq \lambda$. Therefore $\gamma = \lambda = 1/\mu$, as was to be proved.

There remains the case where $f_1 = 0$. Consider then the collection of all integers j for which $f_j > 0$. Among them we can find a finite collection $a, b, c, \cdots, m$ whose greatest common divisor is 1. We know that, when $u_{n_\nu} \to \lambda$, also $u_{n_\nu - xa} \to \lambda$, $u_{n_\nu - yb} \to \lambda$, etc., for every fixed $x > 0, y > 0, \cdots, w > 0$. Hence also $u_{n_\nu - xa - yb - \cdots - wm} \to \lambda$. In other words, if an integer k is of the form $k = xa + yb + \cdots + wm$ with positive integers $x, y, \cdots, w$, then $u_{n_\nu - k} \to \lambda$. Now it is known from elementary number theory that *every* integer k exceeding the product $abc \cdots m$ can be written in this form. This means that for $k > abc \cdots m$ we have $u_{n_\nu - k} \to \lambda$. To get the inequality (7.8) it suffices now to apply (7.4) to $n = n_\nu + ab \cdots m$. The remaining part of the proof requires no change.

8. Problems for Solution

1. Suppose that $F(s)$ is a polynomial. Prove for this case all theorems of section 3, using the partial fraction method of Chapter 11, section 7.

2. Let r coins be tossed repeatedly and let $\mathcal{E}$ be the recurrent event that for each of the r coins the accumulated numbers of heads and tails are equal. Is $\mathcal{E}$ certain or uncertain? For the smallest r for which $\mathcal{E}$ is uncertain, estimate the probability that $\mathcal{E}$ ever occurs.

3. Suppose that Peter and Paul toss a coin for unit stakes and let S_n be Peter's accumulated gain at the conclusion of the nth trial. Prove the following

Theorem. The probability that S_n becomes negative for the first time at the trial number $n = 2k + 1$ equals the probability f_{2k} that $S_n = 0$ for the first time at the trial number $2k$.

Hint: Use generating functions, in particular (3.18).

4. The following theorem accentuates the surprising features of the arc sine law.

In the coin-tossing game of section 5 let Z_n be the fraction of time during which Peter leads. Suppose that at the $2n$th trial the accumulated numbers of heads and tails are equal. *Under this hypothesis the conditional probability that $Z_n = r/n$ equals $1/n$, that is, all possible fractions are equally probable.*

Hint: Follow the proof of theorem 1. The present proof is actually simpler. (Cf. the first paper quoted in footnote 4.)

5. Derive theorem 1 of section 5 from the preceding problem.

6. Let $\mathcal{E}$ be a certain aperiodic recurrent event. Assume that the recurrence time has finite mean μ and variance σ^2. Put $q_n = f_{n+1} + f_{n+2} + \ldots$ and $r_n = q_{n+1} + q_{n+2} + \ldots$. Show that the generating functions $Q(s)$ and $R(s)$ converge for $s = 1$. Prove that

$$(8.1) \qquad \sum \left(u_n - \frac{1}{\mu}\right) s^n = \frac{R(s)}{\mu Q(s)}$$

and hence that

$$(8.2) \qquad \sum \left(u_n - \frac{1}{\mu}\right) = \frac{\sigma^2 - \mu + \mu^2}{2\mu^2}.$$

7. Let $\mathcal{E}$ be a certain recurrent event and N_r the number of occurrences of $\mathcal{E}$ in r trials. Prove that $E(N_r) = u_1 + \cdots + u_r$ and hence

$$(8.3) \qquad E(N_r) \sim \frac{r}{\mu}.$$

8. *Continuation.* Prove that

$$E(N_r^2) = u_1 + \cdots + u_r + 2 \sum_{j=1}^{r-1} u_j(u_1 + \cdots + u_{r-j})$$

and hence that $E(N_r^2)$ is the coefficient of s^r in

$$(8.4) \qquad \frac{F^2(s) + F(s)}{(1-s)\{1-F(s)\}^2}.$$

9. Let $q_{k,n} = Pr\{N_k = n\}$. Show that $q_{k,n}$ is the coefficient of s^k in

$$(8.5) \qquad F^n(s) \frac{\{1 - F(s)\}}{1-s}.$$

Deduce that $E(N_r)$ and $E(N_r^2)$ are the coefficients of s^r in

$$(8.6) \qquad \frac{F(s)}{(1-s)\{1-F(s)\}}$$

and (8.4), respectively.

10. Using the notations of problem 6, show that

(8.7) $$\frac{F(s)}{(1-s)\{1-F(s)\}} = -\frac{1}{1-s} + \frac{1}{\mu(1-s)^2} + \frac{R(s)}{\mu\{1-F(s)\}}.$$

Hence, using the last problem, conclude that

(8.8) $$E(N_r) = \frac{r+1}{\mu} + \frac{\sigma^2 - \mu - \mu^2}{2\mu^2} + \epsilon_r$$

with $\epsilon_r \to 0$.

11. *Continuation.* Using a similar argument, show that

(8.9) $$E(N_r^2) = \frac{(r+2)(r+1)}{\mu^2} + \frac{2\sigma^2 - 2\mu - \mu^2}{\mu^3} r + \alpha_r,$$

where α_r remains bounded. Hence

(8.10) $$\mathrm{Var}(N_r) \sim \frac{\sigma^2}{\mu^3} r.$$

*CHAPTER 13

RECURRENT EVENTS:
APPLICATIONS TO
RUNS AND RENEWAL THEORY

The power of the method of recurrent events is well illustrated by the theory of success runs in Bernoulli trials. This is a classical topic which has found applications in modern statistics. The new approach simplifies the theory and (cf. section 2) applies also to more general recurrent patterns which present special difficulties to older methods. This unification is the main advantage of the methods developed in Chapter 12.

Section 3 illustrates the method of partial fractions by a detailed numerical analysis of an asymptotic formula derived in section 1.

The remainder of the chapter is devoted to the renewal theory, a topic to which an extraordinary number of papers has been devoted. The main theorems are derived in section 4 as simple corollaries to theorems on recurrent events. Applications of renewal theory are described in section 5.

1. Success Runs

The term "success run of length r" has been defined in several ways [cf. also example (1.b) in Chapter 12]. It is largely a matter of convention and convenience whether a sequence of three consecutive successes is said to contain 0, 1, or 2 runs of length 2, and for different purposes different definitions have been adopted. However, if we are to use the theory of recurrent events, then the notion of runs of length r must be defined so that we start from scratch every time a run is completed. This means adopting the following definition. *A sequence of n letters S and F contains as many runs of length r as there are non-overlapping uninterrupted successions of exactly r letters S. In a sequence of Bernoulli trials a run of length r occurs at the nth trial, if the nth trial adds a new run to the sequence.* Thus in $SSS \mid SF \mid SSS \mid SSS$ we have three runs of length 3, and they occur at trials number 3, 8, 11; there are five runs of length 2, and they occur at trials number 2, 4, 7,

---

* Starred chapters treat special topics and may be omitted at first reading.

9, 11. This definition has the advantage of a very considerable simplification of the theory. It amounts to counting successions of at least r consecutive successes except that $2r$ consecutive successes count twice, etc.

In the sequel r is a fixed integer and $\mathcal{E}$ means the occurrence of a success run of length r. We shall suppose that the trials are Bernoulli trials, and $\mathcal{E}$ is then a recurrent event. As before, u_n denotes the *probability of $\mathcal{E}$ at the nth trial* ($u_0 = 1$), and f_n *the probability that the first run of length r occurs at the nth trial*, so that $\{f_j\}$ is the distribution of the recurrence times of $\mathcal{E}$.

The probability that the r trials number n, $n-1$, $n-2$, $\cdots$, $n-r+1$ result in success is obviously p^r. In this case $\mathcal{E}$ occurs at one among these r trials; the probability that $\mathcal{E}$ occurs at the trial number $n-k$ ($k = 0, 1, \cdots, r-1$) and that the following k trials result in success is $u_{n-k}p^k$. Since these r possibilities are mutually exclusive, we get the following recurrence relation:[1]

$$(1.1) \qquad u_n + u_{n-1}p + \cdots + u_{n-r+1}p^{r-1} = p^r.$$

This equation is valid for $n \geq r$. Clearly

$$(1.2) \qquad u_1 = u_2 = \cdots = u_{r-1} = 0, \qquad u_0 = 1.$$

Now multiply (1.1) by s^n and sum over $n = r, r+1, r+2, \cdots$. In view of (1.2) we get on the left side

$$(1.3) \qquad \{U(s) - 1\}(1 + ps + p^2s^2 + \cdots + p^{r-1}s^{r-1})$$

and on the right side $p^r(s^r + s^{r+1} + \cdots)$. Summing the two geometric series, we find

$$(1.4) \qquad \{U(s) - 1\} \cdot \frac{1 - (ps)^r}{1 - ps} = \frac{p^r s^r}{1 - s}$$

or

$$(1.5) \qquad U(s) = \frac{1 - s + qp^r s^{r+1}}{(1-s)(1 - p^r s^r)}.$$

Using (3.5) of Chapter 12, we get for *the generating function of the recurrence times*

$$(1.6) \qquad F(s) = \frac{p^r s^r (1 - ps)}{1 - s + qp^r s^{r+1}}.$$

---

[1] The classical approach consists in deriving a recurrence relation for f_n. This method is more complicated and does not apply to, say, runs of either kind or patterns like *SSFFSS*, to which our method applies without change [cf. example (2.c)]. Our method depends on the modified definition of runs which makes them recurrent events.

It is seen that $F(1) = 1$ so that runs of length r are *certain* recurrent events. Clearing (1.6) of the denominator and differentiating implicitly, it is easy to calculate $F'(1)$ and $F''(1)$. We find that *the mean and variance of the recurrence times of success runs of length r are*

$$(1.7) \qquad \mu = \frac{1 - p^r}{qp^r}, \qquad \sigma^2 = \frac{1}{(qp^r)^2} - \frac{2r + 1}{qp^r} - \frac{p}{q^2},$$

respectively. The theorem of section 4 of Chapter 12 implies that for large n *the number N_n of runs of length r produced in n trials is approximately normally distributed*, that is, for fixed $\alpha < \beta$ the probability that

$$(1.8) \qquad \frac{n}{\mu} + \frac{\alpha \sigma n^{1/2}}{\mu^{3/2}} < N_n < \frac{n}{\mu} + \frac{\beta \sigma n^{1/2}}{\mu^{3/2}}$$

tends to $\Phi(\beta) - \Phi(\alpha)$. This fact was first proved by von Mises, but without the theory of Chapter 12 the proof requires rather lengthy calculations. Table 1 gives a few typical means of recurrence times.

TABLE 1

MEAN RECURRENCE TIMES FOR SUCCESS RUNS IF TRIALS ARE PERFORMED AT THE RATE OF ONE PER SECOND

Length of Run	$p = 0.6$	$p = 0.5$ (Coins)	$p = 1/6$ (Dice)
$r = 5$	30.7 seconds	1 minute	2.6 hours
10	6.9 minutes	34.1 minutes	28.0 months
15	1.5 hours	18.2 hours	18,098 years
20	19 hours	24.3 days	140.7 million years

The method of partial fractions (Chapter 11, section 7) permits us to derive an excellent approximation to the probabilities f_n that the first run occurs at the nth trial. The denominator in (1.6) can be factored:

$$(1.9) \quad 1 - s + qp^r s^{r+1}$$
$$= (1 - ps)\{1 - qs(1 + ps + \cdots + (ps)^{r-1})\}.$$

The first factor has the root $s = 1/p$, which is also a root of the numerator and should be cancelled. We have, therefore, only to consider the roots of the second factor, that is,

$$(1.10) \qquad qs(1 + ps + \cdots + p^{r-1}s^{r-1}) = 1.$$

Obviously this equation has a unique *positive root* $s = x$. For every real or imaginary number with $|s| \leq x$ we have

(1.11) $\quad |qs(1 + ps + \cdots + p^{r-1}s^{r-1})|$
$$\leq qx(1 + px + \cdots + p^{r-1}x^{r-1}) = 1$$

where the equality sign is possible only if all terms on the left have the same argument, that is, if $s = x$. Hence x is smaller in absolute value than any other root of the denominator in (1.6). We can, therefore, apply formula (7.9) of Chapter 11 with $s_1 = x$. The coefficient ρ_1 is easily computed from (7.5) with $U(s) = p^r s^r (1 - ps)$ and $V(s) = 1 - s + qp^r s^{r+1}$. We find, using that $V(x) = 0$,

(1.12) $$f_n \sim \frac{(x-1)(1-px)}{(r+1-rx)q} \cdot \frac{1}{x^{n+1}}.$$

The probability that n trials result in no run is $q_n = f_{n+1} + f_{n+2} + f_{n+3} + \ldots$. We get for it from (1.12), summing a geometric series [2]

(1.13) $$q_n \sim \frac{1-px}{(r+1-rx)q} \cdot \frac{1}{x^{n+1}}.$$

We have thus found that, if x is the unique positive root of (1.10), *the probability that n trials produce no success run of length r is, asymptotically, given by* (1.13). Table 2 shows that the formula gives surprisingly good approximations even for very small n, and the goodness of approximation increases rapidly with n. We have here a typical example of the power of the method of generating functions combined with the method of partial fractions. In section 3 we shall see that the calculation of x is rather easy and that estimates of the error in (1.13) can be obtained.

TABLE 2

PROBABILITY OF HAVING NO SUCCESS RUN OF LENGTH $r = 2$ IN n TRIALS WITH $p = \frac{1}{2}$

n	q_n Exact	Approximation (1.13)	Error	Bound for Error According to (3.10)
2	0.75	0.76631	0.0163	0.0835
3	.625	.61996	.0080	.042
4	.500	.50156	.0016	.021
5	.40625	.40577	.0005	.010

[2] A special case of (1.13) is equation (7.13) of Chapter 11, example (7a).

2. More General Patterns

Our method is applicable to more general problems which have been considered as considerably deeper than the theory of runs. Two special examples of recurrent patterns were treated in Chapter 11, examples (7.a) and (7.b). Here we treat two more interesting problems whose solution is surprisingly simple.

Examples. (a) *Runs of Either Kind.* Let $\mathcal{E}$ stand for "*either a success run of length r or a failure run of length ρ.*" We are dealing with *two* recurrent events $\mathcal{E}_1$ and $\mathcal{E}_2$, where $\mathcal{E}_1$ stands for "success run of length r" and $\mathcal{E}_2$ for "failure run of length ρ" and $\mathcal{E}$ means "either $\mathcal{E}_1$ or $\mathcal{E}_2$." To $\mathcal{E}_1$ there corresponds the generating function (1.5) which will now be denoted by $U_1(s)$. The corresponding generating function $U_2(s)$ for $\mathcal{E}_2$ is obtained from (1.5) by interchanging p and q, and replacing r by ρ. Now the probability u_n that $\mathcal{E}$ occurs at the nth trial is the sum of the corresponding probabilities for $\mathcal{E}_1$ and $\mathcal{E}_2$. An exception occurs for $n = 0$, where $u_0 = 1$. It follows that

$$(2.1) \qquad U(s) = U_1(s) + U_2(s) - 1.$$

The generating function $F(s)$ of the recurrence times of $\mathcal{E}$ is again obtained from (3.5) of Chapter 12. We have $F(s) = 1 - U^{-1}(s)$ or

$$(2.2) \quad F(s) = \frac{(1 - ps)p^r s^r (1 - q^\rho s^\rho) + (1 - qs)q^\rho s^\rho (1 - p^r s^r)}{1 - s + qp^r s^{r+1} + pq^\rho s^{\rho+1} - p^r q^\rho s^{r+\rho}}.$$

The *mean recurrence time* follows by differentiation

$$(2.3) \qquad \mu = \frac{(1 - p^r)(1 - q^\rho)}{qp^r + pq^\rho - p^r q^\rho}.$$

As $\rho \to \infty$, this expression tends to the mean recurrence time of success runs as given in (1.7).

(b) In Chapter 8, section 1, we calculated the probability x that a *success run of length r occurs before a failure run of length ρ*. Suppose now that this event is made the object of a bet and that we are interested in the probability distribution of the *duration of the game*. More precisely, we define two recurrent events $\mathcal{E}_1$ and $\mathcal{E}_2$ as in example (a). Let $x_n =$ probability that $\mathcal{E}_1$ occurs for the first time at the nth trial and no $\mathcal{E}_2$ precedes it; $f_n =$ probability that $\mathcal{E}_1$ occurs for the first time at the nth trial (with no condition on $\mathcal{E}_2$). Define y_n and g_n as x_n and f_n, respectively, but with $\mathcal{E}_1$ and $\mathcal{E}_2$ interchanged.

The generating function for f_n is given in (1.6), and $G(s)$ is obtained by interchanging p and q, and replacing r by ρ. For x_n and y_n we have the obvious recurrence relations

(2.4) $\quad x_n = f_n - (y_1 f_{n-1} + y_2 f_{n-2} + \cdots + y_{n-1} f_1)$

$\quad\quad\quad y_n = g_n - (x_1 g_{n-1} + x_2 g_{n-2} + \cdots + x_{n-1} g_1).$

These equations are of the convolution type, and for the corresponding generating functions we have, therefore,

(2.5) $\quad\quad\quad X(s) = F(s) - Y(s)F(s)$

$\quad\quad\quad\quad\quad Y(s) = G(s) - X(s)G(s).$

These are two linear equations in the unknowns $X(s)$ and $Y(s)$. We get

(2.6) $\quad X(s) = \dfrac{F(s)\{1 - G(s)\}}{1 - F(s)G(s)}, \quad Y(s) = \dfrac{G(s)\{1 - F(s)\}}{1 - F(s)G(s)}.$

Expressions for x_n and y_n can again be obtained by the method of partial fractions. For $s = 1$ we get $X(1) = \Sigma x_n = x$, the probability of $\mathcal{E}_1$ occurring before $\mathcal{E}_2$. Now both numerator and denominator vanish, for $s = 1$, and $X(1)$ is obtained from L'Hospital's rule differentiating numerator and denominator: $X(1) = G'(1)/\{F'(1) + G'(1)\}$. Using the values $F'(1) = (1 - p^r)/qp^r$ and $G'(1) = (1 - q^\rho)/pq^\rho$ from (1.7), we find $X(1)$ as given in equation (1.3) of Chapter 8.

(c) Consider the recurrent event defined by the pattern *SSFFSS*. Repeating the argument of section 1, we find easily that

(2.7) $\quad\quad\quad\quad\quad p^4 q^2 = u_n + p^2 q^2 u_{n-4}.$

From this relation we get the generating function and from it the mean recurrence time $\mu = p^{-4}q^{-2} + p^{-2}$. For $p = q = 1/2$ we find $\mu = 68$, whereas the mean recurrence time for a head run of length 6 in coin tossing is 126. This shows that there is an essential difference between head runs and other patterns of the same length.

3. Numerical Estimates [3]

In this section we propose to show that the positive root x of the algebraic equation (1.10) can be calculated very easily; in addition, we shall derive estimates of the error involved in the asymptotic expansions (1.12) and (1.13).

We remember that equation (1.10) has been obtained from (1.9), so that all roots of (1.10) are also roots of the equation

(3.1) $\quad\quad\quad\quad\quad 1 - s + qp^r s^{r+1} = 0.$

---

[3] This section contains a simplification and slight improvement of the results in Chapter 5 of Uspensky's book, *Introduction to Mathematical Probability*, New York, 1937. The results will not be used in the sequel.

This equation has also the extraneous root $s = 1/p$, but x is the only other positive root (unless $s = 1/p$ is a double root, in which case $x = 1/p$).

For abbreviation we put $1 + qp^r s^{r+1} = f(s)$, so that (3.1) becomes $f(s) = s$. We know that the graph of $y = f(s)$ crosses the bisector $y = s$ at $s = x$ and at $s = 1/p$. Between these two roots the graph of $f(s)$ lies *under* the bisector. At the smaller root we must, therefore, have $f'(s) < 1$, while at the larger root $f'(s) > 1$. Now $f'(1/p) = (r+1)q$. For definiteness we shall assume that

$$(3.2) \qquad q > \frac{1}{r+1}, \quad \text{that is,} \quad p < \frac{r}{r+1}.$$

In this case $x < 1/p$.

We first show that the root x can be found by a simple iteration process. Note that, if $s < 1/p$ and $f(s) > s$, then necessarily $x > s$, so that such a value of s serves as a lower bound for x. However, since $f(s)$ is increasing, the inequality $s < x$ implies $f(s) < f(x) = x$. On the other hand, $f(s) > s$, so that the value $f(s)$ is a better approximation to x than s. In this way we can improve any given approximation. For example, start with the approximation $s_0 = 1$. Then the value $s_1 = f(s_0)$, or

$$(3.3) \qquad s_1 = 1 + qp^r,$$

is a better approximation, and certainly $s_1 < x$. Continuing in this way, we get a sequence s_k defined by $s_{k+1} = f(s_k)$; clearly the s_k increase monotonically, and their limit is a root of the equation $s = f(s)$. Since $s_k < x$ for all k and x is the smallest root, we see that *the root x can be obtained as the limit of the sequence s_k.*

In practice the approximation (3.3) is usually sufficient. In fact, the second approximation s_2 differs from s_1 only by terms of the order of magnitude $rq^2 p^{2r}$, which are usually very small. For an estimate of the error involved in (3.3) it suffices to find an upper bound for x, that is, any number s such that $f(s) < s$. Now put

$$(3.4) \qquad s = 1 + qp^r(1 + \epsilon).$$

Since $1 + t < e^t$ for all positive t, we find that

$$(3.5) \qquad f(s) = 1 + qp^r s^{r+1} < 1 + qp^r e^{(r+1)qp^r(1+\epsilon)}.$$

In order for (3.4) to give an upper bound to x it suffices therefore that the right side of (3.5) be smaller than the right side of (3.4), or that

$$(3.6) \qquad (r+1)qp^r(1+\epsilon) < \log(1+\epsilon).$$

This inequality is certainly satisfied if we choose $\epsilon > 0$ so that the left side becomes $\epsilon/2 < \log(1 + \epsilon)$. Simple arithmetic shows that (3.6) holds if we put

$$(3.7) \qquad \epsilon = \frac{2(r+1)qp^r}{1 - 2(r+1)qp^r}.$$

This quantity is positive, since for fixed r the maximum value of qp^r is attained for $p = r/(r+1)$ so that $(r+1)qp^r < 1/2$. The *true value of the root x lies between the approximations* (3.3) *and* (3.4) *with ϵ given by* (3.7). For $p = 0.5$, $r = 10$ we have $s_1 = 1 + 2^{-11} = 1.000488$, while (3.4) leads to the upper bound 1.000493. In general, the approximation s_1 is better than the rough estimate (3.4).

We now turn to an estimate of the error involved in (1.13). We know from equation (7.8) of Chapter 11 that the exact expression of p_n is the sum of r terms with each root of (1.10) contributing one term. The contribution A_k of the root s_k is, of course, given by the right hand in (1.13) when x is replaced by s_k. Thus

$$(3.8) \qquad A_k = \frac{1 - ps_k}{(r+1-rs_k)q} \frac{1}{s_k^{n+1}}.$$

Now we found that the r roots s_k of (1.10) are also roots of (3.1). Let $z = |s_k|$ be the absolute value of a given non-positive root s_k. Then $z = |1 + qp^r s^{r+1}| \leq 1 + qp^r z^{r+1}$. Hence $f(z) > z$, so that z cannot lie between x and $1/p$. Since we know that $z > x$, this means that all roots of (1.10) except x are larger in absolute value than $1/p$. To estimate the first fraction in (3.8) we observe that $\left|\frac{1}{p} - s_k\right| \div \left|\frac{r+1}{r} - s_k\right|$ is the ratio of the distances of the root s_k to the points $1/p$ and $(r+1)/r$ of the real axis. In view of (3.2) the point $(r+1)/r$ lies inside the circle $|s| = 1/p$, and the point s_k outside. The maximum value of the ratio for all s with $|s| \geq 1/p$ is attained for $s = -1/p$. Using (3.2), it follows then that

$$(3.9) \qquad |A_k| < \frac{2p^{n+1}}{(r+1+r/p)q} < \frac{2p^{n+2}}{rq(1+p)}.$$

Therefore *the error committed in* (1.13) *is smaller in absolute value than*

$$(3.10) \qquad \frac{2(r-1)p^{n+2}}{rq(1+p)}.$$

For numerical values see Table 2 of section 1.

4. The Renewal Equation

We now proceed to extend the basic results of Chapter 12, section 3, to a more general case which frequently occurs in applications. Examples will be given in the next section; here we proceed in a formal manner.

Suppose *there are given two sequences of bounded non-negative numbers* $\{a_n\}$ *and* $\{b_n\}$ $(n = 0, 1, 2, \ldots)$ *with* $a_0 \neq 1$. *A third sequence* u_n *is defined by the recursive relations*

$$(4.1) \qquad u_n = b_n + (a_0 u_n + a_1 u_{n-1} + \cdots + a_n u_0).$$

In the notations of Chapter 11, section 2, this equation becomes

$$(4.2) \qquad \{u_n\} = \{b_n\} + \{a_n\}*\{u_n\}.$$

Solving (4.1) successively, we get

$$u_0 = b_0/(1 - a_0), \; u_1 = (b_1 + a_1 u_0)/(1 - a_0), \text{ etc.,}$$

so that no problem as to the existence of a unique solution $\{u_n\}$ arises. We are interested in the behavior of $\{u_n\}$ as $n \to \infty$, a problem to which a great number of papers (mostly of controversial nature) have been devoted.

Note that (4.1) reduces to (3.3) of Chapter 12 if we put $b_n = 0$, $a_n = f_n$ for $n > 0$, and $a_0 = 0$, and $b_0 = 1$. Thus the renewal equation (4.1) contains the fundamental equation of recurrent events as special case. However, we can derive all properties of the renewal equation from the results on recurrent events. Once more we pass to generating functions. Since the coefficients a_n and b_n are bounded, $A(s) = \Sigma a_n s^n$ and $B(s) = \Sigma b_n s^n$ converge at least for $|s| < 1$. The domain of convergence of $U(s) = \Sigma u_n s^n$ remains to be investigated. From (4.2) we get $U(s) = B(s) + A(s)U(s)$ or

$$(4.3) \qquad U(s) = \frac{B(s)}{1 - A(s)}.$$

For $B(s) \equiv 1$ this reduces to (3.5) of Chapter 12. The essential difference is that now $\{a_n\}$ is not necessarily the distribution of a recurrence time, so that $A(s)$ can be larger as well as smaller than 1. *We shall investigate only the case where*

$$(4.4) \qquad B(1) = \Sigma b_k$$

is finite.

We shall say that we have the *periodic case if there exists an integer* $\lambda > 1$, *such that all* a_k *except, perhaps,* a_λ, $a_{2\lambda}$, $a_{3\lambda}$, ... *vanish.* Then $A(s)$ is a power series in s^λ. The largest integer λ with the said property is called the *period*.

Theorem 1. *If* $\{a_k\}$ *is a probability distribution (that is, if* $A(1) = 1$), *then, except in the periodic case,*

$$(4.5) \qquad u_n \to \frac{B(1)}{\mu},$$

where $\mu = A'(1) = \Sigma n a_n$. *In particular,* $u_n \to 0$ *if* $\Sigma n a_n$ *diverges.*

Proof. Let v_n be the coefficient of s^n in $1/\{1 - A(s)\}$. We know from theorem 3 of Chapter 12 that $v_n \to 1/\mu$. Now

$$(4.6) \qquad u_n = v_n b_0 + v_{n-1} b_1 + \cdots + v_0 b_n.$$

For every fixed k the term $v_{n-k} b_k$ tends to b_k/μ as $n \to \infty$. Moreover, the v_n are bounded. It follows that, for N sufficiently large, u_n differs arbitrarily little from

$$(4.7) \qquad u_n' = v_n b_0 + v_{n-1} b_1 + \cdots + v_{n-N} b_N,$$

and $u_n' \to (b_0 + \cdots + b_N)/\mu$ which in turn differs arbitrarily little from $B(1)/\mu$.

Theorem 2. *If* $\Sigma a_k < 1$, *then* $u_n \to 0$ *so fast that the series* $\Sigma u_n = B(1)/\{1 - A(1)\}$ *converges.*

Proof. From (4.3) we have the expansion

$$U(s) = B(s)\{1 + A(s) + A^2(s) + A^3(s) \cdots\}$$

valid for all s for which $|A(s)| < 1$ and therefore at least for $|s| \leq 1$. The right side can be rearranged into a power series in s, converging at least for $|s| \leq 1$, and this proves the theorem.

Theorem 3. *If* $\Sigma a_k > 1$, *then there exists a unique positive root* $x < 1$ *of the equation* $A(s) = 1$. *Then in the non-periodic case*

$$(4.8) \qquad u_n x^n \to \frac{B(x)}{A'(x)},$$

so that u_n is of the order of magnitude x^{-n} and hence increases exponentially. The value $A'(x)$ is finite, since $A(s)$ converges for $|s| < 1$ and $x < 1$.

Proof. It suffices to apply theorem 1 to the coefficients of the power series in

(4.9) $$U(xs) = \frac{B(xs)}{1 - A(xs)},$$

namely, $\{u_n x^n\}$, $\{b_n x^n\}$, and $\{a_n x^n\}$.

There remains the periodic case where $A(s) = \Sigma a_{n\lambda} s^{n\lambda}$ is a power series in s^λ. In this case the coefficients u_n exhibit a certain periodicity, and we divide them into groups of equal phase, $\{u_0, u_\lambda, u_{2\lambda}, u_{3\lambda}, \ldots\}$, $\{u_1, u_{\lambda+1}, u_{2\lambda+1}, u_{3\lambda+1}, \ldots\}$, $\cdots$, $\{u_{\lambda-1}, u_{2\lambda-1}, u_{3\lambda-1}, \ldots\}$. It is obvious from (4.3) that the coefficients $u_{n\lambda}$ depend only on $b_0, b_\lambda, b_{2\lambda}, \ldots$ but not on the b_k with k not divisible by λ. This leads us to represent $U(s)$ and $B(s)$ as the sum of λ power series in s^λ

(4.10) $$U(s) = U_0(s) + sU_1(s) + \cdots + s^{\lambda-1}U_{\lambda-1}(s)$$

$$B(s) = B_0(s) + sB_1(s) + \cdots + s^{\lambda-1}B_{\lambda-1}(s),$$

where

(4.11) $$U_j(s) = \sum_{n=0}^{\infty} u_{n\lambda+j} s^n, \quad B_j(s) = \sum_{n=0}^{\infty} b_{n\lambda+j} s^n.$$

Then, from (4.3) for $j = 0, 1, \cdots, \lambda - 1$,

(4.12) $$U_j(s) = \frac{B_j(s)}{1 - A(s)}.$$

Here all functions are power series in s^λ, and the preceding theorems apply after the change of variables $s^\lambda = t$. In theorem 1 we had $\mu = A'(1)$, the differentiation being with respect to s. In the present case we must of course replace μ by

(4.13) $$\frac{dA(1)}{dt} = \frac{\mu}{\lambda}.$$

We have thus

Theorem 4. In the periodic case with period λ the sequence $\{u_n\}$ is asymptotically periodic; if $A(1) = 1$, each of the λ subsequences $\{u_{n\lambda+j}\}$ has a limit

(4.14) $$\lim_{n \to \infty} u_{n\lambda+j} = \frac{\lambda B_j(1)}{\mu}$$

where $B_j(1) = b_j + b_{\lambda+j} + b_{2\lambda+j} + b_{3\lambda+j} + \cdots$.

5. Examples

(a) *Self-renewing Aggregates.* We consider an electric bulb, fuse, or other piece of equipment with a finite life span. As soon as the piece fails, it is replaced by a new piece of like kind, which in due time is replaced by a third piece, and so on. We assume that the life span is a random variable which ranges only over multiples of a unit time interval (year, day, or second). Each time unit then represents a trial with possible outcomes "replacement" and "no replacement." The successive replacements may be treated as recurrent events. If a_n is the probability that a new piece will serve for exactly n time units, then $\{a_n\}$ is the distribution of the recurrence times. If it is certain that the life span is finite, then $\Sigma a_n = 1$ and the recurrent event is certain. Usually it is known that the life span cannot exceed a fixed number m, and in this case the generating function $A(s)$ is a polynomial of a degree not exceeding m.

So far we have considered only a single piece and the line of its direct descendants (replacements). We now turn to the study of a whole population of pieces of like kind, each of which is replaced as soon as it expires, so that the replacements keep the population size constant. The term "self-renewing aggregates" is used to describe this situation (although the meaning of the prefix "self" is not apparent). Of course, in special cases a self-renewing aggregate may well consist of people.

Suppose that the initial population (at time 0) contains exactly v_k elements (pieces or people) of age k, so that $N = \Sigma v_k$ is the original population size. Each of these N elements originates a line of descendants, and at any time n there is a certain probability that a replacement is required in this line. The sum of these probabilities for all N elements is the *expected number* u_n of replacements at time n. In the present case it is natural to put $u_0 = 0$. Without loss of generality we may assume that the life span is necessarily positive, that is, $a_0 = 0$.

To see that the renewal equation applies, note that the replacements at time n are of two kinds. First, an element to be replaced may have been installed as a new element at time j $(1 \leq j < n)$. At time n such an element has age $n - j$, and the expected number of replacements of this kind is clearly $u_j a_{n-j}$. Adding over all possible j, we get $u_1 a_{n-1} + u_2 a_{n-2} + \cdots + u_{n-1} a_1$ (remember that we have $u_0 = a_0 = 0$). This accounts for the second term on the right in the renewal equation (4.1). Second, the element to be replaced at time n may be of the initial population and have been of age $k \geq 0$ at time 0 (that is, of age $n + k$ at the moment of expiration). The probability of a life span exceeding k is $r_k = a_{k+1} + a_{k+2} + \cdots$. To find the probability that an element of age k will expire after exactly n years we require the conditional probability of a life span $k + n$ on the hypothesis that the life span exceeds k. This conditional probability is obviously a_{n+k}/r_k. Now at time 0 there were v_k elements of age k. The expected number of those among them which expire at time n is $v_k a_{n+k}/r_k$. The expected number of elements of the original population which expire at time $n \geq 1$ is therefore

$$(5.1) \qquad b_n = \sum_{k=0}^{\infty} \frac{v_k a_{n+k}}{r_k}.$$

Adding this term to the one previously found, we see that for $n \geq 1$ u_n satisfies the renewal equation (4.1). To have (4.1) true for all n we put $b_0 = 0$. We can then apply our theorems [4] to find the asymptotic behavior of the renewal coefficients u_n.

It is also easy to obtain the age distribution at time n. Let $v_k(n)$ be the expected number of elements of age k at time n [which implies $v_k(0) = v_k$]. Then clearly

$$(5.2) \qquad v_k(n) = u_{n-k} r_k \quad \text{if} \quad k < n$$

$$v_k(n) = \frac{v_{k-n} r_k}{r_{k-n}} \quad \text{if} \quad k \geq n.$$

In the non-periodic case we know that $u_n \to B(1)/\mu = N/\mu$ as $n \to \infty$, and it follows from (5.2) that $v_k(n) \to N r_k/\mu$. Hence, in the non-periodic case, there is a stable *limiting age distribution:* in the limit the expected number of elements of age k is $N r_k/\mu$, where N is the (constant) population size, and $\mu = \Sigma r_k$ the mean duration of life (if $\mu = \infty$, then the population ages indefinitely). The basic fact is that the *limiting age distribution is independent of the initial age distribution* and depends only on the mortality distribution $\{a_n\}$ (cf. problems 9 and 10).

As a numerical illustration consider a population of 1000 elements with the age distribution $v_0 = 500$, $v_1 = 320$, $v_2 = 74$, $v_3 = 100$, $v_4 = 6$. Let the life-time distribution be given by $a_1 = 0.20$, $a_2 = 0.43$, $a_3 = 0.17$, $a_4 = 0.17$, $a_5 = 0.03$ (no element can attain an age exceeding 5). Here $r_0 = 1$, $r_1 = 0.80$, $r_2 = 0.37$, $r_3 = 0.20$, $r_4 = 0.03$, $r_5 = 0$, whence $v_0/r_0 = 500$, $v_1/r_1 = 400$, $v_2/r_2 = 200$, $v_3/r_3 = 500$,

---

[4] For further properties cf. W. Feller, Fluctuation Theory of Recurrent Events, *Transactions of the American Mathematical Society,* vol. 67 (1949), pp. 98–119. A great many papers treat special cases. For numerical applications cf. for example N. R. Campbell, The Replacement of Perishable Members of a Continually Operating System, *Supplement, Journal of the Royal Statistical Society,* vol. 7 (1941), pp. 110–130, or D. J. Bishop, The Renewal of Aircraft, *Ministry of Aircraft Production, Aeronautical Research Committee, Report and Memoranda* no. 1907 (6342), 1942.

$v_4/r_4 = 200$. Hence from (5.1) $b_1 = 397$, $b_2 = 332$, $b_3 = 159$, $b_4 = 97$, $b_5 = 15$, and from (4.3) we get

$$(5.3) \quad U(s) = s \frac{397 + 332s + 159s^2 + 97s^3 + 15s^4}{1 - 0.20s - 0.43s^2 - 0.17s^3 - 0.17s^4 - 0.03s^5}.$$

The roots of the denominator are $s_1 = 1$, $s_2 = -5/3$, $s_3 = -5$, $s_4 = 2i$, $s_5 = -2i$, and hence

$$U(s) = \frac{1250s}{3(1-s)} - \frac{972s}{61(1+3s/5)} + \frac{38s}{87(1+s/5)} - \frac{78{,}225s^2 + 22{,}125s}{5307(1+s^2/4)}.$$

Expanding each term into a geometric series, we get exact expressions for u_n.

The age distributions $v_k(n)$ are given in the following table.

k	n								
	0	1	2	3	4	5	6	7	∞
0	500	397	411.4	412	423.8	414.3	417.0	416.0	416.7
1	320	400	317.6	329.1	329.6	339.0	331.5	333.6	333.3
2	74	148	185	146.9	152.2	152.4	156.8	153.3	154.2
3	100	40	80	100	79.4	82.3	82.4	84.8	83.3
4	6	15	6	12	15	11.9	12.3	12.4	12.5

(b) *Population Theory.* This theory is analogous to renewal theory except that the population size is variable and that female births play the role of replacements. The essential novelty is that a mother can have zero, one, or more daughters, so that lines may become extinct or branch. We now define a_n as the probability of a newborn female to survive and at age n give birth to a female child (the dependence on the number and ages of previous children is neglected). Then Σa_n is the expected number of daughters, and hence all three possibilities $\Sigma a_n < 1$, $\Sigma a_n = 1$, $\Sigma a_n > 1$ are now possible. The preceding argument applies with this obvious modification.

6. Problems for Solution

1. Find an approximation to the probability that in 10,000 tossings of a coin the number of head runs of length 3 will lie between 700 and 730.

2. In a sequence of tossings of a coin let $\mathcal{E}$ stand for the pattern HTH. Let r_n be the probability that $\mathcal{E}$ does not occur in n trials. Verify the relation (7.15) and hence (7.14) of Chapter 11.

3. In example (2.b) show that the expected duration of the game is $\mu_1\mu_2/(\mu_1 + \mu_2)$, where μ_1 and μ_2 are the mean recurrence times for success runs of length r and failure runs of length ρ, respectively.

4. The possible outcomes of each trial are A, B, and C; the corresponding probabilities are α, β, γ ($\alpha + \beta + \gamma = 1$). Find the generating function of the probability that in n trials there is no run of length r: (a) of A's, (b) of A's or B's, (c) of any kind.

5. *Continuation.* Find the probability that the first A-run of length r precedes the first B-run of length ρ and terminates at the nth trial. [Note that this problem does *not* reduce to that of example (2.b) with $p = \alpha/(\alpha + \beta)$, $q = \beta/(\alpha + \beta)$.]

6. In a sequence of Bernoulli trials let $q_{k,n}$ be the probability that exactly n success runs of length r occur in k trials. Using problem 9 of Chapter 12 show that the generating function $Q_k(x) = \Sigma q_{k,n} x^n$ is the coefficient of s^k in

$$\frac{1 - p^r s^r}{1 - s + qp^r s^{r+1} - (1 - ps)p^r s^r x}.$$

Show, furthermore, that the root of the denominator which is smallest in absolute value is $s_1 \approx 1 + qp^r(1 - x)$.

7. *Continuation. The Poisson distribution of long runs.*[5] If the number k of trials and the length r of runs both tend to infinity, so that $kqp^r \to \lambda$, then the probability of having exactly n runs of length r tends to $e^{-\lambda}\lambda^n/n!$.

Hint: Using the preceding problem, show that the generating function is asymptotically $\{1 + qp^r(1 - x)\}^{-k} \sim e^{-\lambda(1-x)}$.

The following problems refer to the renewal theory, specifically to example (5.a).

8. *Constancy of the population.* For the quantities (5.2) prove by induction that $\sum_k v_k(n) = N$ for every n.

9. If the mortality distribution is given by $p_k = q^{k-1}p$ (with $p + q = 1$), find u_n and the limiting age distribution, assuming that the original population consists of N elements aged zero.

10. An age distribution is called *stable* if $v_k(n)$ does not depend on n. Show that this is the case if, and only if, $v_k = Cr_k$, where C is a constant.

---

[5] This problem is best solved using the continuity theorem of Chapter 11, section 8. The theorem was proved by different methods by von Mises.

CHAPTER 14

RANDOM WALK AND RUIN PROBLEMS

1. General Orientation

The main part of this chapter is devoted to certain problems connected with Bernoulli trials in which the probabilities of success and failure are p and q, respectively. For simplicity and clarity of language we shall formulate the problems and theorems in terms of two intuitive models.

First, we shall consider the familiar gambler who wins a dollar for each success and loses a dollar for each failure. We shall suppose that the gambler and his adversary own a total of a dollars and start with z and $a - z$ dollars, respectively. The game continues until the gambler's capital either is reduced to zero or has increased to a, that is, until one of the two players is ruined. We are interested in the probability of the gambler's ruin and the probability distribution of the duration of the game. This is the *classical ruin problem*.

Physical applications and analogies suggest another intuitive interpretation. We imagine that the trials are performed at times $t = 1$, 2, 3, ... and interpret their results in terms of the motion of a variable point or *particle* on the x-axis. At time $t = 0$ this particle has the position $x = z$, and at times $t = 1, 2, 3, \ldots$ it moves a unit step to the right or left according to whether the corresponding trial results in success or failure. Thus the position of our particle at time n represents the gambler's capital at the conclusion of the nth trial. The trials terminate when the particle for the first time reaches either $x = 0$ or $x = a$. We say that the particle performs a *random walk*. The limiting positions $x = 0$ and $x = a$ are called *absorbing barriers;* the gambler's ruin is interpreted as *absorption at* $x = 0$. We say that our random walk is *restricted* to the possible positions $x = 0, 1, \cdots, a$; in the absence of absorbing barriers the random walk is called unrestricted. If $p = q = 1/2$, the random walk is called *symmetric*. Physicists use the random-walk model as a crude approximation to one-dimensional diffusion processes and Brownian motion, where a physical particle is exposed to a great number of molecular collisions or shocks which impart to it a random motion. If there is a *drift* to the right, shocks from the left are more probable and we have $p > q$.

So far we have merely described the classical ruin problem in a new terminology. However, the random-walk model leads to new problems which are also suggested by physical analogies. Thus, instead of absorbing barriers we may consider other boundary conditions. For example, we may imagine a reflecting wall at $x = \frac{1}{2}$ with the property that if the particle starts from $x = 1$ and moves to the left, it is reflected at $x = \frac{1}{2}$ and returns to $x = 1$ instead of reaching $x = 0$. In other words, whenever the particle is at the position $x = 1$, then it has probability p to move a unit step to the right and probability q to stay. We describe this condition by referring to a *reflecting barrier at* $x = \frac{1}{2}$. In gambling terminology this corresponds to a convention that whenever the gambler loses his last dollar it is generously replaced by his adversary so that the game can continue. A reflecting barrier at $x = a - \frac{1}{2}$ is defined in a similar way. With two reflecting barriers the random walk never terminates. We may also consider random walks for which one boundary acts as a reflecting, the other as an absorbing barrier. Finally, there exist *elastic barriers*, which are partly absorbing, partly reflecting.

Consider next an unrestricted random walk with the possible positions $x = 0, \pm 1, \pm 2, \ldots$. We may inquire as to the probability that the particle eventually returns to its initial position and, if it does, that the first return occurs at the nth step. This is the problem of *recurrence times* which was solved and discussed at length in Chapter 12 [example (3.a) and section 5]. We found there that the fluctuations of these recurrence times exhibit several unexpected properties and differ from the more familiar type of fluctuation phenomena described by the central limit theorem.

Instead of inquiring as to the return to the initial position we may also ask for the probability that the particle reaches a preassigned position x for the first time at the nth step. This is the problem of *first passage times*, which is related to the ruin problem. In fact, suppose that a gambler starts with an initial capital $z > 0$ and plays against an infinitely rich adversary. (This is the limiting case of the classical ruin problem when $a \to \infty$.) The gambler's capital is represented by a particle performing a random walk, and the gambler is ruined when the particle reaches the position $x = 0$ for the first time. The problem of the *duration* of this game is equivalent to the problem of the first passage time through the origin in an unrestricted random walk starting at z.

Even though we have used various intuitive descriptions, all problems described are obviously concerned with sums of mutually independent random variables $X_1, X_2, \ldots$ which assume the values $+1$

and -1 with probabilities p and q, respectively. Among the many generalizations of random-walk problems we shall here consider only two.

First, instead of Bernoulli trials we may consider arbitrary trials. This means that the gambler's gain X_k at the kth trial is now a random variable with an arbitrary distribution, or, in random-walk terminology, that the particle may change its position in jumps which are not necessarily of magnitude ± 1. We formulate the corresponding *ruin* or *absorption problem* as follows. The particle starts at $z > 0$ and the process ends when for the first time it jumps to a position $x \leq 0$ or $x \geq a$, that is, when for the first time [1] the sum $X_1 + \cdots + X_k$ is either $\leq -z$ or $\geq a - z$. Required are the corresponding probabilities and the probability distribution of the duration of the game. This problem has attracted widespread interest in connection with *sequential sampling*. There the X_k represent certain characteristics of samples or observations. Measurements are taken until a sum $X_1 + \cdots + X_k$ falls outside two preassigned limits (our $-z$ and $a - z$). In the first case the procedure leads to what is technically known as *rejection*, in the second case to *acceptance*. The first sampling procedure of this kind was described by W. Bartky[2]; the general theory was outlined by A. Wald, to whom the above formulation is due.[3] In section 8 the methods of ordinary random walks are adapted to this more general case. However, it is more natural to consider the generalized problem as a special case of Markov chains, and a full treatment is postponed to Chapter 15.[4] It must be understood that all our random walks can be treated as special Markov chains, and that the present chapter serves mostly as an introduction to the next.

A second generalization consists in letting the particle perform a random walk in *two or more dimensions*. For example, in two dimen-

---

[1] For an interpretation in betting language it is, of course, necessary that the gambler's initial capital be at least z plus the maximum possible loss in a single trial; similarly the adversary's initial capital must be at least a plus his maximum single loss.

[2] W. Bartky, Multiple Sampling with Constant Probability, *Annals of Mathematical Statistics*, vol. 14 (1943), pp. 363–377.

[3] A. Wald, On Cumulative Sums of Random Variables, *Annals of Mathematical Statistics*, vol. 15 (1944), pp. 283–296. The methods described in the present book are different from Wald's. Cf. also Wald's book, *Sequential Analysis*, John Wiley & Sons, New York, 1947.

[4] In the theory of sequential sampling it has become usual to consider random walks in which the particle has probability p to move in the direction of the positive x-axis, and probability q to move in the direction of the positive y-axis. Most of the qualitative results in this direction follow from the results of Chapter 15, section 2, example IX.

sions we consider the regular net formed by the lines $x = 0$, ± 1, $\pm 2, \ldots$ and $y = 0, \pm 1, \pm 2, \ldots$. The particle moves in unit steps, but from each position it has the choice of four possible directions. An interesting difference between random walks in two and in three dimensions will be discussed in section 7.

The discussion of the various problems will proceed as follows. In section 2 the probability of the gambler's ruin is derived and various implications of the solution are discussed. In section 3 it is shown how the expected value of the duration of the game can be derived in an elementary way. In sections 4 and 5 we turn to the more delicate problem of the probability distributions of the duration of the game and of first passage times. We use the method of difference equations because of its intrinsic interest and because of its intimate connections with physical diffusion theory. An alternative derivation of the results is outlined in problems 8–13. The random walk with reflecting barriers is not considered in this chapter but will be treated by the more appropriate methods of Markov chains (Chapter 16, section 3; cf. also problem 13).

In section 6 we pass to the limit of a continuous chance process and discuss the connection of random walks with diffusion theory. In section 7 random walks in the plane and space are considered. Finally, section 8 is devoted to the generalized random walk connected with sequential sampling.

2. The Gambler's Ruin

We consider the problem stated at the opening of the present chapter. Let q_z be the probability of the gambler's ultimate[5] ruin, and p_z the probability of his winning. In random-walk terminology q_z and p_z are the probabilities that a particle starting at z will be absorbed at $x = 0$ and $x = a$, respectively. We shall show that $p_z + q_z = 1$, so that we need not consider the possibility of an unending game.

After the first trial the gambler's fortune is either $z - 1$ or $z + 1$, and therefore we must have

(2.1) $$q_z = pq_{z+1} + qq_{z-1}$$

provided $1 < z < a - 1$. For $z = 1$ the first trial may lead to ruin,

[5] Strictly speaking, the probability of ruin is defined in a sample space of infinitely prolonged games. However, we can work with the sample space of n trials. The probability of ruin in less than n trials increases with n and has therefore a limit. We call this *limit* "the probability of ruin." All probabilities in this chapter may be interpreted in this way without referring explicitly to infinite sample spaces (cf. the introduction to Chapter 8).

and (2.1) is to be replaced by $q_1 = pq_2 + q$. Similarly, for $z = a - 1$ the first trial may result in victory, and therefore $q_{a-1} = qq_{a-2}$. To unify our equations we shall define

(2.2) $$q_0 = 1, \quad q_a = 0.$$

With this convention the probability q_z of ruin satisfies (2.1) for $z = 1, 2, \cdots, a - 1$.

Equation (2.1) is a *difference equation*, and (2.2) represents the *boundary conditions* on q_z. We shall derive an explicit expression for q_z by the *method of particular solutions*, which also will be used in more general cases.

Suppose first that $p \neq q$. It is easily verified that the difference equation (2.1) admits of the two particular solutions $q_z = 1$ and $q_z = (q/p)^z$. It follows that for arbitrary constants A and B the sequence

(2.3) $$q_z = A + B \left(\frac{q}{p}\right)^z$$

represents a formal solution of (2.1). We wish to adjust the constants A and B so that the boundary conditions (2.2) will be satisfied. This means that A and B must satisfy the two linear equations $A + B = 1$ and $A + B(q/p)^a = 0$. Thus

(2.4) $$q_z = \frac{(q/p)^a - (q/p)^z}{(q/p)^a - 1}$$

is a formal solution of the difference equation (2.1), satisfying the boundary conditions (2.2). In order to prove that (2.4) is the required probability of ruin it remains to show that the solution is unique. In other words, we have to prove that *all* solutions of (2.1) can be written in the form (2.3). Now, given an arbitrary solution of (2.1), the two constants A and B can be chosen so that (2.3) will agree with it for $z = 0$ and $z = 1$. However, from these two values all other values can be found by substituting in (2.1) successively $z = 1, 2, 3, \ldots$. This means that two solutions which agree for $z = 0$ and $z = 1$ are identical, and hence that every solution is of the form (2.3).

Our argument breaks down if $p = q = 1/2$, for then (2.4) is meaningless. This is due to the fact that in this case the two formal particular solutions $q_z = 1$ and $q_z = (q/p)^z$ are identical. However, we now have a second formal solution in $q_z = z$, and therefore $q_z = A + Bz$ is a solution of (2.1) depending on two constants. In order to satisfy the boundary conditions (2.2) we must put $A = 1$ and $A + Ba = 0$. Hence

(2.5) $$q_z = 1 - \frac{z}{a}.$$

[The same numerical value can be obtained formally from (2.4) by finding the limit as $p \to 1/2$, using L'Hospital's rule.]

We have thus proved that the required *probability of the gambler's ruin is given by* (2.4) *if* $p \neq q$, *and by* (2.5) *if* $p = q = 1/2$. The probability p_z of the gambler's winning the game equals the probability of his adversary's ruin, and is therefore obtained from our formulas on replacing p, q, and z by q, p, and $a - z$, respectively. It is readily seen that $p_z + q_z = 1$, as stated previously.

We can reformulate our result as follows: *Let a gambler with an initial capital z play against an infinitely rich adversary who is always willing to play, while the gambler has the privilege of stopping at his pleasure. The gambler adopts the strategy of playing until he either loses his capital or increases it to a (or a net gain $a - z$). Then q_z is the probability of his losing and $1 - q_z$ the probability of his winning.*

Under this system the gambler's ultimate gain or loss is a random variable G which assumes the values $a - z$ and $-z$ with probabilities $1 - q_z$ and q_z, respectively. The expectation of gain is found to be

(2.6) $$E(G) = a(1 - q_z) - z.$$

Introducing the value q_z from (2.5), it is found that, if $p = q = 1/2$, then $E(G) = 0$. Conversely, it follows from (2.6) that $E(G) = 0$ implies (2.5) and hence that $p = q$. This means that, with the system described, a "fair" game remains fair, and no "unfair" game can be changed into a "fair" one.

From (2.5) we see that in the case $p = q$ a player with initial capital $z = 999$ has a probability $999/1000$ to win a dollar before losing his capital. With $q = 0.6$, $p = 0.4$ the game is unfavorable indeed, but still the probability (2.4) of winning a dollar before losing the capital is about $1/3$. In general, a gambler with a relatively large initial capital z has a reasonable chance to win a small amount $a - z$ before being ruined.[6]

---

[6] A certain man used to visit Monte Carlo year after year and was always successful in recovering the costs of his vacations. He firmly believed in a magic power over chance. Actually his experience is not surprising. Assuming that he started with ten times the ultimate gain, the chances of success are nearly $9/10$. The probability of an unbroken sequence of ten successes is about $(1 - 1/10)^{10} \approx e^{-1} \approx 0.37$. *One* failure would, of course, be blamed on an oversight or momentary indisposition.

Let us now consider the effect of *changing stakes*. If the initial capital of the player and of his adversary are z and $a - z$, respectively, then the probability of the player's ruin is given by (2.4). Suppose now that the unit is changed from a dollar to a half-dollar. This means simply that in (2.4) we must replace z by $2z$ and a by $2a$. The new probability of ruin is therefore

$$(2.7) \qquad q_z^* = \frac{(q/p)^{2a} - (q/p)^{2z}}{(q/p)^{2a} - 1} = q_z \cdot \frac{(q/p)^a + (q/p)^z}{(q/p)^a + 1}.$$

If $q > p$, then the last fraction is greater than unity and hence $q_z^* > q_z$. Hence, *if the stakes are doubled while the initial capitals remain unchanged, then the probability of ruin decreases for the player whose probability of success $p < 1/2$, and increases for the adversary (for whom the game is advantageous since $q > p$).* A similar statement holds true in general when the stakes are increased (not necessarily doubled). Suppose, for example, that our player plays on the unfavorable side ($q > p$) and owns 900 dollars which he is willing to risk in order to win 100 dollars. If he stakes 1 dollar at each trial, the probability of ruin is given by (2.4) with $z = 900$, $a = 1000$. If he stakes 10 dollars at each trial, we must put $z = 90$ and $a = 100$. In general, if k dollars are staked at each trial, we find the probability of ruin from (2.4), replacing z by z/k and a by a/k, and the probability of ruin decreases as k increases. The gambler therefore minimizes the probability of ruin by selecting the stakes as large as is consistent with his goal of gaining an amount fixed in advance. In this sense, the popular *doubling system* is optimal. In fact, suppose a player sets out to win an amount c (which should be reasonably small in comparison with his initial capital). The optimal stake at the first trial is c, for we have just shown that a smaller stake would increase the probability of ruin, and the same is true of a larger stake since the possibility of a larger gain is necessarily compensated by an increased probability of ruin. If the first trial is successful, then his goal is achieved and he leaves the game. Otherwise, he has to recover the loss c and win additional c dollars. His new goal is therefore $2c$, and hence the doubling of the stake.

These results are classical. It has been contended that every "unfair" bet is unreasonable. If this were to be taken seriously, it would mean the end of all insurance business. Actually no theorem of probability suggests that a careful driver who insures against liability at average rates acts unreasonably, but he plays a game which is technically "unfair."

If in our formulas we pass to the limit as $a \to \infty$, we expect to get the *probability of ruin in a game against an infinitely rich adversary.*

With an initial capital z this probability should be 1 if $q \geq p$ and $(q/p)^z$ if $q < p$. Note, however, that the case $a = \infty$ (random walk on a semi-infinite line) is defined on its own merits and not as a limiting case. It will be seen in section 4 that the result of a direct treatment agrees with the described formal passage to the limit.

3. Expected Duration of the Game

The probability distribution of the duration of the game will be deduced in the following sections. However, its expected value can be derived by a much simpler method which is of such wide applicability that it will now be explained at the cost of a slight duplication.

We are still concerned with the classical ruin problem formulated at the beginning of this chapter. We shall assume as known that the duration of the game has a finite expectation D_z. A rigorous proof will be given in the next section.

The argument which led to the difference equation (2.1) and the boundary conditions (2.2) shows directly that the expected duration D_z satisfies the difference equation

$$(3.1) \qquad D_z = pD_{z+1} + qD_{z-1} + 1, \qquad 0 < z < a$$

with the boundary conditions

$$(3.2) \qquad D_0 = 0, \quad D_a = 0.$$

The appearance of the term 1 makes the difference equation (3.1) non-homogeneous. If $p \neq q$, then $D_z = z/(q-p)$ is a formal solution of (3.1). It is readily seen that the difference Δ_z of any two solutions of (3.1) satisfies the homogeneous equations $\Delta_z = p\Delta_{z+1} + q\Delta_{z-1}$, and we know already that all solutions of this equation are of the form $A + B(q/p)^z$. It follows that if $p \neq q$ all solutions of (3.1) are of the form

$$(3.3) \qquad D_z = \frac{z}{q-p} + A + B\left(\frac{q}{p}\right)^z.$$

The values of the constants A and B follow again from the boundary conditions (3.2), according to which we must have $A + B = 0$ and $A + B(q/p)^a = -a/(q-p)$. Solving for A and B, we find

$$(3.4) \qquad D_z = \frac{z}{q-p} - \frac{a}{q-p} \cdot \frac{1 - (q/p)^z}{1 - (q/p)^a}.$$

Again the formula breaks down if $q = p = 1/2$. In this case we must replace $z/(q-p)$ by $-z^2$, which is now a solution of (3.1). It follows that when $p = q = 1/2$ all solutions of (3.1) are of the form $D_z = -z^2 + A + Bz$. The required solution D_z which satisfies the boundary conditions (2.2) is then

(3.5) $$D_z = z(a - z).$$

The expected duration of the game in the classical ruin problem is given by (3.4) or (3.5), according as $p \neq q$ or $p = q = 1/2$.

It should be noted that this duration is considerably longer than one would naively expect. If two players with 500 dollars each toss a coin until one is ruined, the average duration of the game is 250,000 trials. If a gambler has only one dollar and his adversary 1000, the average duration is 1000 trials.

TABLE 1

Illustrating the Classical Ruin Problem

p	q	z	a	Probability of		Expectation of	
				Ruin	Success	Gain	Duration
0.5	0.5	9	10	0.1	0.9	0	9
.5	.5	90	100	.1	.9	0	900
.5	.5	900	1,000	.1	.9	0	90,000
.5	.5	950	1,000	.05	.95	0	47,500
.5	.5	8,000	10,000	.2	.8	0	16,000,000
.45	.55	9	10	.210	.790	-1.1	11
.45	.55	90	100	.866	.134	-76.6	765.6
.45	.55	99	100	.182	.818	-17.2	171.8
.4	.6	90	100	.983	.017	-88.3	441.3
.4	.6	99	100	.333	.667	-32.3	161.7

The initial capital is z. The game terminates with ruin (loss z) or capital a (gain $a - z$).

Most interesting is the passage to the limit as $a \to \infty$, which corresponds to a play against an infinitely rich adversary (cf. the concluding remark to section 2). From (2.4) and (2.5) we concluded that there is probability one of ruin if $q \geq p$, while if $q < p$ (favorable game), the probability of ruin is $(q/p)^z$. If $p = q$ *the duration of the game has infinite expectation.* This is in accordance with the fact discussed in Chapter 12 that, if a coin is tossed until for the first time the number of heads equals the number of tails, the game has infinite expected duration. For an unrestricted symmetric random walk this means that the time until the first return to the initial position has infinite expectation. Our new result states that the first passage time to *any* position (even the adjacent ones) has infinite expectation. We shall see that a similar statement is true in the more refined diffusion theory. The reader is referred to Chapter 12 (section 5) for a discussion of several startling features of recurrence times, in particular the arc sine law.

4. Generating Functions for the Duration of the Game and First Passage Times

We shall use the method of generating functions to study the duration of the game in the classical ruin problem or restricted random walk with absorbing barriers at $x = 0$ and $x = a$. The initial position is z (with $0 < z < a$). Let $u_{z,n}$ denote the probability that the process ends with the nth step at the barrier $x = 0$ (gambler's ruin at the nth trial). After the first step the position is $z + 1$ or $z - 1$, and we conclude that for $1 < z < a - 1$ and $n \geq 1$

$$(4.1) \qquad u_{z,n+1} = pu_{z+1,n} + qu_{z-1,n}.$$

This is a difference equation analogous to (2.1), but depending on the two variables z and n. In analogy with the procedure of section 2 we wish to define boundary values $u_{0,n}$, $u_{a,n}$, and $u_{z,0}$ so that (4.1) becomes valid also for $z = 1$, $z = a - 1$, and $n = 0$. For this purpose we put

$$(4.2) \qquad u_{0,n} = u_{a,n} = 0 \qquad \text{when} \quad n \geq 1$$

and

$$(4.3) \qquad u_{0,0} = 1, \quad u_{z,0} = 0 \qquad \text{when} \quad z > 0.$$

Then (4.1) holds for all z with $0 < z < a$ and all $n \geq 0$.

We now introduce the generating function

$$(4.4) \qquad U_z(s) = \sum_{n=0}^{\infty} u_{z,n} s^n.$$

Multiplying (4.1) by s^{n+1} and adding for $n = 0, 1, 2, \ldots$, we find for $0 < z < a$

$$(4.5) \qquad U_z(s) = psU_{z+1}(s) + qsU_{z-1}(s).$$

Moreover, equations (4.2) and (4.3) lead to the boundary conditions

(4.6) $$U_0(s) = 1, \quad U_a(s) = 0.$$

Equation (4.5) is a difference equation analogous to (2.1), and the boundary conditions (4.6) correspond to (2.2). The novelty lies in the circumstance that the coefficients and the unknown $U_z(s)$ now depend on the variable s, but as far as the difference equation is concerned, s is merely an arbitrary constant. We can again apply the method of section 2 provided we succeed in finding two particular solutions of (4.5). It is natural to inquire whether there exist two solutions $U_z(s)$ of the form $U_z(s) = \lambda^z(s)$. Substituting this expression into (4.5), we find that $\lambda(s)$ must satisfy the quadratic equation

(4.7) $$\lambda(s) = ps\lambda^2(s) + qs,$$

which has the two roots

(4.8) $$\lambda_1(s) = \frac{1 + (1 - 4pqs^2)^{\frac{1}{2}}}{2ps}, \quad \lambda_2(s) = \frac{1 - (1 - 4pqs^2)^{\frac{1}{2}}}{2ps}$$

(we take $0 < s < 1$ and the positive square root).

We have thus found two particular solutions of (4.5) and conclude as in section 2 that for two arbitrary functions $A(s)$ and $B(s)$

(4.9) $$U_z(s) = A(s)\lambda_1^z(s) + B(s)\lambda_2^z(s)$$

is a solution of (4.5). If this solution is to satisfy the boundary conditions (4.6), we must have $A(s) + B(s) = 1$ and $A(s)\lambda_1^a(s) + B(s)\lambda_2^a(s) = 0$. We find in this way

(4.10) $$U_z(s) = \frac{\lambda_1^a(s)\lambda_2^z(s) - \lambda_1^z(s)\lambda_2^a(s)}{\lambda_1^a(s) - \lambda_2^a(s)}.$$

Using the obvious relation $\lambda_1(s)\lambda_2(s) = q/p$, the last formula simplifies to

(4.11) $$U_z(s) = \left(\frac{q}{p}\right)^z \frac{\lambda_1^{a-z}(s) - \lambda_2^{a-z}(s)}{\lambda_1^a(s) - \lambda_2^a(s)}.$$

This is *the required generating function of the probability of ruin at the nth trial (absorption at $x = 0$)*. The corresponding generating function for the probability of absorption at $x = a$ is obtained on replacing p, q, z by q, p and $a - z$, respectively. The generating function of the *duration of game* is, of course, the sum of the two generating functions.

All results of the preceding two sections are contained in formula (4.11). In particular, we have for the probability of ruin

$$(4.12) \qquad q_z = \sum_{n=0}^{\infty} u_{z,n} = U_z(1).$$

Now $1 - 4pq = (p - q)^2$, and from (4.8) we find that when $p \geq q$ we have $\lambda_1(1) = 1$ and $\lambda_2(1) = q/p$, while $\lambda_1(1) = q/p$ and $\lambda_2(1) = 1$ in the case $q \geq p$. Substituting into (4.11), we see that (4.12) reduces to (2.4). Similarly we get the expected duration D_z as given in (3.4) by a simple differentiation. For $p = q$ our expression becomes indeterminate, but the formulas (2.5) and (3.5) follow by a passage to the limit as $s \to 1$, using L'Hospital's rule. In the next section we shall derive from (4.11) an explicit formula for $u_{z,n}$.

Our method applies also when $a = \infty$, which corresponds to the case of a random walk with the single absorbing barrier at $x = 0$ (or playing against an infinitely rich adversary). We have now the sole boundary condition $U_0(s) = 1$. All solutions of (4.5) are of the form (4.9), but since $\lambda_1(s) > 1$ and $\lambda_2(s) < 1$ for $0 < s < 1$, we find that $U_z(s)$ is unbounded unless $A(s) = 0$. Hence the required solution is

$$(4.13) \qquad V_z(s) = \lambda_2^z(s).$$

This is *the generating function of the probability that, starting from $z > 0$, the particle will be absorbed at $x = 0$ exactly at the nth trial* (in the absence of other barriers). It is also the generating function of *the first passage time through $x = 0$ of a free particle* starting at $z > 0$. In particular, for $z = 1$ we find that $\lambda_2(s)$ is the generating function of the first passage time through the neighboring position to the left. The first passage time from z to 0 is the sum of the first passage times from z to $z - 1$, from $z - 1$ to $z - 2$, etc., and is therefore the sum of z mutually independent random variables each having the generating function $\lambda_2(s)$. This explains why $V_z(s)$ is the zth power of a generating function.

Substituting $s = 1$ into (4.13), we find the probability of ruin in the case of an infinitely rich adversary. It is $(q/p)^z$ or 1, according as $q \leq p$ or $q \geq p$.

If $z < 0$, the generating function for the first passage time through the origin is obtained from $\lambda_2^z(s)$ by interchanging p and q. An easy computation shows that this generating function is $\lambda_1^z(s)$.

*5. Explicit Expressions

We shall now derive an explicit formula for $u_{z,n}$ by expanding $U_z(s)$ into partial fractions. Formally, the expression (4.11) for $U_z(s)$ de-

---

* Starred sections treat special topics and may be omitted at first reading.

pends on a square root, but in reality $U_z(s)$ is a rational function. In fact, expanding the expressions (4.8) according to the binomial theorem, we see that the difference $\lambda_1^k(s) - \lambda_2^k(s)$ is a rational function in s multiplied by $(1 - 4pqs^2)^{1/2}$; this root appears as a factor in both the numerator and the denominator of (4.11), and hence $U_z(s)$ is the ratio of two polynomials. The degree of the denominator is $a - 1$ or $a - 2$, according to whether a is odd or even; the degree of the numerator is $a - 1$ or $a - 2$, according to whether $a - z$ is odd or even. In no case can the degree of the numerator exceed the degree of the denominator by more than one. Hence for $n > 1$ we can compute $u_{z,n}$ from formula (7.8) of Chapter 11, provided only that all the roots of the denominator are distinct.

We could calculate the roots of the denominator and the corresponding coefficients ρ_ν directly, but the algebra simplifies if we introduce a new independent variable ϕ by

$$(5.1) \qquad \frac{1}{\cos\phi} = 2(pq)^{1/2}s.$$

From (4.8) we find

$$(5.2) \qquad \lambda_{1,2}(s) = \left(\frac{q}{p}\right)^{1/2}(\cos\phi \pm i\sin\phi) = \left(\frac{q}{p}\right)^{1/2}e^{\pm i\phi},$$

and hence from (4.11)

$$(5.3) \qquad U_z(s) = \left(\frac{q}{p}\right)^{z/2}\frac{\sin(a-z)\phi}{\sin a\phi}.$$

The roots of the denominator are obviously $\phi = 0, \pi/a, 2\pi/a, \ldots$. The corresponding values of s are

$$(5.4) \qquad s_\nu = \frac{1}{2(pq)^{1/2}\cos\nu\pi/a}.$$

We get all possible values for s_ν, putting $\nu = 0, 1, \cdots, a$. However, to $\nu = 0$ and $\nu = a$ there correspond the extraneous values $\phi = 0, \pi$, which are also roots of the numerator in (5.3), and if a is even, no number s_ν corresponds to $\nu = a/2$. Hence, when a is odd, we get all $a - 1$ roots s_ν, putting $\nu = 1, 2, \cdots, a - 1$; when a is even, the value $\nu = a/2$ must be omitted.

We know that

$$(5.5) \qquad \left(\frac{q}{p}\right)^{z/2}\frac{\sin(a-z)\phi}{\sin a\phi} = \frac{\rho_1}{s_1 - s} + \cdots + \frac{\rho_{a-1}}{s_{a-1} - s}.$$

To find ρ_ν multiply both sides by $s_\nu - s$ and let $s \to s_\nu$. We get (putting $\phi_\nu = \pi\nu/a$) as in Chapter 11, formula (7.5),

$$(5.6) \qquad \rho_\nu = -\left(\frac{q}{p}\right)^{z/2} \frac{\sin(a-z)\pi\nu/a}{a \cdot \cos \nu\pi \cdot (d\phi/ds)_{s=s_\nu}}$$

$$= \left(\frac{q}{p}\right)^{z/2} \frac{\sin z\pi\nu/a \cdot \sin \pi\nu/a}{2a(pq)^{1/2} \cos^2 \pi\nu/a}.$$

Hence we get finally from (5.5) for the coefficient $u_{z,n}$ of s^n

$$(5.7) \qquad u_{z,n} = a^{-1} 2^n p^{(n-z)/2} q^{(n+z)/2} \sum_{\nu=1}^{a-1} \cos^{n-1}\frac{\pi\nu}{a} \cdot \sin\frac{\pi\nu}{a} \cdot \sin\frac{\pi z\nu}{a}.$$

(Strictly speaking, the term $\nu = a/2$ should be omitted when a is even, but it is zero anyway and therefore does no harm.)

For $n > 1$ formula (5.7) represents *the probability of ruin (absorption) at the nth trial*. It goes back to Lagrange and has been derived in many different ways.[7] Despite an honorable history and its availability in textbooks, the formula is rediscovered at frequent intervals. For an alternative explicit expression see problem 6; for limiting forms cf. section 6 and problem 7.

If we let $a \to \infty$, the sum in (5.7) may be interpreted as a Riemann sum approximating an integral. In this way we find that *in a game against an infinitely rich adversary (single absorbing barrier at $x = 0$) the probability $w_{z,n}$ that a player with initial capital $z > 0$ will be ruined exactly at the nth step is*

$$(5.8) \qquad w_{z,n} = 2^n p^{(n-z)/2} q^{(n+z)/2} \int_0^1 \cos^{n-1} \pi x \sin \pi x \sin \pi x z \cdot dx.$$

This integral can be expressed in an elementary way [8] as follows

$$(5.9) \qquad w_{z,n} = \frac{z}{n} \binom{n}{\frac{1}{2}(n-z)} p^{(n-z)/2} q^{(n+z)/2};$$

the binomial coefficient is again to be interpreted as zero if $(n-z)/2$ is not an integer of the interval $[0, n]$. The corresponding *generating function* was found to be $\lambda_2{}^z(s)$ (cf. end of section 4).

[7] An elementary derivation using trigonometric interpolation was given by Ellis, *Cambridge Mathematical Journal*, vol. 4 (1844), or *The Mathematical and Other Writings of R. E. Ellis*, Cambridge and London, 1863.

[8] Integrating by parts and observing that $\cos \pi xz = \cos \pi x(z-1) \cos \pi x - \sin \pi x(z-1) \sin \pi x$, we get a recursion formula for $w_{z,n}$ which checks with (5.9). A simpler proof consists in verifying that (5.9) is a solution of the difference equation (4.1) with the appropriate boundary conditions (4.2)–(4.3) at $z = 0$.

6. Passage to the Limit; Diffusion Processes

It has already been pointed out that our random-walk models serve as a first approximation to the theory of diffusion and Brownian motion, where small particles are exposed to a tremendous number of molecular shocks. Each shock has a negligible effect, but the superposition of many small actions produces an observable motion. Accordingly, we now want to study random walks where the individual steps are extremely small and occur in very rapid succession. In the limit the process will appear as a continuous motion. The point of interest is that in passing to this limit our formulas remain meaningful and agree with physically significant formulas of diffusion theory which can be derived under much more general conditions by more streamlined methods.[9] This explains partly why the random-walk model, despite its crudeness, describes diffusion processes reasonably well; only the limiting case is physically significant, and various discrete models lead to the same limiting formulas. The situation is in many ways analogous to the central limit theorem where we saw that under extremely general conditions the cumulative effect of many chance components is practically independent of the nature of the individual components.

Let us begin with an *unrestricted random walk starting at the origin*, and let $v_{x,n}$ be the probability that the nth step takes the particle to the position x. If r among the n steps are directed to the right, $n - r$ are directed to the left, and the total displacement is $r - (n - r) = 2r - n$ units. Since this displacement is to equal x, we must have $2r - n = x$. This is possible only if n and x are either both even or both odd (which means that after an even number of steps the abscissa x is an even integer). Out of n steps r can be selected in $\binom{n}{r}$ ways, and therefore

$$(6.1) \qquad v_{x,n} = \binom{n}{\frac{1}{2}(n+x)} p^{(n+x)/2} q^{(n-x)/2};$$

here the binomial coefficient should be interpreted as 0 whenever $(n + x)/2$ is not an integer in the interval $[0, n]$.

---

[9] The limiting formulas of the present section agree with those of the now classical Einstein-Wiener theory. The newer, more refined theories (Uhlenbeck, Ornstein) are not considered here. Credit for discovering the connection between random walks and diffusion is due principally to L. Bachelier (1870–). His work is frequently of a heuristic nature, but he derived many new results. Kolmogorov's theory of stochastic processes of the Markov type is based largely on Bachelier's ideas. Cf. in particular L. Bachelier, *Calcul des probabilités*, Paris, 1912.

There is an alternative way of deriving (6.1) by using the argument which led to the difference equation (4.1) and the boundary conditions (4.2) and (4.3). One verifies that $v_{x,n}$ must satisfy the difference equation

(6.2) $$v_{x,n+1} = pv_{x-1,n} + qv_{x+1,n}$$

with the boundary conditions

(6.3) $$v_{0,0} = 1, \quad v_{x,0} = 0 \qquad \text{for} \quad x \neq 0.$$

Given (6.3), we put in (6.2) successively $n = 1, 2, \ldots$ and get first all values $v_{x,1}$, and then successively $v_{x,2}$, $v_{x,3}$, ... This shows that the conditions (6.2) and (6.3) uniquely determine $v_{x,n}$. On the other hand, it is readily seen that (6.1) is a solution.

Let us now change the unit of length so that *each step has length Δx and suppose that the time between any two consecutive steps is Δt*. During time t the particle performs about $t/\Delta t$ jumps, and a displacement x is now equivalent to $x/\Delta x$ units. Only multiples of Δx and Δt represent meaningful coordinates, but in the limit $\Delta x \to 0$, $\Delta t \to 0$ every displacement and all times become possible.

We must not expect sensible results if we let Δx and Δt approach zero in an arbitrary manner. It suffices to notice that the maximum possible displacement in time t amounts to $t\Delta x/\Delta t$, so that in the limit no motion exists if $\Delta x/\Delta t \to 0$. Physically speaking, we must keep the x- and t-scales in an appropriate ratio or the process will degenerate in the limit, the velocities tending to zero or infinity. To find the proper ratio we note that the total displacement during time t is the sum of about $t/\Delta t$ mutually independent random variables each having the mean $(p - q)\Delta x$ and variance $\{1 - (p - q)^2\}(\Delta x)^2 = 4pq(\Delta x)^2$. The mean and variance of the total displacement in time t are therefore about $t(p - q)\Delta x/\Delta t$ and $4pqt(\Delta x)^2/\Delta t$, respectively. To obtain reasonable results we must let Δx and Δt approach zero so that the mean and variance remain finite for all t. The finiteness of the variance requires that $(\Delta x)^2/\Delta t$ should remain bounded; the finiteness of the mean implies that $p - q$ must be of the order of magnitude of Δx. This suggests putting

(6.4) $$\frac{(\Delta x)^2}{\Delta t} = 2D, \qquad p = \frac{1}{2} + \frac{c}{2D}\Delta x, \qquad q = \frac{1}{2} - \frac{c}{2D}\Delta x,$$

where D and c are constants. The numerical value of D introduces only a scale factor; for mathematical simplicity it would be best to put $D = 1$, but we keep D unspecified in order to facilitate comparison with physical theories. The constants D and c are, respectively, the *diffusion coefficient* and the *drift*. If $c = 0$, the random walk is symmetric, and, in general, the sign of c determines the direction of the drift. In the limit both p and q approach $1/2$; with any other norming the particle would drift away so fast that the probability of finite displacements would tend to zero.

We shall use the norming (6.4) to pass to the limit $\Delta x \to 0$, $\Delta t \to 0$. The total displacement at time $t \approx n\Delta t$ is determined by n Bernoulli trials, and therefore the limiting form of $v_{x,n}$ is known from Chapter 7 to be given by the normal distribution. The necessary computations were effected there and need not be repeated. For a fixed Δx the displacement is the sum of finitely many independent variables, and its mean is $t(p - q)\Delta x/\Delta t = 2ct$; its variance $4pqt(\Delta x)^2/\Delta t = 2Dt$. We find therefore that *the probability that at time t the displacement lies between x_0 and x_1 ($x_0 < x_1$) tends to*

$$(6.5) \qquad (2\pi)^{-1/2} \int_{y_0}^{y_1} e^{-\lambda^2/2} \, d\lambda$$

where $y_1 = (x_1 - 2ct)/(2Dt)^{1/2}$ and $y_0 = (x_0 - 2ct)/(2Dt)^{1/2}$. (According to the central limit theorem, the same conclusion holds for more general random walks.)

As for equation (6.2), we pass to the usual functional notation and write it in the form $v(x, t + \Delta t) = p \cdot v(x - \Delta x, t) + q \cdot v(x + \Delta x, t)$. Expanding according to Taylor's theorem up to terms of second order, we get formally

$$(6.6) \qquad \Delta t \cdot \frac{\partial v(x, t)}{\partial t} = (q - p)\Delta x \cdot \frac{\partial v(x, t)}{\partial x} + \frac{(\Delta x)^2}{2} \frac{\partial^2 v(x, t)}{\partial x^2} + \cdots$$

Using (6.4), we get in the limit

$$(6.7) \qquad \frac{\partial v(x, t)}{\partial t} = -2c \cdot \frac{\partial v(x, t)}{\partial x} + D \cdot \frac{\partial^2 v(x, t)}{\partial x^2}.$$

This is the well-known *Fokker-Planck* equation for diffusion with drift, which can be derived from more general and more convincing assumptions. In the usual theory, the solution (6.5) is derived from (6.7), while we have obtained both results by the same limiting process. Our procedure is only heuristic but can be justified more rigorously. The fact is that all formulas of the discrete random walk permit a similar passage to the limit.

As a further example, consider the limiting form of the probabilities for the *first passage*. For simplicity let us first consider formula (5.9) which corresponds to a single barrier. Of the two quantities $w_{z,n}$ and $w_{z,n+1}$, one is necessarily zero. The sum $w_{z,n} + w_{z,n+1}$ represents, asymptotically, the probability of absorption during the time interval $(t, t + 2\Delta t)$. We shall show that $w_{z,n} + w_{z,n+1} \sim f(z, t)(2\Delta t)$, where $f(z, t)$ is a continuous function. Then the limiting probability of absorption within any time interval (t_1, t_2) is the integral of $f(z, t)$ extended over that interval. Suppose now that $n - z$ is even. Then $w_{z,n+1} = 0$, and to find $f(z, t)$ we must replace z in (5.9) by $z/\Delta x$ and n by $t/\Delta t$, and apply (6.4). Using the normal approximation to the binomial distribution and the last equation (6.9), we find easily

$$(6.8) \qquad f(z, t) \sim \frac{z}{2(\pi D t^3)^{1/2}} e^{-(z+2ct)^2/(4Dt)}.$$

This is the limiting form of (5.9); again it coincides with the corresponding formula of diffusion theory. In fact, it is easily verified that $f(-x, t)$ is a solution of (6.7). (In the definition of $w_{z,n}$ the variable z plays the role of $-x$ in $v_{x,n}$.)

A similar argument applies to (5.7). An inspection of this formula shows that the contributions of $\nu = k$ and $\nu = a - k$ cancel if $n - z$ is odd and add if $n - z$ is even. Hence we get the limiting form of $f(z, t) \sim (u_{z,n} + u_{z,n+1})/(2\Delta t)$ by extending in (5.7) the sum twice over $1 \leq \nu < a/2$. Replace z, a, n respectively by $z/\Delta x, a/\Delta x, t/\Delta t$ and observe that for fixed ν

$$\sin \frac{\pi \nu \Delta x}{a} \sim \frac{\pi \nu \Delta x}{a}$$

$$(6.9) \qquad \left(\cos \frac{\pi \nu \Delta x}{a} \right)^{t/\Delta t} \sim \left(1 - \frac{D\pi^2 \nu^2 \Delta t}{a^2} \right)^{z/\Delta t} \sim e^{-D\pi^2 \nu^2 t/a^2},$$

$$(4pq)^{t/2\Delta t} \left(\frac{q}{p} \right)^{z/2\Delta x} \sim e^{-c(ct+z)/D}.$$

We obtain formally the limiting form

$$(6.10) \qquad f(z, t) \sim 2\pi D a^{-2} e^{-c(ct+z)/D} \sum_{\nu=1}^{\infty} \nu e^{-D\pi^2 \nu^2 t/a^2} \sin \frac{\pi z \nu}{a}.$$

The formal passage to the limit is justified because of uniform convergence: the contribution of the terms with large ν is negligible both in (6.10) and in the original sum (5.7) (where we have $\nu < a/2$).

In diffusion theory (6.10) is known as Fürth's formula for first passages and is derived directly from the Fokker-Planck equation.

In free diffusion the integral over (6.10), extended over the time interval (t_1, t_2), gives the probability that a particle starting at $z > 0$ will within that time interval for the first time reach the origin and not have previously passed the barrier $x = a$.

*7. Random Walks in the Plane and Space

In a two-dimensional random walk the particle moves in unit steps in one of the four directions parallel to the x- and y-axes. If the particle starts at the origin, the possible positions are all points of the plane with integral-valued coordinates. Each position has four *neighbors*. Similarly, in three dimensions each position has six neighbors. In order to define the random walk the corresponding four or six probabilities must be specified. For simplicity we shall consider only the *symmetric* case where all directions have the same probability. The complexity of problems is considerably greater than in one dimension, for now the domains to which the particle is restricted may have arbitrary shapes so that complicated boundaries take the place of the single-point barriers in the one-dimensional case.

We begin with an interesting theorem due to Polya.[10]

Theorem. In the symmetric random walks in one and two dimensions there is probability one that the particle will sooner or later (and therefore infinitely often) return to its initial position. In three dimensions, however, this probability is only about 0.35 [the expected number of returns is then $0.65 \Sigma k(0.35)^k = 0.35/0.65 \approx 0.53$].

Before proving the theorem let us give two alternative formulations, both due to Polya. First, it is almost obvious that the theorem implies that in *one and two dimensions there is probability* 1 *that the particle will pass infinitely often through every possible point*; in three dimensions this is not true, however. Thus the statement "all roads lead to Rome" is, in a way, justified in two dimensions.

Alternatively, consider *two* particles performing independent symmetric random walks, the steps occurring simultaneously. Will they ever meet? To simplify language let us define the *distance* of two possible positions as the smallest number of steps leading from one position to the other. (Then distance = sum of absolute differences of

---

* Starred sections treat special topics and may be omitted at first reading.

[10] G. Polya, Über eine Aufgabe der Wahrscheinlichkeitsrechnung betreffend die Irrfahrt im Strassennetz, *Mathematische Annalen*, vol. 84 (1921), pp. 149–160. The numerical value 0.35 was calculated by W. H. McCrea and F. J. W. Whipple, Random Paths in Two and Three Dimensions, *Proceedings of the Royal Society of Edinburgh*, vol. 60 (1940), pp. 281–298.

the coordinates). If the two particles move one step each, their mutual distance either remains the same or changes by two units. Accordingly, the distance of our two particles either is even at all times or else is always odd. In the second case the two particles can never occupy the same position. In the first case it is readily seen that the probability of the two particles meeting at the nth step equals the probability of the first particle to reach in $2n$ steps the initial position of the second particle. Hence our theorem states that in two, but not in three, dimensions the two particles are sure infinitely often to occupy the same position. If the initial distance of the two particles is odd, a similar argument shows that they will infinitely often occupy neighboring positions. If this is called meeting, then our theorem asserts that *in one and two dimensions the two particles are certain to meet infinitely often, while in three dimensions there is a positive probability that they never meet.*

Proof. For one dimension the theorem has been proved in Chapter 12, except that there we referred to a coin-tossing game rather than to a symmetric random walk. The proof for two and three dimensions proceeds along the same lines. Let u_n be the probabilite that the nth trial takes the particle to the initial position. According to theorem 2 of Chapter 12, section 3, we have to prove that in the case of two dimensions Σu_n diverges, while in the case of three dimensions $\Sigma u_n \approx 0.53$. In two dimensions a return to the initial position is possible only if the numbers of steps in the positive x- and y-directions equal those in the negative x- and y-directions, respectively. Hence $u_n = 0$ if n is odd while (using the multinomial distribution of Chapter 6, section 7)

$$(7.1) \quad u_{2n} = \frac{1}{4^{2n}} \sum_{k=0}^{2n} \frac{(2n)!}{k!k!(n-k)!(n-k)!} = \frac{1}{4^{2n}} \binom{2n}{n} \sum_{k=0}^{n} \binom{n}{k}^2.$$

The last expression equals $4^{-2n} \binom{2n}{n}^2$, by Chapter 2, formula (9.8).

Stirling's formula shows that u_{2n} is of the order of magnitude $1/n$, so that Σu_{2n} diverges as asserted.

In the case of three dimensions we find similarly

$$(7.2) \quad u_{2n} = \frac{1}{6^{2n}} \sum_{j,k} \frac{(2n)!}{j!j!k!k!(n-j-k)!(n-j-k)!},$$

the summation extending over all j, k with $j + k \leq n$. It is easily verified that

$$(7.3) \quad u_{2n} = \frac{1}{2^{2n}} \binom{2n}{n} \sum_{j,k} \left\{ \frac{1}{3^n} \frac{n!}{j!k!(n-j-k)!} \right\}^2.$$

Within the braces we have the terms of a trinomial distribution, and we know that they add to unity. Hence the sum of the squares is smaller than the maximum term within braces, and this is attained when both j and k are close to $n/3$. Stirling's formula shows that this maximum is of the order of magnitude n^{-1}, and therefore u_{2n} is of the magnitude $n^{-3/2}$ so that Σu_{2n} converges as asserted.

Polya's theorem is analogous to the facts concerning multiple coin tossings discussed in Chapter 12, example (3.c).

We conclude this section with another problem which generalizes the concept of *absorbing barriers*. To fix ideas we consider the case of two dimensions where instead of the interval $0 \leq x \leq a$ we have a plane domain D, that is, a collection of points with integral-valued coordinates. Each point has four neighbors, but for some points of D one or more of the neighbors lie outside D. Such points form the boundary of D, while all other points are called interior points. In the one-dimensional case the two barriers form the boundary, and our problem consisted in finding the probability that, starting from z, the particle will reach the boundary point $x = 0$ before reaching $x = a$. By analogy, we now ask for the probability that the particle will reach a certain section of the boundary before reaching any boundary point which is not in this section. This means that we divide all boundary points into two sets B' and B''. If (x, y) is an interior point, we ask for the probability $u(x, y)$ that, starting from (x, y), the particle will reach a point of B' before reaching a point of B''. In particular, if B' consists of a single point, then $u(x, y)$ is the probability that the particle will, sooner or later, be absorbed at that particular point.

Let (x, y) be an interior point. The first step takes the particle from (x, y) to one of the four neighbors $(x \pm 1, y)$, $(x, y \pm 1)$, and if all four of them are interior points, we must have

(7.4)
$$u(x,y) = \tfrac{1}{4}\{u(x + 1, y) + u(x - 1, y) + u(x, y + 1) + u(x, y - 1)\}.$$

This is a partial difference equation which takes the place of (2.1) (with $p = q = 1/2$). If $(x + 1, y)$ is a boundary point, then its contribution $u(x + 1, y)$ must be replaced by 1 or 0, according to whether $(x + 1, y)$ belongs to B' or B''. Hence (7.4) *will be valid for all interior points if we agree that for a boundary point* (ξ, η) *we put* $u(\xi, \eta) = 1$ *if* (ξ, η) *is in B' and* $u(\xi, \eta) = 0$ *if* (ξ, η) *is in B''. This convention takes the place of the boundary conditions* (2.2).

In (7.4) we now have a system of linear equations for the unknowns $u(x, y)$; to each interior point there correspond one unknown and one equation. The system is non-homogeneous, since in it there appears at least one boundary point (ξ, η) of B' and it gives rise to a contribution $\frac{1}{4}$ on the right side. If the domain D is finite, we have as many equations as unknowns, and it is well known that the system has a unique solution if and only if the corresponding homogeneous system [with $u(\xi, \eta) = 0$ for all boundary points] has no non-vanishing solution. Now $u(x, y)$ is the mean of the four neighboring values $u(x \pm 1, y)$, $u(x, y \pm 1)$ and hence cannot exceed all four. In other words, $u(x, y)$ cannot have either a maximum or a minimum in the strict sense, so that the greatest and the smallest value occur at boundary points. Hence, if all boundary values vanish, so does $u(x, y)$ at all interior points. This proves the existence and uniqueness of the solution of (7.4). Since the boundary values are 0 and 1, all values $u(x, y)$ lie between 0 and 1, as is required for probabilities. These statements are true also for the case of infinite domains, as will be seen from a general theorem on infinite Markov chains.[11]

8. The Generalized One-dimensional Random Walk (Sequential Sampling)

We now return to one dimension but consider the general case where the particle does not necessarily pass to a neighboring point. The possible positions are still $x = 0, \pm 1, \pm 2, \ldots$. However, we shall assume that *at each step the particle has probability p_k to move from x to the point $x + k$;* here the integer k is allowed to be zero, positive, or negative (in the ordinary random walk $p_1 = p$, $p_{-1} = q$). We shall consider the following *ruin problem. The particle starts from a position z with $0 < z < a$; required is the probability u_z that the particle will arrive at some position $x \leq 0$ before reaching any position $x \geq a$.* An interpretation of this problem in the terminology of gambling was discussed at the end of section 1, where it was also stated that our problem is of great importance in Wald's sequential analysis. There the term "rejection" is used instead of ruin. Also, in Wald's terminology the particle always starts from $x = 0$, and the game (i.e., sampling) terminates when it reaches a position to the left of $x = -b$ or to the right of $x = a$; this, however, is only a notational difference.[12]

[11] Explicit solutions are known in only a few cases and are always very complicated. Solutions for the case of rectangular domains, infinite strips, etc., will be found in the paper by McCrea and Whipple cited in footnote 10.

[12] Wald treats also the case of continuous variables and uses different tools. The present methods apply also to general random variables.

Without loss of generality we shall suppose that steps are possible in both the positive and negative directions. Otherwise we would have either $u_z = 0$ or $u_z = 1$ for all z.

The probability of ruin at the *first* step is obviously

(8.1) $$r_z = p_{-z} + p_{-z-1} + p_{-z-2} + \cdots$$

(a quantity which may be zero). After the first step the random walk continues only if the particle moved to a position x with $0 < x < a$. The probability of a jump from z to x is p_{x-z}, and the probability of subsequent ruin is then u_x. Therefore

(8.2) $$u_z = \sum_{x=1}^{a-1} u_x p_{x-z} + r_z.$$

Once more we have here $a - 1$ linear equations for $a - 1$ unknowns u_z. The system is non-homogeneous, since at least for $z = 1$ the probability r_1 is different from zero (steps in the negative direction being possible, which obviously implies $r_1 > 0$). We claim that the corresponding homogeneous system

(8.3) $$u_z = \sum_{x=1}^{a-1} u_x p_{x-z}$$

can have only the solution $u_x = 0$.

In fact, if it had another solution, one of the values u_z would be largest in absolute value, say $u_z = M > 0$. Suppose first that $p_{-1} \neq 0$. Since the coefficients p_{x-z} in (8.3) add to at most unity, the equation is possible only if all those p_{x-z} which actually appear on the right side (with a coefficient different from zero) equal M, and if their coefficients add to 1. Hence $u_{z-1} = M$, and, arguing the same way, $u_{z-2} = u_{z-3} = \cdots = u_1 = M$. However, for $z = 1$ the coefficients p_{x-z} in (8.3) add to less than unity, so that M must be zero. The same argument obviously applies also if $p_{-1} = 0$, since we can replace p_{-1} by some other coefficient p_k with $k < 0$ which is positive.

It follows that (8.2) has a unique solution, and thus our problem is determined. Equation (8.2) plays the role of the difference equation (2.1). Again we can simplify the writing by introducing the boundary conditions

(8.4) $$\begin{aligned} u_x &= 1 \quad \text{if} \quad x \leq 0 \\ u_x &= 0 \quad \text{if} \quad x \geq a. \end{aligned}$$

Then (8.2) can be written in the form

(8.5) $$u_z = \Sigma u_x p_{x-z},$$

the summation now extending over all x [for $x \geq a$ we have no contribution owing to the second condition (8.4); the contributions for $x \leq 0$ add to r_z owing to the first condition].

For large a it is cumbersome to solve $a - 1$ linear equations directly, and it is preferable to use the *method of particular solutions* analogous to the procedure of section 2. It works whenever the probability distribution $\{p_k\}$ has relatively few positive terms. Suppose that only the p_k with $-\nu \leq k \leq \mu$ are different from zero, so that the largest possible jumps in the positive and negative directions are μ and ν, respectively. Consider the *characteristic equation*

$$(8.6) \qquad \Sigma p_k s^k = 1.$$

It is equivalent to an algebraic equation of degree $\nu + \mu$. If s is a root of (8.6), then $u_z = s^z$ is a formal solution of (8.5) for all z, but this solution does not satisfy the boundary conditions (8.4). If (8.6) has $\mu + \nu$ distinct roots $s_1, s_2, \ldots$, then the linear combination

$$(8.7) \qquad u_z = \Sigma A_k s_k^z$$

is again a formal solution of (8.5) for all z. We have to adjust the constants A_k so that the boundary conditions are satisfied. Now for $0 < z < a$ only values x with $-\nu + 1 \leq x \leq a + \mu - 1$ appear in (8.5). It suffices therefore to satisfy the boundary conditions (8.4) for $x = 0, -1, -2, \cdots, -\nu + 1$, and $x = a, a + 1, \cdots, a + \mu - 1$, so that we have $\mu + \nu$ conditions in all. If s_k is a double root of (8.5), we lose one constant, but in this case it is easily seen that $u_z = z s_k^z$ is another formal solution. In every case the $\mu + \nu$ boundary conditions determine the $\mu + \nu$ arbitrary constants.

Example. Suppose that each individual step takes the particle to one of the four nearest positions, and we let $p_{-2} = p_{-1} = p_1 = p_2 = 1/4$. The characteristic equation (8.6) is $s^{-2} + s^{-1} + s + s^{+2} = 4$. To solve it we put $t = s + s^{-1}$: with this substitution our equation becomes $t^2 + t = 6$, which has the roots $t = 2, -3$. Solving $t = s + s^{-1}$ for s, we find the four roots

$$(8.8) \quad s_1 = s_2 = 1, \quad s_3 = \frac{-3 + 5^{1/2}}{2} = s_4^{-1}, \quad s_4 = \frac{-3 - 5^{1/2}}{2} = s_3^{-1}.$$

Since s_1 is a double root, the general solution of (8.5) in our case is

$$(8.9) \qquad u_z = A_1 + A_2 z + A_3 s_3^z + A_4 s_4^z.$$

The boundary conditions are $u_0 = u_{-1} = 1$, and $u_a = u_{a+1} = 0$.

They lead to four linear equations for the coefficients A_j and to the final solution

$$(8.10) \quad u_z = 1 - \frac{z}{a} + \frac{(2z-a)(s_3{}^a - s_4{}^a) - a(s_3{}^{2z-a} - s_4{}^{2z-a})}{a\{(a+2)(s_3{}^a - s_4{}^a) - a(s_3{}^{a+2} - s_4{}^{a+2})\}}$$

with s_3 and s_4 given by (8.8).

Numerical Approximations. If the degree $\mu + \nu$ of the characteristic equation (8.6) is not very small, then it is cumbersome to find all its roots. In practice rather satisfactory approximations can be obtained in a surprisingly simple way. Consider first the case where the probability distribution $\{p_k\}$ has mean zero. Then the characteristic equation (8.6) has a double root at $s = 1$, and hence $A + Bz$ is a formal solution of (8.5). Of course, the two constants A and B do not suffice to satisfy the $\mu + \nu$ boundary conditions (8.4). However, if we determine A and B so that $A + Bz$ vanishes for $z = a + \mu - 1$ and equals 1 for $z = 0$, then we shall have $A + Bx \geq 1$ for $x \leq 0$ and $A + Bx \geq 0$ for $a \leq x < a + \mu$. Our $A + Bz$ then satisfies the boundary conditions (8.4) with the equality sign replaced by "greater than or equal to." The difference $A + Bz - u_z$ is therefore a formal solution of (8.5) with non-negative boundary values whence $A + Bz - u_z \geq 0$. In like manner we can get a lower bound for u_z by determining A and B so that $A + Bz$ vanishes for $z = a$ and equals 1 for $z = -\nu + 1$. Hence we have

$$(8.11) \quad \frac{a-z}{a+\nu-1} \leq u_z \leq \frac{a+\mu-z-1}{a+\mu-1}.$$

If a is large as compared to $\mu + \nu$, we have here an excellent estimate. (Of course, $u_z \approx (1 - z/a)$ is a better approximation, but does not give precise bounds.)

Next, consider the general case where the mean of the distribution $\{p_k\}$ is not zero. The characteristic equation (8.6) has then a simple root at $s = 1$. The left side of (8.6) approaches ∞ as $s \to 0$ and as $s \to \infty$. It is continuous for $s > 0$, and its second derivative is positive; this means that for positive s the curve $y = \Sigma p_k s^k$ is continuous and convex. Since it intersects the line $y = 1$ at $s = 1$, there exists exactly one more intersection. Therefore, the characteristic equation (8.6) has exactly two positive roots, 1 and s_1. As before, we see that $A + Bs_1{}^z$ is a formal solution of (8.5), and we can apply our previous argument, substituting this solution for $A + Bz$. We find in this case

$$(8.12) \quad \frac{s_1{}^a - s_1{}^z}{s_1{}^a - s_1{}^{-\nu+1}} \leq u_z \leq \frac{s_1{}^{a+\mu-1} - s_1{}^z}{s_1{}^{a+\mu-1} - 1}.$$

Hence the

Theorem. *The solution of our ruin problem satisfies the inequalities* (8.11) *if* $\{p_k\}$ *has zero mean, and* (8.12) *otherwise. Here* s_1 *is the unique positive root different from* 1 *of* (8.6), *and* μ *and* $-\nu$ *are defined, respectively, as the largest and smallest subscript for which* $p_k \neq 0$.

Let $m = \Sigma k p_k$ be the *expected gain* in a single trial (or expected length of a single step). It is easily seen from (8.6) that $s_1 > 1$ or $s_1 < 1$ according to whether $m < 0$ or $m > 0$. Letting $a \to \infty$, we conclude from our theorem that *in a game against an infinitely rich adversary the probability of an ultimate ruin is one if and only if* $m \leq 1$.

The *duration of game* can be discussed by similar methods (cf. problem 18).

9. Problems for Solution

1. We modify the ruin problem of section 2 so that the gambler has probability α to win a dollar, probability β to lose a dollar, while with probability γ the trial ends in a tie. Show that the probability of ruin is still given by (2.4) and (2.5) with $p = \alpha/(1 - \gamma)$, $q = \beta/(1 - \gamma)$. The expected duration of the game is $D_z/(1 - \gamma)$ with D_z given by (3.4) and (3.5).

2. In the one-dimensional random walk with absorbing barriers at $x = 0$ and $x = a$, let z be the initial position, and let $w_{z,n}(x)$ be the probability that the nth step takes the particle to x. Show that $w_{z,n}(x)$ satisfies the difference equation $w_{z,n+1}(x) = pw_{z+1,n}(x) + qw_{z-1,n}(x)$ with the boundary conditions (1) $w_{0,n}(x) = w_{a,n}(x) = 0$ for $n \geq 1$; (2) $w_{z,0}(x) = 0$ if $z \neq x$ and $w_{x,0}(x) = 1$.

3. In problem 2 let there be *reflecting barriers* at $x = \frac{1}{2}$ and $x = a - \frac{1}{2}$ (cf. section 1). Show that the statements of the preceding problem hold with the boundary conditions (1) replaced by $w_{0,n}(x) = w_{1,n}(x)$ and $w_{a,n}(x) = w_{a-1,n}(x)$.

In the following problems $v_{x,n}$ is always the probability that in an unrestricted random walk starting at the origin the nth step takes the particle to the position x. This probability is given by (6.1); the symbols $u_{z,n}$, $\lambda_1(s)$, $\lambda_2(s)$, $w_{z,n}$ will be used as defined in sections 4 and 5.

4. Suppose there is a single absorbing barrier at the origin. Let $u_{z,n}(x)$ be the probability that a particle starting at $x > 0$ is after n steps at $z > 0$. If the random walk is *symmetric* ($p = q = 1/2$), show that $u_{z,n}(x) = v_{z-x,n} - v_{z+x,n}$.

Hint: Show that a difference equation similar to (4.1) and the appropriate boundary condition are satisfied.

5. *Continuation*.[13] If there are absorbing barriers at $x = 0$ and $x = a$, show that

$$u_{z,n}(x) = \sum_{k} \{v_{z-x-2ka,n} - v_{z+x-2ka,n}\},$$

the summation extending over all k, positive and negative (only finitely many terms are different from zero).

6. *Alternative formula for the probability of ruin* (5.7). Expanding (4.11) into a geometric series, prove that

$$u_{z,n} = \sum_{k=0}^{\infty} \left(\frac{p}{q}\right)^{ka} w_{z+2ka,n} - \sum_{k=1}^{\infty} \left(\frac{p}{q}\right)^{ka-z} w_{2ka-z,n}$$

with $w_{z,n}$ defined in (5.9).

7. If the passage to the limit of section 6 is applied to the expression for $u_{z,n}$ given in the preceding problem, show that the probability of absorption during a short time interval of length Δt is asymptotically [14]

$$\frac{1}{2} \Delta t (\pi D t^3)^{-1/2} e^{-c(ct+z)/D} \sum_{k=-\infty}^{+\infty} (z + 2ka) e^{-(z+2ka)^2/4Dt}.$$

Hint: Apply the normal approximation to the binomial distribution.

[13] This solution is obtained by the *method of images* used in potential theory and due to Lord Kelvin. The term $v_{z-x,n}$ is the desired probability in the absence of barriers (free random walk); then $v_{z+x,n}$ represents the corresponding probability for an "image" particle which starts at the point $-x$ (the point x mirrored at the left barrier); v_{z+x-2a} corresponds to another "image" starting at the point $2a - x$ (which is x mirrored at the right end); there follow images of images, etc.

[14] The agreement of the new formula with the limiting form (6.10) is a well-known fact of the theory of theta functions.

8.[15] *First passage times.* In an unrestricted random walk starting at $x = 0$, let g_n be the probability that the particle reaches $x = -1$ for the first time at the nth step. Without using any previous results prove directly that the generating function $G(s)$ of $\{g_n\}$ satisfies the equation $G(s) = qs + ps\, G^2(s)$. Hence show that $G(s) = \lambda_2(s)$ with $\lambda_2(s)$ defined in (4.8). Similarly, $1/\lambda_1(s) = p\lambda_2(s)/q$ is the generating function of the first passage time through $x = +1$. Show also that this implies that the first passage time through any position z has the generating function $\lambda_2^{-z}(s)$ if $z < 0$ and $\lambda_1^{-z}(s)$ if $z > 0$ [cf. (4.13) which corresponds to the first passage through $-z$].

9. *Continuation: recurrence times.* Let f_n be the probability that the particle returns to its initial position for the first time at the nth step. If $F(s)$ is the corresponding generating function, we must have $F(s) = ps\lambda_2(s) + qs/\lambda_1(s)$. [*Note:* This is a new derivation of the equation $F(s) = 1 - (1 - 4pqs^2)^{1/2}$ found in Chapter 12, section 3.]

10. Let $V_x(s) = \Sigma v_{x,n} s^n$ (cf. the note preceding problem 4). Using the results of problem 8, prove that $V_x(s) = V_0(s)\lambda_2^{-x}(s)$ if $x < 0$ and $V_x(s) = V_0(s)\lambda_1^{-x}(s)$ if $x > 0$. [*Note:* These relations are almost obvious and should be proved without calculations. It is easily verified that $V_0(s) = (1 - 4pqs^2)^{-1/2}$.]

11. *Renewal method for the ruin problem.* In the random walk with two absorbing barriers of section 4 let $u_{z,n}$ and $u_{z,n}{}^*$ be, respectively, the probabilities of absorption at the left and the right barriers. By a proper interpretation prove the truth of the following two equations:

$$V_{-z}(s) = U_z(s)V_0(s) + U_z{}^*(s)V_{-a}(s),$$

$$V_{a-z}(s) = U_z(s)V_a(s) + U_z{}^*(s)V_0(s).$$

By solving this system for $U_z(s)$, derive (4.11).

12. Let $u_{z,n}(x)$ be the probability that the particle, starting from z, will at the nth step be at x without having previously touched the absorbing barriers. Using the notations of problem 11, show that for the corresponding generating function $U_z(s;x) = \Sigma u_{z,n}(x)s^n$ we have $U_z(s;x) = V_{x-z}(s) - U_z(s)V_x(s) - U_z{}^*(s)V_{x-a}(s)$. (No calculations are required.)

13. *Continuation.* The generating function $U_z(s;x)$ of the preceding problem can be obtained by putting $U_z(s;x) = V_{x-z}(s) - A\lambda_1^z(s) - B\lambda_2^z(s)$ and determining the constants so that the boundary conditions $U_z(s;x) = 0$ for $z = 0$ and $z = a$ are satisfied. If there are *reflecting barriers* at $\frac{1}{2}$ and $a - \frac{1}{2}$, the boundary conditions are $U_0(s;x) = U_1(s;x)$ and $U_a(s;x) = U_{a-1}(s;x)$.

14. A symmetric unrestricted random walk starts at the origin. The probability that the rth return to the origin occurs at the nth step equals the probability that the first passage through $x = r$ occurs at the $(n + r)$th step. (*Hint:* Compare the generating functions.)

15. Prove the formula

$$v_{x,n} = (2\pi)^{-1} 2^n p^{(n+x)/2} q^{(n-x)/2} \int_{-\pi}^{\pi} \cos^n t \cdot \cos tx \cdot dt.$$

by showing that the appropriate difference equation is satisfied. Conclude that [16]

$$V_x(s) = (2\pi)^{-1} \left(\frac{p}{q}\right)^{x/2} \int_{-\pi}^{\pi} \frac{\cos tx}{1 - 2(pq)^{1/2} \cdot s \cdot \cos t}\, dt.$$

[15] Problems 8–13 contain a new and independent derivation of the main results concerning random walks in one dimension.

[16] The formulas of problem 10 now follow easily by the calculus of residues.

16. In a three-dimensional symmetric random walk the particle has probability one to pass infinitely often through any particular line $x = m$, $y = n$. (*Hint:* Cf. problem 1.)

17. In a two-dimensional symmetric random walk starting at the origin the probability that the nth step takes the particle to (x,y) is

$$(2\pi)^{-2} 2^{-n} \int_{-\pi}^{\pi} \int_{-\pi}^{\pi} (\cos \alpha + \cos \beta)^n \cdot \cos x\alpha \cdot \cos y\beta \cdot d\alpha \, d\beta.$$

Verify this formula and find the analogue for three dimensions. (*Hint:* Check that the expression satisfies the proper difference equation.)

18. In the generalized random-walk problem of section 8 put [in analogy with (8.1)] $\rho_z = p_{a-z} + p_{a+1-z} + p_{a+2-z} + \ldots$, and let $d_{z,n}$ be the probability that the game lasts for exactly n steps. Show that for $n \geq 1$

$$d_{z,n+1} = \sum_{x=1}^{a-1} d_{x,n} p_{x-z}$$

with $d_{z,1} = r_z + \rho_z$. Hence prove that the generating function $d_z(s) = \Sigma d_{z,n} s^n$ is the solution of the system of linear equations

$$s^{-1} d_z(s) - \sum_{x=1}^{a-1} d_x(s) \, p_{x-z} = r_z + \rho_z.$$

By differentiation it follows that the expected duration e_z is the solution of

$$e_z - \sum_{x=1}^{a-1} e_x p_{x-z} = 1.$$

CHAPTER 15

MARKOV CHAINS

1. Definition

Up to now we have been concerned mostly with independent trials, which can be described as follows. A set of possible outcomes E_1, E_2, ..., (finite or infinite in number) is given, and with each there is associated a probability p_k; the probabilities of sample sequences are defined by the multiplicative property $Pr\{(E_{j_0}, E_{j_1}, \cdots, E_{j_n})\} = p_{j_0} p_{j_1} \cdots p_{j_n}$. In the theory of Markov[1] chains we consider the simplest generalization which consists in permitting the outcome of any trial to depend on the outcome of the directly preceding trial (and only on it). The outcome E_k is then no longer associated with a fixed probability p_k, but to every pair (E_j, E_k) there corresponds a fixed *conditional probability* p_{jk}: given that E_j has occurred at some trial, the probability of E_k at the next trial is p_{jk}. In addition to the p_{jk} we must be given the probability a_k of the outcome E_k at the *initial* trial. If the p_{jk} are to have the meaning that we attributed to them, then we must define the probabilities of sample sequences corresponding to two, three, or four trials by

$$Pr\{(E_j, E_k)\} = a_j p_{jk}, \quad Pr\{(E_j, E_k, E_r)\} = a_j p_{jk} p_{kr},$$
$$Pr\{(E_j, E_k, E_r, E_s)\} = a_j p_{jk} p_{kr} p_{rs},$$

and generally

(1.1) $\quad Pr\{(E_{j_0}, E_{j_1}, \cdots, E_{j_n})\} = a_{j_0} p_{j_0 j_1} p_{j_1 j_2} \cdots p_{j_{n-2} j_{n-1}} p_{j_{n-1} j_n}.$

Here the initial trial is numbered zero, so that trial number one is the second trial. (This convention is convenient and has been introduced tacitly in the preceding chapter.) Before scrutinizing the legitimacy of the definition (1.1), two simple examples will render the notion more intuitive. The next section contains more interesting illustrations.

Examples. (*a*) Suppose we are given two unbalanced coins with faces marked E_1 and E_2 such that with the first coin these faces have probabilities α and $\beta = 1 - \alpha$, with the second α' and $\beta' = 1 - \alpha'$. One

[1] A. A. Markov (1856–1922).

of the two coins is selected at random and tossed; this is the initial (or zero) trial. Each following trial consists in tossing the first or second coin, according to whether the preceding trial resulted in E_1 or E_2. The probabilities of E_1 and E_2 at the initial (or zeroth) trial are obviously $a_1 = \frac{1}{2}(\alpha + \alpha')$ and $a_2 = \frac{1}{2}(\beta + \beta')$, respectively. Moreover, $p_{11} = \alpha$, $p_{12} = \beta$, $p_{21} = \alpha'$, $p_{22} = \beta'$. The probability of E_1 at the second trial is $Pr\{(E_1, E_1)\} + Pr\{(E_2, E_1)\} = \frac{1}{2}(\alpha + \alpha')\alpha + \frac{1}{2}(\beta + \beta')\alpha'$, etc.

(b) Independent trials are illustrated by drawings with replacement from an urn of fixed composition. Similarly, our new type of trials may be realized by drawings from a sequence of urns. In the jth urn balls of various colors are represented in the proportions $p_{j1} : p_{j2} : p_{j3} : \ldots$. If a drawing has resulted in a ball of jth color, the next drawing is made from the urn numbered j.

It is clear that, if a_k is the probability of E_k at the initial (or zero-th) trial, we must have $a_k \geq 0$ and $\Sigma a_k = 1$. Similarly, since whenever E_j occurs it must be followed by some E_k, we must have for all j and k

(1.2) $$p_{j1} + p_{j2} + p_{j3} + \ldots = 1, \quad p_{jk} \geq 0.$$

We want to show that for any numbers a_k and p_{jk} satisfying these conditions, the assignment (1.1) is a permissible definition of probabilities in the sample space corresponding to $n + 1$ trials. Since the numbers defined in (1.1) are obviously non-negative, we need only prove that they add to unity. Now first fix $j_0, j_1, \cdots, j_{n-1}$ and add the numbers (1.1) for all possible j_n. Using (1.2) with $j = j_{n-1}$, we see immediately that the sum is $a_{j_0} p_{j_0 j_1} \cdots p_{j_{n-2} j_{n-1}}$. Thus the sum over all numbers (1.1) does not depend on n, and since $\Sigma a_{j_0} = 1$, the sum equals unity for all n.

The definition (1.1) depends formally on the number of trials, but our argument proves the mutual consistency of the definitions (1.1) for all n. For example, to obtain the probability of the event "the first two trials result in (E_j, E_k)," we have to fix $j_0 = j$ and $j_1 = k$, and add the probabilities (1.1) for all possible $j_2, j_3, \cdots, j_n$. We have just shown that the sum is $a_j p_{jk}$, and thus independent of n. This means that it is usually not necessary explicitly to refer to the number of trials: the event $(E_{j_0}, \cdots, E_{j_r})$ has the same probability in all sample spaces of more than r trials. In connection with independent trials it has been pointed out repeatedly that, from a mathematical point of view, it is most satisfactory to introduce only the unique sample space of unending sequences of trials and to consider the result of finitely many trials as the beginning of an infinite sequence. This statement holds true also for Markov chains. Unfortunately, sample spaces of infinitely many trials lead beyond the theory of discrete probabilities to which we are restricted in the present volume.

To summarize, our starting point is the following

Definition. A sequence of trials with possible outcomes E_1, E_2, ... will be called a Markov chain[2] *if the probabilities of sample sequences are defined by* (1.1) *in terms of an initial probability distribution* $\{a_k\}$ *for the states E_k at time 0 and fixed conditional probabilities p_{jk} of E_k, given that E_j has occurred at the preceding trial.*

We shall now modify our terminology so as to conform to the usage in physical applications. Instead of saying "the nth trial results in E_k" we shall say that *at time n the system is in state E_k*. The conditional probability p_{jk} will be called the *probability of the transition $E_j \to E_k$* (from state E_j to state E_k).

The transition probabilities p_{jk} will be arranged in a *matrix of transition probabilities*

$$(1.3) \qquad P = \begin{bmatrix} p_{11} & p_{12} & p_{13} & \cdots \\ p_{21} & p_{22} & p_{23} & \cdots \\ p_{31} & p_{32} & p_{33} & \cdots \\ \cdot & \cdot & \cdot & \cdots \\ \cdot & \cdot & \cdot & \cdots \\ \cdot & \cdot & \cdot & \cdots \end{bmatrix}$$

where the first subscript stands for row, the second for column. Clearly P is a square matrix with non-negative elements and unit row sums. Such a matrix (finite or infinite) is called a *stochastic matrix*. *Any stochastic matrix can serve as a matrix of transition probabilities; together with our initial distribution* $\{a_k\}$ *it completely defines a Markov chain.*

In some special cases it is convenient to number the states starting with 0 rather than with 1. A zero row and zero column are then to be added to P.

---

[2] This is not the standard terminology. We are here considering only a special class of Markov chains, and, strictly speaking, here and in the following sections the term Markov chain should always be qualified by adding the clause "with constant transition probabilities." Actually, the general type of Markov chain is rarely studied. It will be defined in section 10, where the Markov property will be discussed in relation to general stochastic processes. There the reader will also find examples of dependent trials that do not form Markov chains.

2. Illustrative Examples

This section contains a list of various special examples which will familiarize the reader with the notion of a Markov chain. To save space we shall repeatedly refer to these examples to illustrate various definitions and theorems. The reader is advised not to attempt to keep these examples continuously in mind but to consider each reference to them independently. For an application of Markov chains to card shuffling cf. section 9.

I. *Independent Trials.* Let $p_{jk} = a_k$ be independent of j, so that all rows of the transition matrix P are identical. Then the trials are independent. For example, to Bernoulli trials there corresponds a 2 by 2 matrix with rows (p, q).

II. *Success Runs.* Consider a sequence of Bernoulli trials, and let us agree to say that at time t the system is in state E_k ($k = 1, 2, \ldots$) if the tth trial results in a success which is the kth success in an uninterrupted sequence [in other words, if the trials numbered t, $t - 1$, $t - 2$, $\cdots$, $t - k + 1$ resulted in success but the $(t - k)$th trial in failure]. Further, we say that at time n the system is in state E_0 if the nth trial resulted in failure; at time 0 the system starts from the state E_0. We have here a Markov chain with states $E_0, E_1, E_2, \ldots$; the initial distribution is $(1, 0, 0, 0, \ldots)$; the transition probabilities are defined by $p_{j,j+1} = p$, $p_{j,0} = q$, and $p_{jk} = 0$ whenever k is not either $j + 1$ or 0. Thus

$$P = \begin{bmatrix} q & p & 0 & 0 & 0 & \cdots \\ q & 0 & p & 0 & 0 & \cdots \\ q & 0 & 0 & p & 0 & \cdots \\ . & . & . & . & . & \cdots \\ . & . & . & . & . & \cdots \\ . & . & . & . & . & \cdots \end{bmatrix}$$

III. *Random Walk with Absorbing Barriers.* The random-walk problems of the preceding chapter are examples of Markov chains with positions playing the roles of states. If there are absorbing barriers at $x = 0$ and $x = a$, the possible states are $E_0, E_1, \cdots, E_a$ with E_k standing for $x = k$. For $1 \leq j \leq a - 1$ the system can pass from E_j either to E_{j-1} or to E_{j+1}, but no further change is possible once the system reaches either E_0 or E_a. Hence $p_{00} = p_{aa} = 1$, and $p_{j,j+1} = p$, $p_{j,j-1} = q$ provided $1 \leq j \leq a - 1$; all other transition probabilities vanish. The matrix P is given by

$$P = \begin{bmatrix} 1 & 0 & 0 & 0 & \cdots & 0 & 0 & 0 \\ q & 0 & p & 0 & \cdots & 0 & 0 & 0 \\ 0 & q & 0 & p & \cdots & 0 & 0 & 0 \\ \cdot & \cdot & \cdot & \cdot & \cdots & \cdot & \cdot & \cdot \\ \cdot & \cdot & \cdot & \cdot & \cdots & \cdot & \cdot & \cdot \\ 0 & 0 & 0 & 0 & \cdots & q & 0 & p \\ 0 & 0 & 0 & 0 & \cdots & 0 & 0 & 1 \end{bmatrix}$$

The initial probabilities are, in principle, arbitrary. In the preceding chapter we have assumed that the particle starts from the state E_z. This corresponds to $a_z = 1$, $a_k = 0$ for $z \neq k$.

IV. *Reflecting Barriers.* We modify the preceding example so that the possible states are $E_1, E_2, \cdots, E_a$ (with E_0 omitted). From the interior states $E_2, E_3, \cdots, E_{a-1}$, the system can pass either to the right or to the left neighbor, exactly as in the ordinary random walk. However, from E_1, the system has probability p to pass to E_2 and probability q to stay in E_1. Similarly, from E_a only the transitions $E_a \to E_a$ and $E_a \to E_{a-1}$ are possible, and the corresponding probabilities are p and q. In the terminology of random walks this means reflecting barriers at $x = \frac{1}{2}$ and $x = a + \frac{1}{2}$ (cf. Chapter 14, section 1). Alternatively, the state of the system may stand for a gambler's fortune if the familiar gambling for unit stakes is amended by an agreement that whenever a player loses his last dollar he is given one dollar by his adversary. A continuation of the game is then always possible; the combined capital of the two gamblers is $a + 1$ and remains constant. The matrix of transition probabilities is now

$$P = \begin{bmatrix} q & p & 0 & 0 & \cdots & 0 & 0 & 0 \\ q & 0 & p & 0 & \cdots & 0 & 0 & 0 \\ 0 & q & 0 & p & \cdots & 0 & 0 & 0 \\ \cdot & \cdot & \cdot & \cdot & \cdots & \cdot & \cdot & \cdot \\ \cdot & \cdot & \cdot & \cdot & \cdots & \cdot & \cdot & \cdot \\ 0 & 0 & 0 & 0 & \cdots & q & 0 & p \\ 0 & 0 & 0 & 0 & \cdots & 0 & q & p \end{bmatrix}.$$

[Continued in example (6.b), problem 10, and Chapter 16, section 3.]

V. *Cyclical Random Walks.* Again let the possible states be $E_1, E_2, \cdots, E_a$ but order them cyclically so that E_a has the neighbors E_{a-1} and E_1. If, as before, the system always passes either to the right or to the left neighbor, the rows of the matrix P are as in the preceding example, except that the first row is $(0, p, 0, 0, \cdots, 0, q)$ and the last $(p, 0, 0, 0, \cdots, 0, q, 0)$.

More generally, we may permit transitions between any two states. Let $q_0, q_1, \cdots, q_{a-1}$ be, respectively, the probability of staying fixed or moving $1, 2, \cdots, a - 1$ units to the right (where k units to the right is the same as $a - k$ units to the left). Then P is the cyclical matrix

$$P = \begin{bmatrix} q_0 & q_1 & q_2 & \cdots & q_{a-2} & q_{a-1} \\ q_{a-1} & q_0 & q_1 & \cdots & q_{a-3} & q_{a-2} \\ q_{a-2} & q_{a-1} & q_0 & \cdots & q_{a-4} & q_{a-3} \\ \cdot & \cdot & \cdot & \cdots & \cdot & \cdot \\ \cdot & \cdot & \cdot & \cdots & \cdot & \cdot \\ q_1 & q_2 & q_3 & \cdots & q_{a-1} & q_0 \end{bmatrix}.$$

If $q_1 = p$, $q_{a-1} = p$, and $q_k = 0$ for $1 < k < a - 1$, then this random walk reduces to the simple case discussed at the beginning of this example. [The discussion is continued in Chapter 16, example (2.c).]

VI. *Unrestricted Random Walks.* An unrestricted one-dimensional random walk is a Markov chain where it is most natural to order the states in a doubly infinite sequence $(\ldots E_{-2}, E_{-1}, E_0, E_1, E_2, \ldots)$. In order to write the matrix of transition probabilities in the familiar form, we must rearrange the states. For example, we may write them in the order $(E_0, E_1, E_{-1}, E_2, E_{-2}, \ldots)$; the first row of P is then $(0, p, q, 0, 0, \ldots)$, the second $(q, 0, 0, p, 0, 0, \ldots)$, etc. Unfortunately, the natural symmetry is lost, and the formulas become unpleasant. The situation grows even worse in two dimensions. In such cases the methods of this chapter are not convenient for deriving explicit formulas, but the general theorems apply and contain pertinent information.

VII. *The Ehrenfest Model of Diffusion.* Once more we consider a chain with the $a + 1$ states $E_0, E_1, \cdots, E_a$ and transitions possible only to the right and the left neighbor; however, this time we put $p_{j,j+1} = 1 - j/a$ and $p_{j,j-1} = j/a$, so that

$$P = \begin{bmatrix} 0 & 1 & 0 & 0 & \cdots & 0 & 0 \\ a^{-1} & 0 & 1 - a^{-1} & 0 & \cdots & 0 & 0 \\ 0 & 2a^{-1} & 0 & 1 - 2a^{-1} & \cdots & 0 & 0 \\ \cdot & \cdot & \cdot & \cdot & \cdots & \cdot & \cdot \\ 0 & 0 & 0 & 0 & \cdots & 0 & a^{-1} \\ 0 & 0 & 0 & 0 & \cdots & 1 & 0 \end{bmatrix}.$$

This chain has two interesting physical interpretations. For a discussion of various recurrence problems in statistical mechanics P. and T. Ehrenfest [3] described a conceptual experiment where a molecules are distributed in two containers A and B. At time n a molecule is chosen at random and removed from its container to the other. Let the state of the system be determined by the number of molecules in A. Suppose that at a certain moment there are exactly k molecules in the container A. At the next trial the system passes into E_{k-1} or E_{k+1} according to whether a molecule in A or B is chosen; the corresponding probabilities are k/a and $(a-k)/a$, and therefore our chain describes Ehrenfest's experiment. However, our chain can also be interpreted as *diffusion with a central force*,[4] that is, a random walk in which the probability of a step to the right varies with the position. From $x = j$ the particle is more likely to move to the right or to the left according as $j < a/2$ or $j > a/2$; this means that the particle has a tendency to move towards $x = a/2$, which corresponds to an attractive elastic force increasing in direct proportion to the distance. [Discussion continued in example (6.c) and problem 6.[5]]

VIII. *Occupancy Problems.* In Chapter 3 we considered random placements of balls into a cells. Let the number of occupied cells determine the state of the system. If j cells are occupied, the probability that the next ball is placed into an empty cell is $(a-j)/a$. Hence the experiment is described by a chain with transition probabilities $p_{jj} = j/a$, $p_{j,j+1} = (a-j)/a$, and $p_{j,k} = 0$ for all other combinations of j and k. The initial distribution (all cells empty) is given by $p_0 = 1$, $p_k = 0$ for $1 \leq k \leq a$. [Cf. Chapter 16, example (2.e).]

IX. *Sequential Sampling.* In Chapter 14 (end of section 1 and section 8) we considered the following generalized ruin problem connected with sequential sampling. Given a sequence of mutually independent random variables X_ν which assume only integral values (positive and negative) and have a common distribution $\{p_k\}$ ($k = 0$, $\pm 1, \pm 2, \ldots$). Put $\mathbf{S}_n = \mathbf{X}_1 + \cdots + \mathbf{X}_n$. There exists a smallest subscript n for which either $\mathbf{S}_n \geq b$ or $\mathbf{S}_n \leq -z$; here b and z are preassigned postive numbers and n is, of course, a random variable. A

---

[3] P. and T. Ehrenfest, Über zwei bekannte Einwände gegen das Boltzmannsche H-Theorem, *Physikalische Zeitschrift*, vol. 8 (1907), pp. 311–314.

[4] Ming Chen Wang and G. E. Uhlenbeck, On the Theory of the Brownian Motion II, *Reviews of Modern Physics*, vol. 17 (1945), pp. 323–342.

[5] For a more complete discussion (by methods essentially equivalent to those of Chapter 16) cf. M. Kac, Random Walk and the Theory of Brownian Motion, *American Mathematical Monthly*, vol. 54 (1947), pp. 369–391. See also B. Friedman, A Simple Urn Model, *Communications on Pure and Applied Mathematics*, vol. 2 (1949), pp. 59–70.

general problem of sequential sampling according to Wald consists in finding the distribution of n and the probabilities of the two contingencies $S_n \leq -z$ and $S_n \geq b$.

We interpret this problem as follows. Put $a = b + z$ and consider a Markov chain with possible states $x = 0, 1, 2, \cdots, a$. The initial state of the system is z. If S_1 has a value $-z < S_1 < b$, then we say that the first step takes the system into the state $x = S_1 + z$; if $S_1 \leq -z$, the first step takes the system into $x = 0$, while if $S_1 \geq b$, then a transition into $x = a$ occurs. If one of the two limiting states $x = 0$ and $x = a$ is reached, the system remains in it for all future time (which is a way of expressing that the process stops). Otherwise the process continues as described: at time n the system is in state $x = S_n + z$ provided that all partial sums $S_1, S_2, \cdots, S_{n-1}$ lie in the interval $-z < S_k < b$. Otherwise the system is in $x = 0$ or $x = a$ according to whether the first sum S_k which falls outside this interval is negative or positive. The matrix of transition probabilities is then

$$P = \begin{bmatrix} 1 & 0 & 0 & 0 & \cdots & 0 & 0 \\ r_1 & p_0 & p_1 & p_2 & \cdots & p_{a-1} & \rho_1 \\ r_2 & p_{-1} & p_0 & p_1 & \cdots & p_{a-2} & \rho_2 \\ r_3 & p_{-2} & p_{-1} & p_0 & \cdots & p_{a-3} & \rho_3 \\ \cdot & \cdot & \cdot & \cdot & \cdots & \cdot & \cdot \\ \cdot & \cdot & \cdot & \cdot & \cdots & \cdot & \cdot \\ \cdot & \cdot & \cdot & \cdot & \cdots & \cdot & \cdot \\ r_a & p_{-a+1} & p_{-a+2} & p_{-a+3} & \cdots & p_0 & \rho_a \\ 0 & 0 & 0 & 0 & \cdots & 0 & 1 \end{bmatrix}$$

where
$$r_k = p_{-k} + p_{-k-1} + p_{-k-2} + p_{-k-3} + \cdots$$
and
$$\rho_k = p_{a-k+1} + p_{a-k+2} + \cdots.$$

As an example consider Bartky's original sampling scheme (mentioned in section 1 of Chapter 14) which was the first sequential scheme to be proposed. The integer a, the so-called rejection level, is fixed. The lot of items which is subjected to sampling inspection must be large, and, for theoretical purposes, we shall assume it infinitely large. A preliminary sample is drawn and the lot is accepted if the sample contains no defective item and rejected if it contains at least a defective items. In either case the process terminates. The Markov chain starts only if the number j of defective items lies between the limits 0 and a, and in this case j is the initial state (so that the initial distribution depends on the manner in which the preliminary sample is taken). The process consists in drawing successive independent samples of fixed size N, counting each time the number of defectives. Allowance is made for one defective per lot, that is, whenever the new sample contains exactly one defective, the state remains unchanged. If no defective is found, the system moves to the next lower state, $j \to j - 1$. If $r + 1$ defectives are found, the system moves from j to $j + r$, except that it moves to a if $j + r \geq a$. In practice passing to 0 means acceptance, and passing to a rejection; sampling is continued until one of the two alternatives occurs.

In our previous notation X_n is the number of defectives in the nth sample minus one. Assuming that the number of defectives has a Bernoulli distribution, we have for $k \geq 0$

$$(2.1) \qquad p_k = \binom{N}{k+1} p^{k+1} q^{N-k-1},$$

and $p_{-1} = q^N$, $p_{-2} = p_{-3} \ldots = 0$.

X. *An Example from Genetics.*[6] Consider a population which is kept constant in size by the selection of N individuals in each successive generation. We classify individuals with respect to a particular gene pair (A, a). There are $2N$ genes in the population, and if in the nth generation A occurs j times, then a occurs $2N - j$ times. In this case we say that the population is at time n in state j ($0 \leq j \leq 2N$). Assuming random mating, the composition of the following generation is determined by $2N$ Bernoulli trials in which the A-gene has probability $j/2N$. We have therefore a Markov chain with

$$(2.2) \qquad p_{jk} = \binom{2N}{k} \left(\frac{j}{2N}\right)^k \left(1 - \frac{j}{2N}\right)^{2N-k}.$$

[Cf. example (8.c).]

XI. *A Breeding Problem.* In the so-called brother-sister mating two individuals are mated, and among their direct descendants two individuals of opposite sex are selected at random. These are again mated, and the process continues indefinitely. If there are the three genotypes AA, Aa, aa for each parent, then we have to distinguish six combina-

[6] This problem was discussed at length by R. A. Fisher and S. Wright. The formulation in terms of Markov chains is due to G. Malécot, Sur un problème de probabilités en chaine que pose la génétique, *Comptes rendus de l'Académie des Sciences*, vol. 219 (1944), pp. 379–381.

tions of parents. We order these six possible states of our system as follows: $E_1 = AA \times AA$, $E_2 = AA \times Aa$, $E_3 = Aa \times Aa$, $E_4 = Aa \times aa$, $E_5 = aa \times aa$, $E_6 = AA \times aa$. Using the rules of Chapter 5, it is easily seen that the matrix of transition probabilities is in this case

$$\begin{bmatrix} 1 & 0 & 0 & 0 & 0 & 0 \\ 1/4 & 1/2 & 1/4 & 0 & 0 & 0 \\ 1/16 & 1/4 & 1/4 & 1/4 & 1/16 & 1/8 \\ 0 & 0 & 1/4 & 1/2 & 1/4 & 0 \\ 0 & 0 & 0 & 0 & 1 & 0 \\ 0 & 0 & 1 & 0 & 0 & 0 \end{bmatrix}.$$

[The discussion is continued in example (4.c) and problem 3; a complete treatment is given in Chapter 16, example (4.b).]

XII. *Decomposable Chains.* The following is a rather artificial example designed to illustrate certain points of the theory.

Given a coin with faces E_1 and E_2, and a die with faces $E_3, \cdots, E_8$. We select one of the two pieces at random and perform independent trials with it. In other words, the entire process consists either in tossing of a coin or in throwing a die, each alternative having probability 1/2. The matrix of transition probabilities can be exhibited schematically in the form of a *partitioned matrix*,

$$P = \begin{pmatrix} A & 0 \\ 0 & B \end{pmatrix},$$

where A stands for the 2 by 2 matrix with elements 1/2, and B for the 6 by 6 matrix with elements 1/6; the zeros indicate that the remaining 24 elements vanish.

Obviously our chain is an artificial combination of the two chains representing coin tossing and die throwing. The matrices of transition probabilities corresponding to these two chains are A and B. It is more natural to study the two chains separately, and, at any rate, all properties of the combined chain can be obtained from a study of the two component chains. We have here a typical example of a *decomposition* (for the definition cf. section 4), and a similar procedure can be used for more complicated artificial combinations of several chains.

XIII. *Periodic Chains.* Let trials consist in throwing alternately a coin and a die and number the states as in the preceding example.

We have now for the transition probabilities the partitioned matrix

$$P = \begin{pmatrix} 0 & U \\ V & 0 \end{pmatrix}$$

where U is a 2 by 6, and V a 6 by 2 matrix. (The first two rows are 0, 0, 1/6, 1/6, 1/6, 1/6, 1/6, 1/6. The last six are 1/2, 1/2, 0, 0, 0, 0, 0, 0.) If we consider the process only at times 2, 4, 6, ..., we have a simple die-throwing experiment, whereas at times 1, 3, 5, ..., we are concerned with coin tossing. This chain has period 2.

3. Higher Transition Probabilities

A transition from E_j to E_k in exactly n steps can occur via different paths $E_j \to E_{j_1} \to E_{j_2} \to \cdots \to E_{j_{n-1}} \to E_k$. The conditional probability that the system passes through this particular path if it is at E_j at a certain time is $p_{jj_1}p_{j_1j_2} \cdots p_{j_{n-1}k}$. The sum of the corresponding expressions for all possible paths is the *probability of finding the system at time $r + n$ in state E_k, given that at time r it was in state E_j*. We shall denote this probability by $p_{jk}^{(n)}$.

We have, in particular, $p_{jk}^{(1)} = p_{jk}$, and

(3.1) $$p_{jk}^{(2)} = \sum_\nu p_{j\nu} p_{\nu k}.$$

By induction we find easily the *recursion formula*

(3.2) $$p_{jk}^{(n+1)} = \sum_\nu p_{j\nu} p_{\nu k}^{(n)};$$

a further induction on m shows that more generally

(3.3) $$p_{jk}^{(m+n)} = \sum_\nu p_{j\nu}^{(m)} p_{\nu k}^{(n)}.$$

This equation reflects the simple fact that the first m steps lead the system from E_j to some intermediate state E_ν, and the last n steps from E_ν to E_k. It characterizes Markov chains. For more general processes (cf. section 10) a similar equation holds, but the last factor depends not only on ν and k but also on j.

In the same way as the p_{jk} form the matrix P, we arrange the $p_{jk}^{(n)}$ in a matrix to be denoted by P^n. Equation (3.2) states that to obtain the element $p_{jk}^{(n+1)}$ of P^{n+1} we have to multiply the elements of the jth row of P by the corresponding elements of the kth column of P^n and add all products. This operation is called row into column multiplication of the matrices P and P^n, and is expressed symbolically by the equation $P^{n+1} = PP^n$. This suggests calling P^n the nth power of P; equation (3.3) expresses the associative law $P^{m+n} = P^m P^n$.

Examples. (a) In example I we have $P^n = P$ for all n.

(b) In example II the numbering of rows and columns starts with zero. In the zero column of P^n all elements equal q; in the first column they equal qp; and, generally, for $k \leq n-1$ all elements of column number k equal $p^k q$. Moreover, we have $p_{0,n}^{(n)} = p_{1,n+1}^{(n)} = p_{2,n+2}^{(n)} = \cdots = p^n$. These elements are on a line parallel to the main diagonal. All other elements of P^n vanish.

(c) In example XII all powers of P are identical with P.

(d) In example XIII the square of P equals the matrix of example XII; it follows that also $P^2 = P^4 = \ldots$. On the other hand, $P = P^3 = P^5 = \ldots$.

Absolute Probabilities. If the initial probability of the state E_j is a_j, then the (unconditional) probability of finding the system at time n in state E_k is obviously

$$(3.4) \qquad a_k^{(n)} = \sum_j a_j p_{jk}^{(n)}.$$

The most important properties of Markov chains depend on the asymptotic behavior of $p_{jk}^{(n)}$ as $n \to \infty$. Intuitively one would expect that the influence of the initial stage gradually wears off, so that for large n the probability of finding the system at time n in state E_k should be independent of the state at time 0. We mean by this that $p_{jk}^{(n)}$ tends to a limit u_k which is independent of j (a property called ergodicity). We shall show that our intuitive surmise is generally true, but exceptions exist.

4. Irreducible Chains.

We shall say that *the state E_k can be reached from E_j if there exists some n such that $p_{jk}^{(n)} > 0$*, that is, if there is a positive probability of reaching E_k from E_j, in n steps. It is not necessary that E_k can be reached from E_j in one step. For example, in an unrestricted random walk one-step transitions are possible only to the neighboring states, but every state can be reached from every other state. On the other hand, in example XII, only E_1 and E_2 can be reached from E_1 or E_2, while from E_3 only the states $E_3, \cdots, E_8$ can be reached. The following definition is designed to cope with such situations.

Definition. A set C of states is called closed if no one-step transition is possible from any state of C to any state outside C, that is, if $p_{jk} = 0$ whenever E_j is in C and E_k outside. A chain is called irreducible if there are no closed sets other than the set of all states.

If E_j is in the closed set C and E_k outside, then $p_{jk} = 0$, and it follows from (3.1) that also $p_{jk}^{(2)} = 0$. More generally we conclude from (3.2) that $p_{jk}^{(n)} = 0$ for every n. This expresses the intuitively obvious fact that no escape is possible from a closed set: *no state outside a closed set C can be reached from any state in C*. It follows that, if E_j is in C, then the sum of $p_{j\nu}^{(n)}$ extended over all those ν for which E_ν is also in C is unity. In the jth row of P^n the elements corresponding to states in C add to unity, while all others vanish. In other words:

If in the matrices P^n all rows and all columns corresponding to states outside the closed set C are deleted, there remain matrices for which the fundamental relations (3.2) *and* (3.3) *again hold.* This means that we have a Markov chain defined on C, and this subchain can be studied independently of other states.

Examples. (a) In example XII we have two closed sets; they are formed by E_1, E_2 and by $E_3, \cdots, E_8$, respectively.

(b) With the matrix of transition probabilities

$$P = \begin{bmatrix} 1/3 & 0 & 2/3 & 0 \\ 0 & 1/4 & 0 & 3/4 \\ 1/2 & 0 & 1/2 & 0 \\ 0 & 1/2 & 0 & 1/2 \end{bmatrix}$$

we have the two closed sets (E_1, E_3) and (E_2, E_4). Their matrices are

$$P_1 = \begin{pmatrix} 1/3 & 2/3 \\ 1/2 & 1/2 \end{pmatrix}, \quad P_2 = \begin{pmatrix} 1/4 & 3/4 \\ 1/2 & 1/2 \end{pmatrix},$$

respectively. If we reorder the states into the sequence E_1, E_3, E_2, E_4, then the matrix of the composite chain can be written in the form of a partitioned matrix

(4.1) $$P = \begin{pmatrix} P_1 & 0 \\ 0 & P_2 \end{pmatrix},$$

and it is easily verified that

(4.2) $$P^n = \begin{pmatrix} P_1^n & 0 \\ 0 & P_2^n \end{pmatrix}.$$

The last statement obviously does not depend on the particular form of P_1 and P_2.

(c) In the random walk with absorbing barriers (example III) each barrier is a closed set consisting of a single state. The same is true of the states E_0 and E_a in the sequential sampling example IX, of E_0

and E_{2N} in the example X, taken from genetics, and of E_1 and E_5 in the breeding example XI.

(d) In example V let a be even and $q_1 = q_3 = \cdots = q_{a-1} = 0$. Then the even states form one closed set, the odd states another.

If a single state E_k forms a closed set, it will be called an *absorbing state*. A necessary and sufficient condition for E_k to be an absorbing state is that $p_{kk} = 1$. (In this case all elements in the kth row of P except the diagonal element vanish.) A chain in which there exist two or more closed sets is called *decomposable*.

5. Classification of States

Consider an arbitrary, but fixed, state E_j and suppose that at time 0 the system is in E_j. Let $f_j^{(n)}$ be the probability that the *first return* to E_j occurs at time n. In particular,

$$f_j^{(1)} = p_{jj}, \quad f_j^{(2)} = p_{jj}^{(2)} - f_j^{(1)} p_{jj};$$

generally, the $f_j^{(n)}$ can be calculated from the obvious recurrence relation

$$(5.1) \quad f_j^{(n)} = p_{jj}^{(n)} - f_j^{(1)} p_{jj}^{(n-1)} - f_j^{(2)} p_{jj}^{(n-2)} - \cdots - f_j^{(n-1)} p_{jj}$$

which will be used only in an indirect way.

The sum

$$(5.2) \quad f_j = \sum_{n=1}^{\infty} f_j^{(n)}$$

may be interpreted as *the probability that the system ever returns to E_j*. If $f_j = 0$, a return is impossible, while $f_j = 1$ is interpreted as certainty of return. Once the system is back at E_j, the initial situation is reestablished, and the process starts from the beginning as a replica of the preceding trials. Hence the return to the state E_j is a *recurrent event* as defined in Chapter 12. If $f_j = 1$, this recurrent event is certain, and

$$(5.3) \quad \mu_j = \sum_{n=1}^{\infty} n f_j^{(n)}$$

is the *mean recurrence* time. [Note that equation (5.1) is, except for notation, identical with equation (3.1) of Chapter 12.]

From the theory of recurrent events we have the following double

Classification of States. (1) *The state E_j is called recurrent or transient according to whether a return to E_j is certain or uncertain (that is, accord-*

ing to whether $f_j = 1$ or $f_j < 1$). *A recurrent state E_j with infinite mean recurrence time is called a null state.*

(2) *The state E_j is called periodic with period t if a return to E_j is impossible except, perhaps, in t, $2t$, $3t$, ... steps, and $t > 1$ is the greatest integer with this property.* (In this case $p_{jj}^{(n)} = 0$ whenever n is not divisible by t.)

A recurrent state which is neither a null state nor periodic will be called ergodic.

Examples. In examples I and II all states are recurrent. A random walk with absorbing barriers necessarily ends after finitely many steps at one of the barriers; hence in example III the states E_j with $1 \leq j \leq a - 1$ are transient. Moreover, they have period 2 since a return can obviously occur only after an even number of steps. The two limiting states E_0 and E_a are recurrent and non-periodic, since for them a return after one step has probability one. In the case of of reflecting barriers (example IV) it is intuitively clear that all states are recurrent. However, they are non-periodic, since the system can stay at E_1 for an arbitrary time and then return to E_j. From Chapter 14, section 7, we know that in an unrestricted symmetric random walk in one or two dimensions all states are recurrent null states, whereas in three dimensions all states are transient. In either case all states have period 2.

From the fundamental theorem of Chapter 12, we get directly the

Criterion. *For a transient state E_j the series*

$$(5.4) \qquad \sum_{n=1}^{\infty} p_{jj}^{(n)}$$

converges. For a recurrent null state E_j this series diverges, but $p_{jj}^{(n)} \to 0$ as $n \to \infty$. If E_j is ergodic (that is, recurrent, but neither null state nor periodic), then $\mu_j < \infty$ and

$$(5.5) \qquad p_{jj}^{(n)} \to \frac{1}{\mu_j}.$$

If E_j has period t and is a recurrent non-null state, then $\mu_j < \infty$, and

$$(5.6) \qquad p_{jj}^{(nt)} \to \frac{t}{\mu_j}.$$

(while, of course, $p_{jj}^{(n)} = 0$ for all n not divisible by t).

The salient fact revealed by this criterion is that, except in the periodic case, $p_{jj}^{(n)}$ has a unique limit; the latter is zero if E_j is a null state or transient, and otherwise is given by (5.5). In the periodic case a limit exists for the subsequence $n = t, 2t, 3t, \ldots$.

Now let E_j be a fixed *recurrent* state and E_k some other state which can be reached from it. Furthermore, let N be the length of the *shortest* possible path from E_j to E_k, so that $p_{jk}^{(N)} = \alpha > 0$. A return from E_k to E_j must have positive probability, for otherwise the probability of the system not returning to E_j would be at least α, and $f_j \leq 1 - \alpha < 1$ contrary to the assumption that E_j is recurrent. It follows that there exists an index M such that $p_{kj}^{(M)} = \beta > 0$. Now for any n we have obviously

$$(5.7) \qquad p_{jj}^{(n+N+M)} \geq p_{jk}^{(N)} p_{kk}^{(n)} p_{kj}^{(M)} = \alpha\beta \cdot p_{kk}^{(n)}$$

and

$$(5.8) \qquad p_{kk}^{(n+N+M)} \geq p_{kj}^{(M)} p_{jj}^{(n)} p_{jk}^{(N)} = \alpha\beta \cdot p_{jj}^{(n)}.$$

These relations imply that the sequences $p_{jj}^{(n)}$ and $p_{kk}^{(n)}$ have the same asymptotic behavior, and from this we can draw important conclusions. To begin with, E_j was assumed recurrent, and therefore the series $\Sigma p_{jj}^{(n)}$ diverges. From (5.8) it follows that also $\Sigma p_{kk}^{(n)}$ diverges, so that E_k must be recurrent. If $p_{jj}^{(n)} \to 0$, then also $p_{kk}^{(n)} \to 0$, and vice versa. Finally, suppose that E_j has period t. A return to E_j is possible in $N + M$ steps, so that $N + M$ must be a multiple of t. It follows then from (5.7) and (5.8) that E_j and E_k must have the same period.

We see thus that *from a recurrent state only recurrent states can be reached, and they are all of the same type:* either they are all null states, or all ergodic, or all periodic non-null states with the same period. Now the set of all states that can be reached from E_j is obviously closed and is therefore the smallest closed set containing E_j. It follows that in an *irreducible* chain every state can be reached from every other state, and hence if one state is transient so are necessarily all others. We have thus proved the important

Theorem. *In an irreducible Markov chain all states belong to the same class: they are all transient, all recurrent null states, or all recurrent non-null states. In every case they have the same period. Moreover, every state can be reached from every other state.*

In every chain the recurrent states can, in a unique manner, be divided into closed sets $C_1, C_2, \ldots$ *such that from any state of a given set all states of that set and no other can be reached.* Since each C_ν can be treated *independently as a Markov chain, all states belonging to the same closed set C_ν are necessarily of the same class.*

In addition to the closed sets C_ν the chain will in general contain transient states from which states of the closed sets C_ν can be reached (but not vice versa).

Examples. (a) In the one-dimensional symmetric random walk (coin tossing) all states are recurrent null states of period 2; if the random walk is unsymmetric, all states are transient (cf. Chapter 12, section 3). In the random walk with absorbing barriers (example III) E_0 forms one closed set, and E_a another closed set. All other states are transient. From each transient state all other states can be reached.

(b) Consider the chain with states $E_1, \cdots, E_6$ and matrix

$$P = \begin{bmatrix} 1/2 & 1/2 & 0 & 0 & 0 & 0 \\ 1/2 & 1/2 & 0 & 0 & 0 & 0 \\ 0 & 0 & 1/3 & 2/3 & 0 & 0 \\ 0 & 0 & 2/3 & 1/3 & 0 & 0 \\ 1/6 & 1/6 & 1/6 & 1/6 & 1/6 & 1/6 \\ 1/6 & 1/6 & 1/6 & 1/6 & 1/6 & 1/6 \end{bmatrix}.$$

Here we have two closed sets; C_1 consists of E_1 and E_2, while C_2 is formed by E_3 and E_4. From E_5 and E_6 the system can pass either to the closed set C_1 or to C_2 and then no return is possible. Hence E_5 and E_6 are transient. Clearly each of the sets C_1 and C_2 can be studied in itself as a complete Markov chain. The transient states connect C_1 and C_2 inasmuch as either closed set can be reached from them. The situation is analogous to the case of a random walk with absorbing barriers, where the two closed sets contained only one state each, but there were many more transient states.

In general, the matrix P corresponding to a chain with two closed sets C_1 and C_2 and additional transient states can be written schematically in the form of a partitioned matrix

$$(5.9) \qquad P = \begin{bmatrix} P_1 & 0 & 0 \\ 0 & P_2 & 0 \\ A & B & C \end{bmatrix}$$

where P_1 and P_2 are the matrices of transition probabilities within the two closed sets. The matrix P^n is then of the same type with P_1, P_2, C replaced by P_1^n, P_2^n, C^n (and A and B by more complicated matrices to be studied in section 8). Note that P_1, P_2, and C are square matrices, but that A and B may be rectangular matrices. Thus, in example III, the two corner elements p_{00} and p_{aa} represent P_1 and P_2. The matrix C is the $a - 1$ by $a - 1$ matrix obtained by deleting the first and last rows and columns. Finally, A and B are single-column matrices with elements $(q, 0, 0, \cdots, 0)$ and $(0, 0, \cdots, 0, p)$, respectively.

It will help to clarify ideas if we mention here that theorem 1 of the next section has the following

Corollary. A finite chain can contain no null states, and it is impossible that all its states are transient.

6. Ergodic Properties of Aperiodic Chains; Stationary Distributions

In the preceding section we have described the asymptotic behavior of the diagonal terms $p_{jj}^{(n)}$. These results will now be used for a discussion of the behavior of $p_{jk}^{(n)}$ for an arbitrary pair j, k. In this section we consider mainly aperiodic and irreducible chains. For them we shall establish the fact stated at the end of section 3, namely, that the probability of finding the system at time n in state E_k is in the limit independent of the initial state.

Let $f_{jk}^{(n)}$ be the probability that, starting from E_j, the system reaches E_k for the first time at the nth step (first passage through E_k starting from E_j). We have clearly

$$p_{jk}^{(n)} = f_{jk}^{(n)} + f_{jk}^{(n-1)} p_{kk}$$
(6.1)
$$+ f_{jk}^{(n-2)} p_{kk}^{(2)} + \cdots + f_{jk}^{(1)} p_{kk}^{(n-1)}.$$

This relation is a direct generalization of (5.1) and enables us to calculate recursively the $f_{jk}^{(n)}$ in terms of the $p_{jk}^{(n)}$ or vice versa.

Theorem 1. If the state E_k is either transient or a recurrent null state, then $p_{jk}^{(n)} \to 0$ for every j.

(Note that here periodic chains are not excluded.)

Proof. The criterion of section 5 assures us that $p_{kk}^{(n)} \to 0$. Hence for every fixed N the last N terms in (6.1) tend to zero. The first $n - N$ terms add to at most $f_{jk}^{(n)} + f_{jk}^{(n-1)} + \cdots + f_{jk}^{(N+1)}$; their sum is therefore less than the Nth remainder of a convergent series and can be made arbitrarily small by choosing N large enough.

This proves the theorem. If the chain contains only a finite number, a, of states, then for at least one k we must have $p_{jk}^{(n)} \geq 1/a$. It is then impossible that all $p_{jk}^{(n)}$ should tend to zero, and this proves the *corollary* stated at the end of section 5.

Theorem 2. *Suppose that the states of an irreducible chain are aperiodic and neither transient nor null states. Then for every pair j, k the limit*

(6.2) $$\lim_{n\to\infty} p_{jk}^{(n)} = u_k > 0$$

exists and is independent of j. The reciprocal of u_k is the mean recurrence time μ_k of E_k. Moreover, $\{u_k\}$ is a probability distribution with positive elements, that is,

(6.3) $$u_k > 0, \quad \Sigma u_k = 1.$$

Finally, the u_k satisfy the system of linear equations:

(6.4) $$u_k = \sum_\nu u_\nu p_{\nu k}.$$

The distribution $\{u_k\}$ is uniquely determined by (6.4) and (6.3) or, more precisely, if $\{v_k\}$ is any other sequence satisfying the conditions

(6.5) $$v_k = \sum_\nu v_\nu p_{\nu k}, \quad \sum |v_k| < \infty,$$

then $u_k = cv_k$ with a constant c.

Proof. By assumption each state E_k has a finite mean recurrence time μ_k. Put $u_k = 1/\mu_k$. We know from (5.5) that $p_{kk}^{(n)} \to u_k$, so that (6.2) holds when $j = k$. Now for any fixed j we have $\Sigma f_{jk}^{(n)} = 1$, for otherwise the system would have a positive probability of passing from E_k to E_j and never to return, which contradicts the hypothesis that E_k is recurrent. Hence we can to every $\epsilon > 0$ select an N so that $f_{jk}^{(1)} + \cdots + f_{jk}^{(N)} > 1 - \epsilon$. The last N terms on the right side in (6.1) differ arbitrarily little from $u_k\{f_{jk}^{(1)} + \cdots + f_{jk}^{(N)}\}$, and hence from u_k; the sum of the first $n - N$ terms is less than the Nth remainder in the series $\Sigma f_{jk}^{(n)}$, and hence less than ϵ. This proves (6.2).

To prove (6.4) we first note that

(6.6) $$\Sigma u_k \leq 1.$$

This follows directly from the fact that for fixed j and n the quantities $p_{jk}^{(n)}$ ($k = 1, 2, \ldots$) add to unity, so that $u_1 + u_2 + \cdots + u_N \leq 1$ for every N. Now put $n = 1$ in (3.3) and let $m \to \infty$. The left side tends to u_k, and the general term of the sum on the right side tends to $u_\nu p_{\nu k}$. Adding an arbitrary finite number of terms, we see that

(6.7) $$u_k \geq \sum_\nu u_\nu p_{\nu k}.$$

Summing these inequalities over all k, we obtain the finite quantity

Σu_k on each side. This shows that in (6.7) the inequality is impossible, and thus (6.4) is proved.

If the sequence $\{v_k\}$ satisfies (6.5), then we may multiply the equation in (6.5) by p_{kr} and sum over all k. We get

$$(6.8) \qquad v_r = \sum_\nu v_\nu p_{\nu r}^{(2)}.$$

Repeating the same operation, it follows that for every n

$$(6.9) \qquad v_r = \sum_\nu v_\nu p_{\nu r}^{(n)}.$$

Now we have assumed that the series of the coefficients v_ν converges absolutely. We can, therefore, in (6.9) let $n \to \infty$ and obtain in the limit

$$(6.10) \qquad v_r = (v_1 + v_2 + v_3 + \ldots) u_r.$$

The sum in the parentheses is a constant independent of r, and this proves that the ratio v_r/u_r is constant. Finally, putting $v_k = u_k$, we find that $\Sigma u_k = 1$. This accomplishes the proof.

Examples. (a) In example II, the system (6.4) reduces to $u_0 = (u_0 + u_1 + \ldots)q = q$, and $u_k = u_{k-1} p$ for $k \geq 1$. The solution is obviously $u_k = p^k q$, in agreement with the fact found in example (3.b) that $p_{jk}^{(n)} \to p^k q$.

(b) *Random Walk with Reflecting Barriers.* In example IV the system (6.4) reduces to

$$qu_1 + qu_2 = u_1$$
$$(6.11) \qquad pu_{k-1} + qu_{k+1} = u_k \qquad (k = 2, 3, \cdots, a-1)$$
$$pu_{a-1} + pu_a = u_a$$

Then $u_k = u_{k-1}(p/q)$ and hence $u_k = u_1(p/q)^{k-1}$. The value of u_1 follows from the condition $\Sigma u_k = 1$. The final result is

$$(6.12) \qquad u_k = \frac{1 - p/q}{1 - (p/q)^a} \cdot (p/q)^{k-1}$$

if $p \neq q$, and $u_k = 1/a$ if $p = q = 1/2$. Wherever the system starts at time 0, the probability of finding the system at time n in state E_k is, in the limit, given by u_k. If $p = q$, all states become equally likely, while in the case $p < q$ the states near the left barrier are more probable.

(c) *The Ehrenfest Model.* In example VII the equations (6.4) take on the form

(6.13)
$$u_k = \left(1 - \frac{k-1}{a}\right) u_{k-1} + \frac{k+1}{a} u_{k+1} \qquad (k = 1, \cdots, a-1)$$

$$u_0 = \frac{u_1}{a}, \quad u_a = \frac{u_{a-1}}{a}.$$

It is easily verified that the required solution is

(6.14)
$$u_k = \binom{a}{k} 2^{-a}.$$

This is a binomial distribution, and the result can be interpreted as follows: whatever the initial number of molecules in the first container, after a long time the probability of finding exactly k molecules in it is the same as if the a molecules had been distributed at random, each molecule having probability $1/2$ to be in the first container. This is a typical example of how our result gains physical significance.

The normal approximation to the binomial distribution shows that, if a is large, then, once the limiting distribution (6.14) is established, we are practically certain to find about one-half of the molecules in each container. To the physicist $a = 10^6$ is a small number, indeed. But even with $a = 10^6$ molecules the probability of finding more than 505,000 molecules in one container (density fluctuation of about 1 per cent) is of the order of magnitude 10^{-23}. With $a = 10^8$ a density fluctuation of one in a thousand has the same negligible probability. It is true that the system will occasionally pass into very improbable states, but their recurrence times are fantastically large as compared with the recurrence times of states near the equilibrium. Physical irreversibility manifests itself in the fact that, whenever the system is in a state far removed from equilibrium, it is much more likely to move towards equilibrium than in the opposite direction.

(d) *Doubly Stochastic Matrices.* The matrix P is called doubly stochastic if not only the row sums, but also the column sums, are unity. Suppose that the chain contains only a finite number, a, of states. The system (6.4) has then obviously the solution $u_k = 1/a$. It follows that, *if a finite irreducible aperiodic chain has a doubly stochastic matrix P, then $u_k = 1/a$ (that is, in the limit all states become equally probable).* It is easily seen that the condition that P be doubly stochastic is not only sufficient but also necessary. Clearly P^n is again a doubly stochastic matrix and, hence, if the finite matrix P is doubly stochastic, *no state can be transient* (cf. problem 9).

It should be remembered that our theorems apply also to *reducible* chains, since each closed set can be treated separately. Suppose that E_j is a recurrent state of an aperiodic irreducible subchain C. The behavior of $p_{jk}^{(n)}$ for all states E_k in C is described in theorems 1 and 2. If E_k is outside C, then $p_{jk}^{(n)} = 0$ for all n. We lack only information concerning $p_{jk}^{(n)}$ in the periodic case, and if E_j is transient and E_k recurrent. The periodic case will be dealt with in the next section, the transient case in section 8.

We can reformulate our theorems in terms of the absolute probabilities $\{a_k^{(n)}\}$ introduced at the end of section 3. It follows from (3.4) that (6.2) implies

$$(6.15) \qquad a_k^{(n)} \to u_k.$$

Our theorems admit

Corollary I. In every aperiodic irreducible chain the probability $a_k^{(n)}$ of finding the system at time n in state E_k tends to a uniquely determined limit which is independent of the initial distribution. If all states are transient or null states, $a_k^{(n)} \to 0$ for all k. Otherwise the probability of E_k is, in the limit, the reciprocal of the (finite) mean recurrence time of E_k.

Stationary Distributions. The initial probability distribution $\{a_k\}$ is called *stationary* if the probabilities $\{a_k^{(n)}\}$ are independent of n, that is, if $a_k^{(n)} = a_k$. The physical significance of stationarity becomes apparent if we imagine a large number of processes going on simultaneously. Let, for example, N particles perform independently the same type of random walk. At time n the expected number of particles in state E_k is $Na_k^{(n)}$. With a stationary distribution these expected numbers remain constant, and we observe (if N is large so that the law of large numbers applies) a state of *macroscopic equilibrium* maintained by a large number of transitions in opposite directions. Most statistical equilibria in physics are of this kind.

Corollary I asserts that $a_k^{(n)}$ has a limit as $n \to \infty$; for a stationary distribution a_k must coincide with this limit. On the other hand, under the conditions of theorem 2 the limits $\{u_k\}$ form a probability distribution, and (6.4) shows that this distribution is stationary. Hence it is the unique stationary distribution, and we have

Corollary II. *If all states of an aperiodic irreducible chain are transient or null states, there exists no stationary distribution; otherwise there exists a unique stationary distribution* $\{u_k\}$ *and the probability distribution* $\{a_k^{(n)}\}$ *necessarily converges towards it.* (For finite chains only the second alternative is possible.)

We have seen that in physics the convergence of $a_k^{(n)}$ to u_k may be interpreted as *tendency towards a state of equilibrium*. A typical example where no state of equilibrium exists is the one-dimensional unrestricted random walk. If $p \neq q$, there exists a drift, and all states are transient; whatever the number of particles, after a long time they will have drifted away towards infinity. If $p = q = 1/2$, all states are recurrent, but the tendency towards equilibrium requires all states to become, in the limit, equally probable, so that the probability of each state tends to zero.

* 7. Periodic Chains

In the preceding section we have excluded the case of periodic chains, but this was done only in order not to obscure salient facts by complicated descriptions. A characterization of the asymptotic behavior of $p_{jk}^{(n)}$ in irreducible periodic chains can be easily derived from the theorems of the preceding sections. We give such a derivation for the sake of completeness, but the results of this section will not be used in the sequel.

By the theorem of section 5 all states of an irreducible chain have the same period t. Consider now any two states E_j and E_k of an irreducible chain with period t. Since every state can be reached from every other, there exist integers a, b such that $p_{jk}^{(a)} > 0$ and $p_{kj}^{(b)} > 0$. Now $p_{jj}^{(a+b)} \geq p_{jk}^{(a)} p_{kj}^{(b)}$, which shows that a return to E_j in $a + b$ steps is possible, so that $a + b$ is necessarily divisible by the period t. It follows that, if E_k can be reached from E_j in a_1 and in a_2 steps, then $a_2 - a_1$ must be divisible by t, and hence a division of a_1 and a_2 by t will leave the same remainder.

Accordingly, for each fixed state E_j there corresponds to every state E_k a certain remainder ν (with $\nu = 0, 1, \cdots, t - 1$), so that a transition from E_j to E_k is possible only in $\nu, \nu + t, \nu + 2t, \nu + 3t, \ldots$ steps. We put $j = 1$ and have then a classification of all states into t groups $G_0, G_1, \cdots, G_{t-1}$, in the following way. If $p_{1k}^{(a)} > 0$ and $a = nt + \nu$ (where ν is the remainder so that $0 \leq \nu < t$), then E_k belongs to G_ν.

* Starred sections treat special topics and may be omitted at first reading.

We imagine the G_ν ordered cyclically so that G_0 and G_{t-1} become neighbors.

It follows in particular that a *one-step* transition from a state in G_ν will always lead to a state in the next following group $G_{\nu+1}$ (or G_0 in case $\nu = t - 1$); a two-step transition will lead to a state in $G_{\nu+2}$ (from G_{t-2} it leads to G_0, from G_{t-1} to G_1), etc. Finally, a t-step transition leads necessarily to a state belonging to the same group. This means that, in a Markov chain whose matrix of transition probabilities is P^t, each group G_ν forms a closed set. Since the original chain is irreducible, each state can be reached from every other. This implies that in the chain with transition probabilities P^t each G_ν forms an irreducible closed set. We have thus the

Theorem. *In an irreducible periodic Markov chain all states can be divided into t groups $G_0, \cdots, G_{t-1}$, so that a one-step transition from a state of G_ν always leads to a state of $G_{\nu+1}$ (to G_0 if $\nu = t - 1$). If we consider the chain only at times $t, 2t, 3t, \ldots$, then we get a new chain whose matrix of transition probabilities is P^t. In it each G_ν forms an irreducible closed set.*

Examples. (a) In an unrestricted random walk all states have period 2. In one dimension the group G_0 is formed by all even positions, G_1 by all odd ones. In more dimensions the same statement is true if a position is called even or odd according to whether the sum of its coordinates is even or odd.

(b) In example XIII the states E_1 and E_2 form G_0, and the remaining six states G_1.

(c) Consider six states $E_1, \cdots, E_6$ with the matrix

$$(7.1) \qquad P = \begin{bmatrix} 0 & 1/2 & 1/2 & 0 & 0 & 0 \\ 0 & 0 & 0 & 1/3 & 1/3 & 1/3 \\ 0 & 0 & 0 & 1/3 & 1/3 & 1/3 \\ 1 & 0 & 0 & 0 & 0 & 0 \\ 1 & 0 & 0 & 0 & 0 & 0 \\ 1 & 0 & 0 & 0 & 0 & 0 \end{bmatrix}.$$

From E_1 the system necessarily passes to E_2 or E_3. From E_2 and E_3 transitions are possible only to the states E_4, E_5, E_6, and from any of these the system necessarily passes to E_1. Consequently, the chain has period 3. The group G_0 consists of only E_1. The group G_1 is formed by E_2 and E_3, the group G_2 by E_4, E_5, E_6. We have

$$P^2 = \begin{bmatrix} 0 & 0 & 0 & 1/3 & 1/3 & 1/3 \\ 1 & 0 & 0 & 0 & 0 & 0 \\ 1 & 0 & 0 & 0 & 0 & 0 \\ 0 & 1/2 & 1/2 & 0 & 0 & 0 \\ 0 & 1/2 & 1/2 & 0 & 0 & 0 \\ 0 & 1/2 & 1/2 & 0 & 0 & 0 \end{bmatrix}, \quad P^3 = \begin{bmatrix} 1 & 0 & 0 & 0 & 0 & 0 \\ 0 & 1/2 & 1/2 & 0 & 0 & 0 \\ 0 & 1/2 & 1/2 & 0 & 0 & 0 \\ 0 & 0 & 0 & 1/3 & 1/3 & 1/3 \\ 0 & 0 & 0 & 1/3 & 1/3 & 1/3 \\ 0 & 0 & 0 & 1/3 & 1/3 & 1/3 \end{bmatrix}$$

and then periodically $P^4 = P$, $P^5 = P^2$, etc.

Our theorem contains complete information concerning the *asymptotic behavior* of $p_{jk}{}^{(n)}$. If all states are transient or null states, then $p_{jk}{}^{(n)} \to 0$ for every pair j, k (theorem 1 of section 6). Otherwise each state E_k has a finite mean recurrence time μ_k. Suppose that E_j belongs to G_ν. On G_ν we have an irreducible non-periodic Markov chain with transition probabilities $p_{jk}{}^{(t)}$, and hence (by theorem 2 of section 6) there exist the limits

(7.2) $$\lim_{n \to \infty} p_{jk}{}^{(nt)} = \begin{cases} u_k & \text{if } E_k \text{ is in } G_\nu \\ 0 & \text{otherwise.} \end{cases}$$

Here u_k is the reciprocal mean recurrence time of E_k in the new chain, one step of which corresponds to t steps of the original chain. Hence

(7.3) $$u_k = \frac{t}{\mu_k}.$$

Using (3.2), we find from (7.2),

(7.4) $$\lim_{n \to \infty} p_{jk}{}^{(nt+1)} = \begin{cases} u_k & \text{if } E_k \text{ is in } G_{\nu+1} \\ 0 & \text{otherwise.} \end{cases}$$

Similarly, $p_{jk}{}^{(nt+2)} \to u_k$ if E_k is in $G_{\nu+2}$, etc. In other words, *for fixed E_j and E_k the sequence $p_{jk}{}^{(n)}$ is asymptotically periodic; in it blocks of $t-1$ consecutive zeros alternate with a positive element which converges to $u_k = t/\mu_k$.*

By theorem 2 of section 6, the u_k within each group G_ν add to unity. Since there are t blocks, it follows from (7.3) that the sequence $\{1/\mu_k\}$ represents a probability distribution. The argument of section 6 shows directly that *this distribution is stationary and that no other stationary distributions exist.*

8. Transient States; Absorption Probabilities

In the two preceding sections we have completely described the asymptotic behavior of $p_{jk}^{(n)}$ for the case where E_j is recurrent. If E_k is transient, then $p_{jk}^{(n)} \to 0$ for all j (theorem 1, section 6). It remains to investigate the case where E_j is transient and E_k recurrent. This is a direct generalization of the classical ruin problem or random walk with two absorbing barriers (example III). In that particular case the two absorbing states E_0 and E_a are the only recurrent states, and $p_{j0}^{(n)}$ and $p_{ja}^{(n)}$ are, respectively, the probabilities that the gambler or his adversary will be ruined at the nth step or before, assuming that their initial capitals are j and $a - j$. In example IX (sequential sampling) we have a similar situation.

In the general case the recurrent state E_k will belong to a closed set C containing more than one state. Once the system is in C, it will remain there and continue occasionally to pass through E_k. We seek the probability x_j that the system, starting from the transient state E_j, will ultimately land in the closed set C.

Suppose that *the system is initially in the transient state E_j and let $x_j^{(n)}$ be the probability that at time n, and not sooner, the system reaches the closed set C. Then*

$$(8.1) \qquad x_j = \sum_{n=1}^{\infty} x_j^{(n)}$$

defines the probability that the system will ultimately reach and stay in C. By analogy with the simple random walk *we shall call x_j the probability of absorption in C.* The difference $1 - x_j$ accounts for the possibility of absorption in other closed sets and (in the case of some infinite chains) of an indefinite continuation in transient states.

It is clear that

$$(8.2) \qquad x_j^{(1)} = \sum_C p_{jk},$$

the summation extending over those k for which E_k is contained in C. If the system reaches C at the $(n+1)$th step, then the first step must lead from E_j to another transient state. It is therefore clear that

$$(8.3) \qquad x_j^{(n+1)} = \sum_T p_{j\nu} x_\nu^{(n)},$$

the summation now extending over those ν for which E_ν is transient. *Equations (8.2) and (8.3) are recurrence relations which uniquely determine the $x_j^{(n)}$.* Adding (8.3) for $n = 1, 2, 3, \ldots$, we find that *the* *absorption probabilities* x_j are solutions of the system of linear equations

(8.4) $$x_j - \sum_T p_{j\nu} x_\nu = x_j^{(1)}.$$

Examples. (a) *Random Walk with Absorbing Barriers* (example III). Take for C the absorbing state E_0. Then $x_1^{(1)} = q$ and $x_j^{(1)} = 0$ if $j > 1$. The system (8.4) therefore reduces to

$$x_1 - px_2 = q,$$

(8.5) $\quad x_j - qx_{j-1} - px_{j+1} = 0 \qquad (j = 2, 3, \cdots, a-2),$

$$x_{a-1} - qx_a = 0.$$

This is the same as the system (2.1)–(2.2) of Chapter 14, and the solution is given in (2.4).

(b) *Sequential Sampling* (example IX). Again let C be the state E_0. Then $x_j^{(1)} = r_j$, and the equations (8.4) reduce to (8.2) of Chapter 14 (where u_x stands for the present x_j; cf. also problem 18 of Chapter 14).

(c) *Genetics* (example X). Here each of the two states E_0 and E_{2N} forms a closed set. Absorption in E_0 and in E_{2N} signifies, respectively, that the population ultimately consists only of aa- or only of AA-individuals. For the absorption in E_0 we have $x_j^{(1)} = p_{j0} = (1 - j/2N)^{2N}$, and hence (8.4) assumes the form

(8.6) $$x_j - \sum_{\nu=1}^{2N-1} \binom{2N}{\nu} \left(\frac{j}{2N}\right)^\nu \left(1 - \frac{j}{2N}\right)^{2N-\nu} x_\nu = \left(1 - \frac{j}{2N}\right)^{2N}.$$

It is plausible that at a moment when the A- and a-genes are in the proportion $j:2N - j$ their survival chances should be in the same ratio. This suggests that the solution to (8.6) should be $x_j = 1 - j/2N$. This is easily verified, using the fact that (8.6) contains the terms of the binomial distribution with mean $2N(j/2N) = j$.

(d) *Example* (5.b). Let C consist of E_1 and E_2. Then $x_5^{(1)} = x_6^{(1)} = \frac{1}{3}$, and equations (8.4) take on the form $x_5 - (x_5 + x_6)/6 = \frac{1}{3}$ and $x_6 - (x_5 + x_6)/6 = \frac{1}{3}$. The solution is $x_5 = x_6 = \frac{1}{2}$, as should be expected for reasons of symmetry.

Once the system is within the closed set C, the process continues as described in the preceding two sections. In particular, if C is not periodic and the system is known to be in C, then the probability of finding it in the particular state E_k of C tends to $1/\mu_k$, where μ_k is the mean recurrence time of E_k. It follows easily that, if E_j is transient,

(8.7) $$p_{jk}^{(n)} \to \frac{x_j}{\mu_k}.$$

Similarly, if C is periodic, then also $p_{jk}^{(n)}$ will be asymptotically periodic. We have thus completed the description of the asymptotic behavior of $p_{jk}^{(n)}$ for all cases.

Two questions have been left open: (1) Is the solution of the system of linear equations (8.4) unique? (2) What is the probability that the system will continue to pass from transient state to transient state without ever reaching a recurrent state? The two questions are closely related.

Let E_j be transient and let $y_j^{(n)}$ be *the probability that the system is at time n in a transient state, given that it started from E_j at time* 0. Obviously

(8.8) $$y_j^{(1)} = \sum_T p_{j\nu},$$
$$y_j^{(n+1)} = \sum_T p_{j\nu} y_\nu^{(n)},$$

the summations again extending over all ν for which E_ν is transient. It follows from (8.8) that $y_j^{(1)} \leq 1$ and hence $y_j^{(2)} \leq y_j^{(1)}$, and generally $y_j^{(n+1)} \leq y_j^{(n)}$. Therefore a limit

(8.9) $$y_j = \lim_{n \to \infty} y_j^{(n)}$$

exists; y_j is the probability of the system forever staying in transient states. From (8.8) we have

(8.10) $$y_j = \sum_T p_{j\nu} y_\nu.$$

If y_j is any solution of (8.10) with $|y_j| \leq 1$, then a comparison of (8.10) and (8.8) shows first that $|y_j| \leq |y_j^{(1)}|$, and then by induction that $|y_j| \leq y_j^{(n)}$ for all n. Hence we have the

Theorem: *the probability y_j of the system forever remaining in the set of transient states satisfies the system of linear equations* (8.10). *This probability is zero for all j if and only if the system* (8.10) *has no bounded solution except $y_j \equiv 0$.* This is always the case for *finite* chains. For, if there are only finitely many y_j, let M be their maximum. From (8.10) we have

(8.11) $$M \leq \sum_T p_{j\nu} M,$$

and the equality sign can hold only if (j being fixed) the $p_{j\nu}$ for which $y_\nu = M$ add to unity. In this case, however, these y_ν would form a closed set, and since the chain is finite, not all of them could be transient.

If now (8.4) has two different bounded solutions, then their difference $\{y_j\}$ is a solution of (8.10). Conversely, if (8.10) has a solution y_j, then $x_j + y_j$ is a new solution of (8.4). Hence:

For the solution x_j of (8.4) *to be unique it is necessary and sufficient that the probability y_j of the system forever remaining in transient states is zero for every initial transient E_j. This is always the case if the chain is finite.*

Duration of the Game. For a given initial transient state E_j let Y_j be the time when the system for the first time passes into a recurrent state (so that $Y_j - 1$ is the number of steps preceding absorption in some closed set). In some of our examples the random variable Y_j is the duration of the game, and we use this term generally. Clearly $Pr\{Y_j = n\} = y_j^{(n-1)} - y_j^{(n)}$; these probabilities add to unity if, and only if, $y_j = 0$. In this case

(8.12) $$d_j = \Sigma n Pr\{Y_j = n\} = \sum_{n=0}^{\infty} y_j^{(n)}$$

is the mean duration of the game. From (8.8) it follows that *the mean duration is the solution of the system of linear equations,*

(8.13) $$d_j - \sum_T p_{j\nu} d_\nu = 1.$$

The solution is *uniquely determined* whenever $y_j = 0$, that is, whenever there is certainty that the game will end after finitely many steps. (If this is not the case, there is no finite mean duration.) (Cf. problem 18 of Chapter 14.)

9. Application to Card Shuffling

A deck of N cards numbered $1, 2, \cdots, N$ can be arranged in $N!$ different orders, and each represents a possible state of the system. Every particular shuffling operation effects a transition from the existing state into some other state. For example, "cutting" will change the order $(1, 2, \cdots, N)$ into one of the N cyclically equivalent orders $(r, r+1, \cdots, N, 1, 2, \cdots, r-1)$. The same operation applied to the inverse order $(N, N-1, \cdots, 1)$ will produce $(N-r+1, N-r+2, \cdots, 1, N, N-1, \cdots, N-r)$. In other words, we con-conceive of each particular shuffling operation as a transformation $E_j \to E_k$. If *exactly* the same operation is repeated, the system will pass (starting from the given state E_j) through a well-defined succession of states, and after a finite number of steps the original order will be re-established. From then on the same succession of states will recur periodically. For most operations the period will be rather small, and in *no* case can all states be reached by this procedure.[7] For example, a perfect "lacing" would change a deck of $2m$ cards from $(1, \cdots, 2m)$ into $(1, m+1, 2, m+2, \cdots, m, 2m)$. With 6 cards four applications of this operation will re-establish the original order. With 10 cards the initial order will reappear after six operations, so that repeated perfect lacing of a deck of 10 cards can produce only six out of the $10! = 3{,}628{,}800$ possible orders.

In practice the player may vary the operations, and certainly the play of chance will introduce variations, so that even a player attempt-

---

[7] In the language of group theory this amounts to saying that the permutation group is not cyclic and can therefore not be generated by a simple operation.

ing to achieve identical operations will not always be successful. We shall assume that we can account for the player's habits and the influence of chance variations by assuming that every particular operation has a certain probability (possibly zero). We need assume nothing about the numerical values of these probabilities, but shall suppose that the player operates without regard to the past and does not know the order of the cards.[8] This implies that the successive operations correspond to independent trials with fixed probabilities: for the actual deck of cards we then have a Markov chain.

We now show that the matrix P of transition probabilities is *doubly stochastic* [example (6.d)]. In fact, if an operation changes a state (order of cards) E_j to E_k, then there exists another state E_r which it will change into E_j. This means that the elements of the jth column of P are identical with the elements of the jth row, except that they appear in a different order. All column sums are therefore unity.

It follows [example (6.d)] that no state can be transient. *If the chain is irreducible and aperiodic, then in the limit all states become equally probable.* In other words, *any* kind of shuffling will do, provided only that it produces an irreducible and aperiodic chain. It is safe to assume that this usually is the case. Suppose, however, that the deck contains an even number of cards and the procedure consists in dividing them equally into two parts and shuffling them separately by any method. If the two parts are put together in their original order, then the Markov chain is reducible (since not every state can be reached from every other state). If the order is inverted, the chain will have period 2. Thus both contingencies can arise in theory, but hardly in practice, when action of chance precludes perfect regularity.

It is seen that continued shuffling may reasonably be expected to produce perfect "randomness" and to eliminate all traces of the original order. It should be noted, however, that the number of operations required for this purpose is extremely large.[9]

---

[8] This assumption corresponds to the usual situation at bridge. It is easy to devise more complicated shuffling techniques in which the operations depend on previous operations and the final outcome is not a Markov chain [cf. example (10.e)].

[9] For an analysis of unbelievably poor results of shuffling in records of extrasensory perception experiments cf. W. Feller, Statistical Aspects of ESP, *Journal of Parapsychology*, vol. 4 (1940), pp. 271–298. In their amusing A Review of Dr. Feller's Critique, *ibid.*, pp. 299–319, J. A. Greenwood and C. E. Stuart try to show that these results are due to chance. Both their arithmetic and their experiments have a distinct tinge of the supernatural (cf. Chapter 2, problem 16 of section 8).

10. The General Markov Process

In applications it is usually convenient to describe Markov chains in terms of random variables. This can be done by the simple device of replacing in the preceding sections the symbol E_k by the integer k. The state of the system at time n then is a random variable $X^{(n)}$, which assumes the value k with probability $a_k^{(n)}$; the joint distribution of $X^{(n)}$ and $X^{(n+1)}$ is given by $Pr\{X^{(n)} = j, X^{(n+1)} = k\} = a_j^{(n)} p_{jk}$, and the joint distribution of $(X^{(1)}, \cdots, X^{(n)})$ is given by (1.1). It is also possible and sometimes preferable to assign to E_k a numerical value e_k different from k. With this notation a Markov chain becomes a special stochastic process,[10] that is, a sequence of (dependent) random variables $(X^{(0)}, X^{(1)}, \ldots)$, every finite collection of which has a well-defined joint probability distribution.[11] The superscript n plays the role of time. In Chapter 17 we shall get a glimpse of more general stochastic processes in which the time parameter is permitted to vary continuously. The term "Markov process" is applied to a very large and important class of stochastic processes (with both discrete and continuous time parameters). Even in the discrete case there exist more general Markov processes than the simple chains which we have studied so far. It will, therefore, be useful to give a definition of the Markov property, to point out the special condition characterizing our Markov chains, and, finally, to give a few examples of non-Markovian processes.

Conceptually, a Markov process is the probabilistic analogue of the processes of classical mechanics, where the future development is completely determined by the present state and is independent of the way in which the present state has developed. The processes of mechanics are in contrast to processes with aftereffect (or hereditary processes), such as occur in the theory of plasticity, where the whole past history of the system influences its future. In stochastic processes the future is never uniquely determined, but we have at least probability relations enabling us to make predictions. For the Markov chains studied in this chapter it is clear that probability relations relating to the future depend on the present state, but not on the manner in which the present state has emerged from the past. In other words, if two independent systems subject to the same transition probabilities happen to be in the same state, then all probabilities

---

[10] The terms "stochastic process" and "random process" are synonyms and cover practically all the theory of probability from coin tossing to harmonic analysis. In practice, the term "stochastic process" is used mostly when a time parameter is introduced.

[11] It is clear that these joint distributions must be mutually consistent.

relating to their future developments are identical. This is a rather vague description which is formalized in the following

Definition. A sequence of discrete-valued random variables is a Markov process if for every finite collection of integers $n_1 < n_2 < \cdots < n_r < n$ the joint distribution of $(X^{(n_1)}, X^{(n_2)}, \cdots, X^{(n_r)}, X^n)$ is defined in such a way that the conditional probability of the relation $X^{(n)} = x$ on the hypothesis that $X^{(n_1)} = x_1, \cdots, X^{(n_r)} = x_r$ is identical with the conditional probability of $X^{(n)} = x$ on the single hypothesis $X^{(n_r)} = x_r$. Here $x_1, \cdots, x_r, x$ are arbitrary numbers for which the hypothesis has a positive probability.

Reduced to simpler terms, this definition states that, given the state x_r at time n_r, no additional data concerning states of the system at previous times can alter the (conditional) probability of the state x at a future time n.

The Markov chains studied in this chapter are obviously Markov processes, but they have the following additional property not implied by the definition. *For the Markov chains studied in the preceding sections the transition probabilities $p_{jk} = Pr\{X^{(m+1)} = k \mid X^{(m)} = j\}$ are independent of m.* The more general transition probabilities

$$(10.1) \qquad p_{jk}{}^{(n-m)} = Pr\{X^{(n)} = k \mid X^{(m)} = j\} \qquad (m < n)$$

then depend only on the difference $n - m$. One says in this case that the transition probabilities are *stationary* (or constant). For a general integral-valued Markov chain the right side in (10.1) depends on m and n. We shall denote it by $p_{jk}(m, n)$ so that $p_{jk}(n, n + 1)$ is the one-step transition probability at time n. Instead of (1.1) we get now for the probability of the path $(j_0, j_1, \cdots, j_n)$ the expression

$$(10.2) \qquad a_{j_0}{}^{(0)} p_{j_0 j_1}(0, 1)\, p_{j_1 j_2}(1, 2) \cdots p_{j_{n-1} j_n}(n - 1, n).$$

The proper generalization of (3.3) is obviously the identity

$$(10.3) \qquad p_{jk}(m, n) = \sum_\nu p_{j\nu}(m, r)\, p_{\nu k}(r, n)$$

which is valid for all r with $m < r < n$. This identity follows directly from the definition of a Markov process and also from (10.2); it is called the *Chapman-Kolmogorov* equation.

In the present chapter we have dealt mostly with the asymptotic behavior of the higher transition probabilities, and few of the established properties are common to the most general discrete Markov process. We shall, therefore, not dwell on the general theory.

Examples of Non-Markovian Processes. (a) *The Polya Urn Scheme* [Chapter 5, example (2.c)]. Let $X^{(n)}$ equal 1 or 0 according to whether the nth drawing results in a black or red ball. The sequence $\{X^{(n)}\}$ is *not* a Markov process. For example,

$$Pr\{X^{(3)} = 1 \mid X^{(2)} = 1\} = (b+c)/(b+r+c),$$

but

$$Pr\{X^{(3)} = 1 \mid X^{(2)} = 1, \quad X^{(1)} = 1\} = (b+2c)/(b+r+2c).$$

(Cf. Chapter 5, problems 16–17.) On the other hand, if $Y^{(n)}$ is the number of black balls in the urn at time n, then $\{Y^{(n)}\}$ is an ordinary Markov chain with constant transition probabilities.

(b) *Higher Sums.* Let $Y_0, Y_1, \ldots$ be mutually independent random variables, and put $S_n = Y_0 + \cdots + Y_n$. The difference $S_n - S_m$ (with $m < n$) depends only on $Y_{m+1}, \cdots, Y_n$, and it is therefore easily seen that the sequence $\{S_n\}$ is a Markov process. Now let us go one step further, and define a new sequence of random variables U_n by $U_n = S_0 + S_1 + \cdots + S_n$ (which means that

$$U_n = Y_n + 2Y_{n-1} + 3Y_{n-2} + \cdots).$$

The sequence $\{U_n\}$ forms a stochastic process whose probability relations can, in principle, be expressed in terms of the distributions of the Y_k. The $\{U_n\}$ process is in general not of the Markov type, since there is no reason why, for example, $Pr\{U_n = 0 \mid U_{n-1} = a\}$ should be the same as $Pr\{U_n = 0 \mid U_{n-1} = a, U_{n-2} = b\}$; the knowledge of U_{n-1} and U_{n-2} permits better predictions than the sole knowledge of U_{n-1}.

In the case of a continuous time parameter the preceding summations are replaced by integrations. In diffusion theory the Y_n play the role of accelerations; the S_n are then velocities, and the U_n positions. If only positions can be measured, we are compelled to study a non-Markovian process even though it is indirectly defined in terms of a Markov process.

(c) *Moving Averages.* Again let $\{Y_n\}$ be a sequence of mutually independent random variables. Moving averages of order r are defined by $X^{(n)} = (Y_n + Y_{n+1} + \cdots + Y_{n+r-1})/r$. It is easily seen that the $X^{(n)}$ are not a Markov process. Processes of this type are common in many applications. (Cf. problem 18.)

(d) *A Traffic Problem.* For an empirical example of a non-Markovian process R. Fürth [12] made extensive observations on the number of pedestrians on a certain segment of a street. An idealized mathematical model of this process can be obtained in the following way. For

---

[12] R. Fürth, Schwankungserscheinungen in der Physik, *Sammlung Vieweg*, Braunschweig, 1920, pp. 17ff. The original observations appeared in *Physikalische Zeitschrift*, vols. 19 (1918) and 20 (1919).

simplicity we assume that all pedestrians have the same speed v; also, we consider only pedestrians moving in one direction. At time $t = 0$ we divide the positive x-axis into segments of fixed length δ, each of which may or may not contain a pedestrian. We suppose that the distribution of pedestrians in our segments is determined by a sequence of Bernoulli trials. In other words, we have a sequence of independent random variables Y_k, each of which assumes the values 1 or 0 with probabilities p and q, respectively. The segment $(k - 1)\delta \leq x < k\delta$ contains a pedestrian if $Y_k = 1$. Let now the whole axis move with velocity v in the negative direction, and let us observe the number of pedestrians in the fixed interval of length $N\delta$, which at time $t = 0$ is covered by the interval $0 \leq x < N\delta$ of the moving x-axis. At time t this fixed interval is covered by the interval $vt \leq x < vt + N\delta$ of the x-axis. Let observations be made at times $n\delta/v$ and let $X^{(n)}$ be the number of pedestrians in our fixed interval observed at time n. Then $X^{(n)} = Y_n + Y_{n+1} + \cdots + Y_{n+N-1}$, so that our process is, except for the factor $1/N$, a moving average process. It is therefore non-Markovian. (Passing to the limit $\delta \to 0$, we obtain a continuous model, in which a Poisson distribution takes over the role of the binomial distribution.)

(e) *Superposition of Markov Processes* (*Composite Shuffling*). There exist many technical devices (such as groups of selectors in telephone exchanges, counters, filters) whose action can be described as a superposition of two Markov processes with an output which is non-Markovian. A fair idea of such mechanisms may be obtained from the study of the following method of card shuffling.

In addition to the target deck of N cards we have a similar auxiliary deck, and the usual shuffling technique is applied to this auxiliary deck. If its cards appear in the order $(a_1, a_2, \cdots, a_N)$, then we permute the cards of the target deck so that the first, second, $\cdots$, Nth cards are transferred to the places number $a_1, a_2, \cdots, a_N$. Thus the shuffling of the auxiliary deck indirectly determines the successive orderings of the target deck. The latter form *a stochastic process which is not of the Markov type*. To prove this, it suffices to show that the knowledge of two successive orderings of the target deck conveys in general more clues as to the future than the sole knowledge of the last ordering. We show this in a simple special case.

Let $N = 4$, and suppose that the auxiliary deck is initially in the order (2431). Suppose, furthermore, that the shuffling operation always consists of a true "cutting," that is, the ordering (a_1, a_2, a_3, a_4) is changed into one of the three orderings (a_2, a_3, a_4, a_1), (a_3, a_4, a_1, a_2),

(a_4, a_1, a_2, a_3); we attribute to each of these three possibilities probability 1/3. With these conventions the auxiliary deck will at any time be in one of the four orderings (2431), (4312), (3124), (1243). On the other hand, a little experimentation will show that the target deck will gradually pass through all 24 possible orderings and that each of them will appear in combination with each of the four possible orderings of the auxiliary deck. This means that the ordering (1234) of the target deck will recur infinitely often, and it will always be succeeded by one of the four orderings (2431), (4312), (3124), (1243). Now the auxiliary deck can never remain in the same ordering, and hence the target deck cannot twice in succession undergo the same permutation. Hence, if at times $n-1$ and n the orderings are (1234) and (1243), respectively, then at time $n+1$ the state (1234) is impossible. Thus the knowledge of the state at times $(n-1)$ and n conveys more information than the sole knowledge of the state at time n.

* 11. Miscellany

1. *Inverse Probabilities.* Although it is most natural to investigate the future development of a system, it is occasionally necessary to study its past. Consider a Markov chain with states E_k and constant transition probabilities p_{jk}, whose absolute probabilities at time n are $a_k^{(n)} = \Sigma a_\nu^{(0)} p_{\nu k}^{(n)}$. The conditional probability that the system was at time $m < n$ in state E_j, given that at time n it is in E_k, is

$$(11.1) \qquad q_{kj}(n, m) = \frac{a_j^{(m)}}{a_k^{(n)}} p_{jk}^{(n-m)}, \qquad m < n.$$

This formula makes sense only if $a_k^{(n)} > 0$; otherwise the conditional probability in question is not defined. If all $a_k^{(n)}$ are positive, then (11.1) defines a system of transition probabilities with all the properties required for a Markov process. In particular, the $q_{kj}(m, n)$ satisfy the Chapman-Kolmogorov identity (10.3) with the time direction reversed, namely,

$$(11.2) \qquad q_{kj}(n, m) = \sum_\nu q_{k\nu}(n, r) \, q_{\nu j}(r, m)$$

($m < r < n$). The $q_{kj}(n, m)$ are called *inverse probabilities*.[13] Consider, in particular, an irreducible chain with stationary probabilities

---

* Starred sections may be omitted at first reading since they treat special topics.

[13] A. Kolmogoroff, Zur Theorie der Markoffschen Ketten, *Mathematische Annalen*, vol. 112 (1935), pp. 155–160.

$\{u_k\}$. Then $a_k{}^{(n)} = u_k$ for all n, and $u_k > 0$ (cf. sections 6 and 7). In this case the *one-step* transitions $q_{k,j}(n+1, n)$ are independent of n and reduce to

$$(11.3) \qquad q_{kj} = \frac{u_j}{u_k} p_{jk}.$$

The matrix $\{q_{kj}\}$ is stochastic, so that here the inverse probabilities define a Markov chain with constant transition probabilities. If $q_{kj} = p_{kj}$ the original chain is called *reversible;* its probability relations are then symmetric in time.

2. *The Central Limit Theorem.* The theory of recurrent events contains further information concerning Markov chains. Let E_k be a fixed recurrent state whose recurrence time has finite variance σ_k^2 (this condition is always satisfied if the chain is finite). Let N_n denote the number of passages up to time n of the system through E_k. Then we know from Chapter 12, section 4, that the variable N_n is asymptotically normally distributed. In the notations of the present chapter we have $E(N_n) = 1/\mu_k = u_k$; a way to calculate the variance in the case of finite chains will be indicated in the next chapter. It follows in particular that for every $\epsilon > 0$ as $n \to \infty$ the probability tends to one that $\left|\dfrac{N_n}{n} - u_k\right| < \epsilon$. This is the *weak law of large numbers* for the number of passages through E_k. Similarly, the strong law of large numbers and the law of the iterated logarithm hold and require no special proof. In the case of an infinite chain, the recurrence time of E_k need not have a finite variance even if its mean is finite. However, the general limit theorems for recurrent events apply in this case.

The random variable N_n may be defined as follows. Define a sequence of random variables X_n so that X_n equals 1 or 0 according to whether the system is or is not at time n in state E_k. Then $N_n = X_1 + \cdots + X_n$. This suggests the following generalization. We assign to the state E_k an arbitrary number x_k and let the random variable X_n equal x_k if at time n the system is in state E_k. As usual, we put $S_n = X_1 + \cdots + X_n$. For *finite* Markov chains Doeblin [14] has shown that in general the central limit theorem and the law of the iterated logarithm hold for S_n. An exception occurs only if the numbers x_k are chosen so that for every shortest path leading from E_k back to E_k the sum of the x_ν equals a constant c independent of the path.

[14] W. Doeblin, Sur les propriétés asymptotiques de mouvements régis par certains types de chaines simples, Thesis, Paris, 1937.

3. *Non-stochastic Matrices.* The theorems of this chapter describe the asymptotic behavior of the powers P^n of an arbitrary stochastic matrix P, that is, of a matrix whose elements satisfy the conditions (1.2). It is easy to generalize these theorems to a more general class of matrices. Let P be an arbitrary (*finite or infinite*) *matrix with non-negative elements and denote its row sums by* S_j *so that* $S_j = \Sigma_k p_{jk}$. *We assume that the sequence* S_j *is bounded,* that is, that there exists a constant M such that $S_j \leq M$. Under these conditions the asymptotic behavior of P^n is still described by our theorems, inasmuch as P can be reduced to a stochastic matrix.

To fix ideas suppose that the rows and columns of P are numbered starting with 1, and consider first the case where $S_j \leq 1$ for all j. In this case we enlarge (border) the matrix P by adding a row and a column number zero whose elements are defined by $p_{00} = 1$, $p_{01} = p_{02} = \cdots = 0$, and $p_{j0} = 1 - S_j$ for $j \geq 1$. The new matrix Q is stochastic, and its asymptotic behavior is given by our theorems. On the other hand, P^n is the submatrix of the corner element $p_{00}^{(n)}$ of Q^n. In the general case the row sums S_j may exceed unity, but we may replace the matrix P by the matrix P^* whose elements are p_{jk}/M. The row sums S_j^* of P^* satisfy the condition $S_j^* \leq 1$, and we are able to describe the asymptotic behavior of the powers P^{*n}. However, the matrices P^n and P^{*n} differ only by the factor M^n, so that our theorems actually describe the asymptotic behavior of $p_{jk}^{(n)}$ in all cases.

Matrices of the described type occur in the theory of generalized random walks with creation or destruction of masses.

4. *Literature.* There exists a huge literature on *finite* Markov chains. A detailed account of the various methods of attack and references to earlier work will be found in the comprehensive treatise by M. Fréchet.[15] An algebraic treatment of finite chains will be described in the next chapter. The entire theory of finite chains can be derived from Frobenius' theory of matrices with positive elements. This method has been exploited in particular by V. Romanovsky. Unfortunately these methods do not carry over to the more interesting case of infinite chains, first considered by A. Kolmogorov.[16] His work was continued

[15] *Recherches théoriques modernes sur le calcul des probabilités*, vol. 2 (théorie des événements en chaine dans le cas d'un nombre fini d'états possibles), Paris, 1938. Another monograph on Markov chains is due to B. Hostinsky, *Méthodes générales du calcul des probabilités*, fasc. 52 of the *Mémorial des sciences mathématiques*, Paris, 1931.

[16] Anfangsgründe der Theorie der Markoffschen Ketten mit unendlich vielen möglichen Zuständen, *Matematičeskii Sbornik*, N.S., vol. 1 (1936), pp. 607–610. This paper contains no proofs. A complete exposition was given only in Russian, in *Bulletin de l'Université d'État à Moscou*, Sect. A, vol. 1 (1937), pp. 1–15.

by W. Doeblin [17] and J. L. Doob.[18] The latter derived the ergodic properties from general group theory. The method used in this chapter, based on the general theory of recurrent events, is new.[19] It permits a uniform treatment of finite and infinite chains and represents a simplification even in the case of finite chains. The states which we call transient are usually called *unessential* and the interesting problem of absorption probabilities is neglected. This is explained by a predominantly abstract attitude. In practical cases the interest often centers on transient states.

12. Problems for Solution

1. Classify the states for the three chains whose matrices P have the rows given below. Find in each case P^2 and the asymptotic behavior of $p_{jk}^{(n)}$.

   (a) (0, 1/2, 1/2), (1/2, 0, 1/2), (1/2, 1/2, 0);

   (b) (0, 0, 0, 1), (0, 0, 0, 1), (1/2, 1/2, 0, 0), (0, 0, 1, 0);

   (c) (1/2, 0, 1/2, 0, 0), (1/4, 1/2, 1/4, 0, 0), (1/2, 0, 1/2, 0, 0),

   (0, 0, 0, 1/2, 1/2), (0, 0, 0, 1/2, 1/2).

2. We consider throws of a true die and agree to say that at time n the system is in state E_j if j is the highest number appearing in the first n throws. Find the matrix P^n and verify that (3.3) holds.

3. In example XI find the (absorption) probabilities x_k and y_k that, starting from E_k, the system will end in E_1 or E_5, respectively ($k = 2, 3, 4, 6$).

4. N black and N white balls are placed in two urns so that each urn contains N balls. The number of black balls in the first urn is the state of the system. At each step one ball is selected at random from each urn, and the two balls thus selected are interchanged. Find the p_{jk}. Show that in the limiting distribution the term u_k equals the probability of getting exactly k black balls if N balls are selected at random out of a collection of N black and N white balls.[20]

5. A chain with states $E_0, E_1, \cdots$ has transition probabilities

$$p_{jk} = e^{-\lambda} \sum_{\nu=0}^{j} \binom{j}{\nu} p^\nu q^{j-\nu} \frac{\lambda^{k-\nu}}{(k-\nu)!}$$

[17] Sur deux problèmes de M. Kolmogoroff concernant les chaines dénombrables, *Bulletin Société Mathématique de France*, vol. 66 (1939), pp. 1–11.

[18] Topics in the Theory of Markoff Chains, and also Markoff Chains—Denumerable Case, *Transactions of the American Mathematical Society*, vol. 52 (1942), pp. 37–64, and vol. 58 (1945), pp. 455–473.

[19] A short outline is given in W. Feller, Fluctuation Theory of Recurrent Events, *Transactions of the American Mathematical Society*, vol. 67 (1949), pp. 98–119.

[20] This problem goes back to Laplace; cf. Fréchet's book (cited in footnote 15), p. 49.

where the terms in the sum should be replaced by zero if $\nu > k$. Show that

$$p_{jk}^{(n)} \to e^{-\lambda/q} \frac{(\lambda/q)^k}{k!}.$$

Note: This chain occurs in statistical mechanics [21] and can be interpreted as follows. The state of the system is defined by the number of particles in a certain volume of space. During each time interval of unit length each particle has probability q to leave the volume, and the particles are statistically independent. Moreover, new particles may enter the volume, and the probability of r entrants is given by the Poisson expression $e^{-\lambda} \lambda^r / r!$. The stationary distribution is then a Poisson distribution with parameter λ/q.

6. *Ehrenfest model.* In example VII let there initially be j molecules in the first container, and let $X^{(n)} = 2k - a$ if at time n the system is in state k [so that $X^{(n)}$ is the difference of the number of molecules in the two containers]. Let $e_n = E(X^{(n)})$. Prove that $e_{n+1} = (a-2)e_n/a$, whence $e_n = (1-2/a)^n(2j-a)$. (Note that $e_n \to 0$ as $n \to \infty$.)

7. If the number of states is $a < \infty$ and if E_k can be reached from E_j, then it can be reached in a steps or less.

8. Let the chain contain a states and let E_j be recurrent. There exists a number $q < 1$ such that for $n \geq a$ the probability of the recurrence time of E_j exceeding n is smaller than q^n. (*Hint:* Use problem 7.)

9. In an infinite chain with doubly stochastic matrix every state is either transient or a recurrent null state [cf. example (6.*d*)].

10. *Random walk with reflecting barriers.* Consider a *symmetric* random walk in a bounded region of the plane. The boundary is reflecting in the sense that, whenever in an unrestricted random walk the particle would leave the region, it is now forced to return to the last position. Show that, if every point of the region can be reached from every other point, there exists a stationary distribution and that $u_k = 1/a$, where a is the number of positions in the region.

11. In a finite chain E_j is transient if and only if there exists an E_k such that E_k can be reached from E_j but not E_j from E_k. (For infinite chains this is false, as shown by the following problem.)

12. Suppose that in an infinite chain only the transitions $E_j \to E_{j+1}$ and $E_j \to E_0$ are possible, and that their probabilities are $1 - p_j$ and p_j. Show that all states are transient or recurrent according to whether Σp_j converges or diverges.

13. An irreducible chain for which *one* diagonal element p_{jj} is positive cannot be periodic.

14. A finite irreducible chain is non-periodic if and only if there exists an n such that $p_{jk}^{(n)} > 0$ for all j and k.

15. In a chain with a states let $(x_1, \cdots, x_a)$ be a solution of the system of linear equations $x_j = \Sigma p_{j\nu} x_\nu$. Prove: (a) the states E_r for which $x_r > 0$ form a closed (not necessarily irreducible) set; (b) if E_j and E_k belong to the same irreducible set, then $x_j = x_k$.

16. *Continuation.* If $(x_1, \cdots, x_a)$ is a solution of $x_j = s\Sigma p_{j\nu} x_\nu$ with $|s| = 1$ but $s \neq 1$, then there exists an integer $t > 1$ such that $s^t = 1$. If the chain is irreducible, then the smallest integer of this kind is the period of the chain.

[21] S. Chandrasekhar, Stochastic Problems in Physics and Astronomy, *Reviews of Modern Physics*, vol. 15 (1943), pp. 1–89, in particular p. 45.

17. *Mean ergodic theorem.*[22] In an arbitrary chain let

$$A_{jk}^{(n)} = \frac{1}{n} \sum_{\nu=1}^{n} p_{jk}^{(\nu)}.$$

If E_j and E_k belong to the same irreducible closed set, then $A_{jk}^{(n)}$ tends to a limit which is independent of j and equals the stationary probability u_k, whenever the latter exists. If E_j and E_k belong to different closed sets, then $A_{jk}^{(n)} = 0$ for all n. If E_k is transient, then $A_{jk}^{(n)} \to 0$ for all j.

18. *Moving averages.* Let $\{Y_k\}$ be a sequence of mutually independent random variables, each assuming the values ± 1 with probability $1/2$. Put $X^{(n)} = (Y_n + Y_{n+1})/2$. Find the transition probabilities

$$p_{jk}(m, n) = Pr\{X^{(n)} = k \mid X^{(m)} = j\},$$

where $m < n$ and $j, k = -1, 0, 1$. Conclude that $\{X^{(n)}\}$ is not a Markov process and that (10.3) does not hold.

[22] This theorem is a simple consequence of the results of the present chapter. However, it is much weaker and can therefore be proved by simpler methods; cf. K. Yosida and S. Kakutani, Markoff Processes with an Enumerable Infinite Number of Possible States, *Japanese Journal of Mathematics*, vol. 16 (1939), pp. 47–55.

*CHAPTER 16

ALGEBRAIC TREATMENT OF FINITE MARKOV CHAINS

In this chapter we consider a Markov chain with finitely many states $E_1, \cdots, E_a$ and a given matrix of transition probabilities p_{jk}. Our main aim is to derive explicit formulas for the n-step transition probabilities $p_{jk}^{(n)}$. We shall not require the results of the preceding chapter, except the general concepts and notations of section 3.

We shall make use of the method of generating functions and shall obtain the desired results from the partial fraction expansions of Chapter 11, section 7. Our results can also be obtained directly from the theory of canonical decompositions of matrices [1] (which in turn can be derived from our results). Also, for *finite* chains the ergodic properties proved in Chapter 15 follow from the results of the present chapter. However, for simplicity, we shall slightly restrict the generality and disregard exceptional cases which complicate the general theory and do not occur in practical examples.

The general method is outlined in section 1 and illustrated in sections 2 and 3. In section 4 special attention is paid to transient states and absorption probabilities. In section 5 the theory is applied to finding the variances of the recurrence times of the states E_j.

1. General Theory

For every fixed j, k we define a generating function

(1.1) $$P_{jk}(s) = \sum_{n=1}^{\infty} p_{jk}^{(n)} s^{n-1}.$$

Multiplying this equation by $sp_{\nu j}$ and adding over all j, we get

(1.2) $$s \sum_{j=1}^{a} p_{\nu j} P_{jk}(s) = P_{\nu k}(s) - p_{\nu k}.$$

For every fixed k we have here a system of a non-homogeneous linear equations for the a unknowns $P_{1k}(s), \cdots, P_{ak}(s)$. Theoretically, this

---

* Starred chapters treat special topics and may be omitted at first reading.
[1] Cf. the treatise by Fréchet cited in Chapter 15, section 11.

system can be solved by means of determinants or by successive eliminations of unknowns. We use only the fact that the determinant $D(s)$ of the system is a polynomial of degree not exceeding a, and that the $P_{\nu k}(s)$ are *rational functions* of s with the common denominator $D(s)$. We shall consider only the case where the equation $D(s) = 0$ has no multiple roots; this is a slight restriction of generality, but the theory will cover most cases of practical interest.

Since the $P_{\nu k}(s)$ are rational functions, we can use the partial fraction expansion of Chapter 11, section 7. It follows that there exist coefficients $\rho_{\nu k}{}^{(1)}, \cdots, \rho_{\nu k}{}^{(a)}$ such that

$$(1.3) \qquad p_{\nu k}{}^{(n)} = \frac{\rho_{\nu k}{}^{(1)}}{s_1{}^n} + \frac{\rho_{\nu k}{}^{(2)}}{s_2{}^n} + \cdots + \frac{\rho_{\nu k}{}^{(a)}}{s_a{}^n}$$

where $s_1, s_2, \ldots$ are the roots of $D(s) = 0$. If the degree of $D(s)$ is smaller than a, then (1.3) will contain fewer than a terms. It is also possible that for some particular values of ν and k one or more roots s_r are common to the numerator and denominator and hence cancel. We can take care of such cases by letting the corresponding coefficients $\rho_{\nu k}{}^{(r)}$ be zero.

We could calculate the roots s_r and the coefficients $\rho_{\nu k}{}^{(r)}$ by the methods of Chapter 11, but it is preferable to take advantage of certain particular properties of Markov chains. Multiply (1.3) by $p_{j\nu}$ and sum over $\nu = 1, 2, \ldots$. The result is

$$(1.4) \qquad p_{jk}{}^{(n+1)} = \sum_{\nu=1}^{a} p_{j\nu} \left\{ \frac{\rho_{\nu k}{}^{(1)}}{s_1{}^n} + \cdots + \frac{\rho_{\nu k}{}^{(a)}}{s_a{}^n} \right\}.$$

If the left side is expressed by means of (1.3), one gets an identity which can hold for all n only if the coefficients of $s_1{}^{-n}, \cdots, s_a{}^{-n}$ on both sides are equal. This means that for every fixed r (with $1 \leq r \leq a$) we must have

$$(1.5a) \qquad \rho_{jk}{}^{(r)} = s_r \sum_{\nu=1}^{a} p_{j\nu} \rho_{\nu k}{}^{(r)}.$$

In like manner we get, on multiplying (1.3) by p_{km} and adding over all k,

$$(1.5b) \qquad \rho_{\nu m}{}^{(r)} = s_r \sum_{k=1}^{a} \rho_{\nu k}{}^{(r)} p_{km}.$$

The relations (1.5a) show that for k and r fixed the a quantities $\rho_{1k}{}^{(r)}, \cdots, \rho_{ak}{}^{(r)}$ are a solution of the system of a linear equations

$$(1.6a) \qquad x_j{}^{(r)} = s_r \sum_{\nu=1}^{a} p_{j\nu} x_\nu{}^{(r)} \qquad (j = 1, \cdots, a)$$

16.1]   GENERAL THEORY   349

Similarly, (1.5b) implies that for ν and r fixed, the $\rho_{\nu 1}^{(r)}, \cdots, \rho_{\nu a}^{(r)}$ satisfy the a linear equations

$$\text{(1.6b)} \qquad y_m^{(r)} = s_r \sum_{k=1}^{a} y_k^{(r)} p_{km} \qquad (m = 1, \cdots, a).$$

For a better understanding let us replace s_r by an arbitrary s and study the two more general systems

$$\text{(1.7a)} \qquad x_j = s \sum_{\nu=1}^{a} p_{j\nu} x_\nu \qquad (j = 1, \cdots, a)$$

and

$$\text{(1.7b)} \qquad y_m = s \sum_{k=1}^{a} y_k p_{km} \qquad (m = 1, \cdots, a).$$

A system of a homogeneous equations in a unknowns can have a non-trivial[2] solution only if its determinant vanishes. Now the matrices of the two systems (1.7a) and (1.7b) are the same except that rows and columns are interchanged. Their determinants are therefore equal. Moreover, the determinant of (1.7a) obviously equals the determinant of the system (1.2), which means that the determinants of the two systems (1.7a) and (1.7b) vanish for $s = s_1, s_2, \cdots, s_a$.

We can now forget about the generating functions $P_{jk}(s)$ and define the roots s_r as those numbers (real or complex) for which the systems (1.7a) and (1.7b) admit of non-trivial solutions. The assumption that s_r is a simple root means that for every fixed r the solutions $(x_1^{(r)}, \cdots, x_a^{(r)})$ and $(y_1^{(r)}, \cdots, y_a^{(r)})$ are uniquely determined except, of course, for a numerical factor. However, our starting point was the discovery that, for k and r fixed, $(\rho_{1k}^{(r)}, \cdots, \rho_{ak}^{(r)})$ is a solution of (1.7a), while for ν and r fixed $(\rho_{\nu 1}^{(r)}, \cdots, \rho_{\nu a}^{(r)})$ is a solution of (1.7b). Since these solutions are determined up to a numerical factor, we must have

$$\text{(1.8)} \qquad \rho_{jk}^{(r)} = c_r x_j^{(r)} y_k^{(r)}.$$

There remains only the calculation of the constants $c_1, \cdots, c_a$.

From (1.8) and (1.3) we have

$$\text{(1.9)} \qquad p_{jk}^{(n)} = \sum_{r=1}^{a} c_r x_j^{(r)} y_k^{(r)} s_r^{-n}.$$

Now

$$\text{(1.10)} \qquad p_{jk}^{(2n)} = \sum_{\nu=1}^{a} p_{j\nu}^{(n)} p_{\nu k}^{(n)},$$

[2] As usual we call an identically vanishing solution trivial and disregard it.

and we can express all quantities in (1.10) by means of (1.9). Equating the coefficients of s_r^{-2n} on both sides, we find

$$(1.11) \qquad 1 = c_r \sum_{\nu=1}^{a} x_\nu^{(r)} y_\nu^{(r)},$$

and thus we have found c_r. It is true that the solutions $x_\nu^{(r)}$ and $y_\nu^{(r)}$ are determined only up to a numerical factor. However, if we replace the $x_j^{(r)}$ by $Ax_j^{(r)}$, and the $y_k^{(r)}$ by $By_k^{(r)}$, then c_r will be changed into c_r/AB and the quantity $\rho_{jk}^{(r)}$ of (1.8) remains unchanged.

Summarizing, we have the following procedure to calculate $p_{jk}^{(n)}$.

Write down the two systems of linear equations (1.7a) and (1.7b). They have a common determinant and admit of non-trivial solutions only for values of s for which this determinant vanishes. We suppose that the roots $s_1, s_2, \ldots$ (of which there are at most a) are simple; then for each r, the solutions $(x_1^{(r)}, \cdots, x_a^{(r)})$ and $(y_1^{(r)}, \cdots, y_a^{(r)})$ are determined up to an arbitrary multiplicative constant. Find these solutions and then the constants c_r from (1.11). Form the quantities $\rho_{jk}^{(r)}$ in accordance with (1.8); then $p_{jk}^{(n)}$ is given by (1.3).

For every fixed r the $\rho_{jk}^{(r)}$ form a matrix which may be constructed in the following way. Form a multiplication table with the $x_j^{(r)}$ heading the rows and the $y_k^{(r)}$ heading the columns. Multiplying all a^2 elements of this square table by c_r, we get the matrix $\rho_{jk}^{(r)}$. To construct the matrix $(p_{jk}^{(n)})$ we have to divide all elements of $\rho_{jk}^{(r)}$ by s_r^n and add the matrices thus obtained for $r = 1, 2, \ldots$. Note that the roots s_r may be simple even if there are fewer than a roots; however, if there are a distinct roots, $s_1, s_2, \cdots, s_a$, they are certainly simple.

The case of multiple roots requires certain changes, but may be treated by similar methods. The case of greatest interest will be discussed in section 4.

In algebra the reciprocals $\lambda_r = 1/s_r$ are called *characteristic values* (or eigen values or latent roots) of the matrix P. Zero is a possible characteristic value, but to it there corresponds no root s_r. This explains why there may be fewer than a roots s_r even though there are always a characteristic values. The use of s_r rather than of their reciprocals is more convenient for the method of generating functions. Moreover, it corresponds to the general usage in the theory of integral equations and is therefore more natural in probability theory.

The value $s = 1$ always occurs among the s_r, and to it there corresponds the solution $(1, 1, \cdots, 1)$ of (1.7a). For all r we have $|s_r| \geq 1$. In fact,[3] a root s_r with $|s_r| < 1$

---

[3] A direct proof is as follows. Let M be the largest term in the sequence $|x_1^{(r)}|, \cdots, |x_a^{(r)}|$ (r fixed). Then from (1.6a) $M \leq |s_r| \Sigma p_{j\nu} M = |s_r| M$ or $|s_r| \geq 1$.

would lead to a divergent development in (1.3). If $s_1 = 1$ is the only root with $|s_r| = 1$, then $p_{jk}^{(n)} \to c_1 x_j^{(1)} y_k^{(1)}$. It is not difficult to show that if there exist other roots with $|s_r| = 1$, then they are necessarily tth roots of unity, where t is an integer; in this case the chain has period t. For details the reader is referred to Fréchet's treatise quoted in Chapter 15, section 11.

Often it is cumbersome or impossible to find *all* roots s_r. However, it is clear that the asymptotic behavior of $p_{jk}^{(n)}$ is determined in first approximation by the s_r with $|s_r| = 1$, and in second approximation by the roots s_r with the next smallest absolute value.

2. Examples

(a) Consider first a chain with only two states. The matrix of transition probabilities assumes the simple form

$$P = \begin{pmatrix} 1-p & p \\ \alpha & 1-\alpha \end{pmatrix}$$

where $0 < p < 1$ and $0 < \alpha < 1$. The equations (1.7a) reduce to $s(1-p)x_1 + spx_2 = x_1$ and $s\alpha x_1 + s(1-\alpha)x_2 = x_2$. Equating the two ratios x_1/x_2, it is found that a solution exists only if either $s = 1$ or $s = 1/(1-\alpha-p)$. The solution corresponding to $s_1 = 1$ is $(1,1)$; the solution corresponding to $s_2 = 1/(1-\alpha-p)$ is $(p, -\alpha)$. Next take the system (1.7b) which now reduces to $s(1-p)y_1 + s\alpha y_2 = y_1$ and $spy_1 + s(1-\alpha)y_2 = y_2$. We know that it can be solved only when $s = s_1$ or $s = s_2$. The corresponding solutions are (α, p) and $(1, -1)$. From (1.11) we get $c_1 = c_2 = 1/(\alpha+p)$. Equations (1.8) and (1.3) now enable us to write down explicit formulas for the quantities $p_{jk}^{(n)}$. The final result can be written in matrix form

$$P^n = \frac{1}{\alpha+p} \begin{pmatrix} \alpha & p \\ \alpha & p \end{pmatrix} + \frac{(1-\alpha-p)^n}{\alpha+p} \begin{pmatrix} p & -p \\ -\alpha & \alpha \end{pmatrix}$$

(where factors common to all four elements have been taken out as factors to the matrices). Since $|1-\alpha-p| < 1$, the second matrix tends to zero as $n \to \infty$, and the first matrix represents the limiting form of P^n.

(b) Let

(2.1) $$P = \begin{bmatrix} 0 & 0 & 0 & 1 \\ 0 & 0 & 0 & 1 \\ 1/2 & 1/2 & 0 & 0 \\ 0 & 0 & 1 & 0 \end{bmatrix}$$

[this is the matrix of problem 1(b), Chapter 15]. The system (1.7a) reduces to

(2.2) $\quad x_1 = sx_4, \quad x_2 = sx_4, \quad x_3 = \dfrac{s(x_1 + x_2)}{2}, \quad x_4 = sx_3.$

Since a multiplicative constant remains arbitrary, we may put $x_4 = 1$. Then $x_1 = s$, $x_2 = s$, $x_3 = s^2$, $x_4 = s^3$, and therefore we must have $s^3 = 1$. Now if we put

(2.3) $\quad \theta = e^{2\pi i/3} = \cos\dfrac{2\pi}{3} + i\sin\dfrac{2\pi}{3},$

then the three roots of $s^3 = 1$ are $s_1 = 1$, $s_2 = \theta$, $s_3 = \theta^2$. (Note that we have only three roots, even though there are four states.) The solutions $x_j^{(r)}$ corresponding to the three roots are $(1, 1, 1, 1)$, $(\theta, \theta, \theta^2, 1)$, $(\theta^2, \theta^2, \theta, 1)$.

From (1.7b) we get $y_1 = sy_3/2$, $y_2 = sy_3/2$, $y_3 = sy_4$, $y_4 = s(y_1 + y_2)$. The three sets of solutions corresponding to $s_1 = 1$, $s_2 = \theta$, and $s_3 = \theta^2$ are $(1, 1, 2, 2)$, $(\theta, \theta, 2, 2\theta^2)$, $(\theta^2, \theta^2, 2, 2\theta)$. Therefore from (1.11) $c_1 = 1/6$, $c_2 = 1/(6\theta^2) = \theta/6$, $c_3 = 1/(6\theta) = \theta^2/6$. We are now able to express all $p_{jk}^{(n)}$. For example,

$$p_{11}^{(n)} = p_{22}^{(n)} = \dfrac{1 + \theta^n + \theta^{2n}}{6}$$

(2.4) $\quad p_{13}^{(n)} = \dfrac{1 + \theta^{2n+2} + \theta^{n+1}}{3}$

$$p_{14}^{(n)} = \dfrac{1 + \theta^{2n+1} + \theta^{n+2}}{3}$$

etc. The chain is obviously periodic with **period 3**.

(c) Let $p + q = 1$, and

(2.5) $\quad P = \begin{bmatrix} 0 & p & 0 & q \\ q & 0 & p & 0 \\ 0 & q & 0 & p \\ p & 0 & q & 0 \end{bmatrix}.$

(This matrix describes a cyclical random walk; cf. example V of Chapter 15.) The equations (1.7a) reduce to $x_1 = s(px_2 + qx_4)$, $x_2 = s(qx_1 + px_3)$, $x_3 = s(qx_2 + px_4)$, $x_4 = s(px_1 + qx_3)$. Suppose that $p \neq q$. From the first and the third equations we find $x_1 + x_3 = s(x_2 + x_4)$, and from the remaining equations $x_2 + x_4 = s(x_1 + x_3)$.

Hence we have either $s^2 = 1$ or $x_1 + x_3 = x_2 + x_4 = 0$. The first alternative leads to the two roots $s_1 = 1$, $s_2 = -1$. On the other hand, substituting $x_3 = -x_1$, $x_4 = -x_2$ into the first two equations, we find $s^2(p-q)^2 = -1$, which yields the remaining two roots s_3 and s_4. Thus

(2.6) $\qquad s_1 = 1, \quad s_2 = -1, \quad s_3 = \dfrac{i}{q-p}, \quad s_4 = -\dfrac{i}{q-p},$

(where $i^2 = -1$). The corresponding solutions $x_j^{(r)}$ contain an arbitrary factor, and we are free to put $x_4^{(r)} = 1$. Then the four sets of solutions are easily found to be $(1, 1, 1, 1)$, $(-1, 1, -1, 1)$, $(i, -1, -i, 1)$, $(-i, -1, i, 1)$. The system (1.7b) reduces in our case to $y_1 = s(qy_2 + py_4)$, $y_2 = s(py_1 + qy_3)$, $y_3 = s(py_2 + qy_4)$, $y_4 = s(qy_1 + pq_3)$. To the four roots (2.6) there correspond the solutions $(1, 1, 1, 1)$, $(-1, 1, -1, 1)$, $(-i, -1, i, 1)$, $(i, -1, -i, 1)$. For the constants c_r we find from (1.11) $c_1 = c_2 = c_3 = c_4 = 1/4$. Using (1.3) and (1.8), we can now write an explicit formula for each sequence $p_{jk}^{(n)}$ ($n = 1, 2, 3, \ldots$). In the present case the solutions $x_j^{(r)}$ and $y_j^{(r)}$ are of the simple form $(\alpha, \alpha^2, \alpha^3, \alpha^4)$, where α is one of the four numbers 1, -1, i, or $-i$. This enables us to express the $p_{jk}^{(n)}$ by the single formula

(2.7) $\qquad p_{jk}^{(n)} = \tfrac{1}{4}\{1 + (q-p)^n (i)^{j-k-n}\} \{1 + (-1)^{k+j-n}\}.$

This formula is valid also for $p = q = 1/2$.

It is seen that the term involving $(q-p)^n$ tends to zero, and that the other term has period 2.

(d) *General Cyclical Random Walk* (example V of Chapter 15). In the preceding example we were able to express the $x_j^{(r)}$ and $y_k^{(r)}$ as powers of the four fourth roots of unity. This suggests trying a similar procedure for the general matrix of example V. It is convenient to number the states from 0 to $a - 1$. For brevity we put

(2.8) $\qquad\qquad\qquad \theta = e^{2i\pi/a}.$

This is an ath root of unity, and all ath roots are represented by the sequence $1, \theta, \theta^2, \cdots, \theta^{a-1}$. It is easily seen that the systems (1.7a) and (1.7b) are satisfied by the a sets of solutions

(2.9) $\qquad\qquad x_j^{(r)} = \theta^{rj}, \quad y_k^{(r)} = \theta^{-rk}$

with $r = 0, 1, 2, \cdots, a - 1$; they correspond to

(2.10) $\qquad\qquad s_r = \left\{ \displaystyle\sum_{\nu=0}^{a-1} q_\nu \theta^{\nu r} \right\}^{-1}.$

From (1.11) and (2.9) we find $c_r = 1/a$ for all r, and thus finally

$$(2.11) \qquad p_{jk}^{(n)} = \frac{1}{a} \sum_{r=0}^{a-1} \theta^{r(j-k)} \left(\sum_{\nu=0}^{a-1} q_\nu \theta^{\nu r} \right)^n.$$

It is interesting to verify this formula for $n = 1$. The factor of q_ν is

$$(2.12) \qquad \sum_{r=0}^{a-1} \theta^{r(j-k+\nu)}.$$

This sum is zero except when $j - k + \nu = 0$ or a, in which case each term equals one. Hence $p_{jk}^{(1)}$ reduces to q_{k-j} if $k \geq j$ and to q_{a+k-j} if $k < j$, and this is the given matrix (p_{jk}).

(e) *The Occupancy Problem.* Example VIII of Chapter 15 shows that the classical occupancy problem can be treated by the method of Markov chains. The system is in state j if there are j occupied and $a - j$ empty cells. If this is the initial situation and n additional balls are placed at random, then $p_{jk}^{(n)}$ is the probability that there will be k occupied and $a - k$ empty cells (so that $p_{jk}^{(n)} = 0$ if $k < j$). For $j = 0$ this probability follows from formula (5.4) of Chapter 4. We now derive a formula for $p_{jk}^{(n)}$, thus generalizing the result of Chapter 4.

Since $p_{jj} = j/a$ and $p_{j,j+1} = (a-j)/a$, it is easily seen that the system of equations (1.7a) reduces to

$$(2.13) \qquad (a - sj)x_j = s(a-j)x_{j+1}, \qquad j = 0, \cdots, a.$$

For $s = 1$ we get the solution $x_j = 1$. It is clear that if $s \neq 1$ then $x_a = 0$, so that $s = 1$ is the only value of s for which all x_j are different from zero. If s is any other value for which (2.13) has a solution, then there must exist some index $r < a$ such that $x_{r+1} = 0$ but $-x_r \neq 0$; from (2.13) it then follows that $sr = a$. Thus the roots s_r for which (2.13) has solutions are $s_r = a/r$ with $r = 1, 2, \cdots, a$. The corresponding solutions of (2.13) are obtained successively, putting $x_0^{(r)} = 1$, and $j = 0, 1, \ldots$. We find

$$(2.14) \qquad x_j^{(r)} = \binom{r}{j} \div \binom{a}{j}$$

so that $x_j^{(r)} = 0$ when $j > r$.

For $s = s_r$ the system (1.7b) reduces to

$$(2.15) \qquad (r-j)y_j^{(r)} = (a-j+1)y_{j-1}^{(r)}$$

and has the solution

$$(2.16) \qquad y_j^{(r)} = \binom{a-r}{j-r}(-1)^{j-r}$$

where, of course, $y_j^{(r)} = 0$ if $j < r$. Since $x_j^{(r)} = 0$ for $j > r$ and $y_j^{(r)} = 0$ for $j < r$, we find from (1.11) easily $c_r = (x_r^{(r)} y_r^{(r)})^{-1} = \binom{a}{r}$, and hence

(2.17)
$$p_{jk}^{(n)} = \sum_{r=j}^{k} \left(\frac{r}{a}\right)^n \binom{a}{r}\binom{r}{j}\binom{a-r}{k-r}(-1)^{k-r} \div \binom{a}{j}.$$

On expressing the binomial coefficients in terms of factorials, this formula simplifies to

(2.18) $$p_{jk}^{(n)} = \binom{a-j}{a-k} \sum_{\nu=0}^{k-j} \left(\frac{\nu+j}{a}\right)^n (-1)^{k-j-\nu} \binom{k-j}{\nu},$$

with $p_{jk}^{(n)} = 0$ if $k < j$.

(Further examples are found in the following two sections.)

3. Random Walk with Reflecting Barriers

The application of Markov chains [4] will now be illustrated by a complete discussion of example IV of Chapter 15. In the terminology of random walks, $p_{jk}^{(n)}$ is the probability that the particle which starts from $x = j$ is at time n at $x = k$.

The equations (1.7a) take on the form

(3.1)
$$x_1 = s(qx_1 + px_2)$$
$$x_j = s(qx_{j-1} + px_{j+1}) \qquad (j = 2, 3, \cdots, a-1)$$
$$x_a = s(qx_{a-1} + px_a).$$

This system admits the solution $x_j \equiv 1$ corresponding to the root $s = 1$. To find all other solutions we apply the method of particular solutions (which we have used for similar equations in Chapter 14, section 4). The middle equation in (3.1) is satisfied by $x_j = \lambda^j$ provided that λ is a root of the quadratic equation $\lambda = qs + \lambda^2 ps$. The two roots of this equation are

(3.2) $$\lambda_1(s) = \frac{1 + (1 - 4pqs^2)^{1/2}}{2ps}, \quad \lambda_2(s) = \frac{1 - (1 - 4pqs^2)^{1/2}}{2ps},$$

---

[4] Part of what follows is a repetition of the theory of Chapter 14. Our quadratic equation occurs there as (4.7); the quantities $\lambda_1(s)$ and $\lambda_2(s)$ of the text were given in (4.8), and the general solution (3.3) appears in Chapter 14 as (4.9). The two methods are related, but in many cases the computational details will differ radically.

and the most general solution of the middle equation in (3.1) is therefore

(3.3) $$x_j = A(s)\lambda_1^j(s) + B(s)\lambda_2^j(s),$$

where $A(s)$ and $B(s)$ are arbitrary. The first and the last equation in (3.1) will be satisfied by (3.3) if and only if $x_0 = x_1$ and $x_a = x_{a+1}$. This requires that $A(s)$ and $B(s)$ satisfy the conditions

(3.4)
$$A(s)\{1 - \lambda_1(s)\} + B(s)\{1 - \lambda_2(s)\} = 0$$
$$A(s)\lambda_1^a(s)\{1 - \lambda_1(s)\} + B(s)\lambda_2^a(s)\{1 - \lambda_2(s)\} = 0.$$

However, these two equations are compatible only if

(3.5) $$\lambda_1^a(s) = \lambda_2^a(s),$$

and we have to determine the values of s for which (3.5) is possible.

From the definition (3.2) we have $\lambda_1(s)\lambda_2(s) = q/p$, and (3.5) implies that $\lambda_1(s)(p/q)^{1/2}$ and $\lambda_2(s)(p/q)^{1/2}$ must be $(2a)$th roots of unity. These roots can be written in the form

(3.6) $$e^{i\pi r/a} = \cos\frac{\pi r}{a} + i\sin\frac{\pi r}{a},$$

where $i^2 = -1$ and $r = 0, 1, 2, \cdots, 2a - 1$. Thus all solutions of (3.5) are among

(3.7) $$\lambda_1(s) = \left(\frac{q}{p}\right)^{1/2} e^{i\pi r/a}, \quad \lambda_2(s) = \left(\frac{q}{p}\right)^{1/2} e^{-i\pi r/a}.$$

The value $r = a$ must be disregarded, since for it $\lambda_1(s) = \lambda_2(s)$, $A(s) = -B(s)$, so that it leads only to the trivial solution $x_j \equiv 0$. To $r = 0$ there corresponds the solution $x_j \equiv 1$, which we have already considered. To $r = 1, 2, \cdots, a - 1$ there correspond $a - 1$ distinct solutions; if we let $r = a + 1, a + 2, \cdots, 2a - 1$, we get the same solutions with $\lambda_1(s)$ and $\lambda_2(s)$ interchanged. Thus we have found a distinct sets of solutions of (3.1), and we know that there can be no more. To each value r we can find the root s_r from (3.7); it is

$$s_r = \{2(pq)^{1/2} \cos \pi r/a\}^{-1}.$$

For $r = 1, 2, \cdots, a - 1$ we get the values of $\lambda_1(s)$ and $\lambda_2(s)$ from (3.7), and then from (3.4) $A(s) = 1 - \lambda_2(s)$ and $B(s) = -\{1 - \lambda_1(s)\}$. (Remember that a multiplicative constant remains arbitrary.) Substituting into (3.3), we find the $a - 1$ sets of solutions

(3.8) $$x_j^{(r)} = \left(\frac{q}{p}\right)^{j/2} \sin\frac{\pi r j}{a} - \left(\frac{q}{p}\right)^{(j+1)/2} \sin\frac{\pi r(j-1)}{a}$$

($r = 1, 2, \cdots, a-1$). To this we add the solution previously found

(3.9) $$x_j^{(0)} = 1.$$

It is easy to verify that (3.8) and (3.9) represent solutions of the given system (3.1).

We have now to find solutions of the second system of linear equations. In the present case (1.7b) takes on the form

(3.10) $$\begin{aligned} y_1 &= sq(y_1 + y_2), \\ y_k &= s(py_{k-1} + qy_{k+1}), \quad (k = 2, \cdots, a-1) \\ y_a &= sp(y_{a-1} + y_a). \end{aligned}$$

The middle equation is the same as (3.1) with p and q interchanged, and its general solution is therefore obtained from (3.3) simply by interchanging p and q. The first and the last equations can be satisfied if (3.7) holds, and a simple calculation shows that for $r = 1, 2, \cdots, a-1$ the solution of (3.10) is

(3.11) $$y_k^{(r)} = \left(\frac{p}{q}\right)^{k/2} \sin \frac{\pi r k}{a} - \left(\frac{p}{q}\right)^{(k-1)/2} \sin \frac{\pi r(k-1)}{a}.$$

For $s = 1$ we find similarly

(3.12) $$y_k^{(0)} = \left(\frac{p}{q}\right)^k.$$

The next step consists in evaluating the coefficients c_r in (1.11). The sum simplifies if $\sin^2 \pi r j/a$ is expressed in terms of the cosine of the double angle, and this in turn by means of complex exponentials. Then we have only to sum finite geometric series and find easily

(3.13) $$c_r = \frac{2p}{a}\left\{1 - 2(pq)^{1/2} \cos \frac{\pi r}{a}\right\}^{-1} \quad (r = 1, 2, \cdots, a-1).$$

For $r = 0$ we get,

(3.14) $$c_0 = \frac{q}{p} \frac{(p/q) - 1}{(p/q)^a - 1},$$

provided that $p \neq q$. If $p = q = 1/2$, then (3.13) remains valid, but (3.14) is to be replaced by $c_0 = 1/a$.

These formulas lead to the *final result*

(3.15) $$p_{jk}^{(n)} = \frac{(p/q) - 1}{(p/q)^a - 1}\left(\frac{p}{q}\right)^{k-1} + 2^{n+1} p^{1+(n-j+k)/2} q^{(n+j-k)/2} a^{-1} \sum_{r=1}^{a-1} S_r$$

where S_r stands for

$$\frac{\cos^n \dfrac{\pi r}{a} \left\{ \sin \dfrac{\pi r j}{a} - \left(\dfrac{q}{p}\right)^{\!1/2} \sin \dfrac{\pi r(j-1)}{a} \right\} \left\{ \sin \dfrac{\pi r k}{a} - \left(\dfrac{q}{p}\right)^{\!1/2} \sin \dfrac{\pi r(k-1)}{a} \right\}}{1 - 2(pq)^{1/2} \cos \dfrac{\pi r}{a}}.$$

As $n \to \infty$, the second term in (3.15) tends to zero, and we find again that $p_{jk}^{(n)}$ tends to a stationary distribution independent of j. [This limiting distribution was derived by other methods in Chapter 15, example (6.b).] Passing to the limit $a \to \infty$, we get the formula for a random walk with a single reflecting barrier; in the limit, the sum is replaced by an integral.[5]

4. Transient States; Absorption Probabilities

The theorem of section 1 was derived under the assumption that the roots $s_1, s_2, \ldots$ are distinct. The presence of multiple roots does not require essential modifications, but we shall discuss only a particular case of special importance. The root $s_1 = 1$ is multiple whenever the chain contains two or more closed subchains, and this is a frequent situation in problems connected with absorption probabilities. It is easy to adapt the method of section 1 to this case. For conciseness and clarity, we shall explain the procedure by means of examples which will reveal the main features of the general case.

Examples. (a) Consider the matrix of transition probabilities

$$(4.1) \qquad P = \begin{bmatrix} 1/3 & 2/3 & 0 & 0 & 0 & 0 \\ 2/3 & 1/3 & 0 & 0 & 0 & 0 \\ 0 & 0 & 1/4 & 3/4 & 0 & 0 \\ 0 & 0 & 1/5 & 4/5 & 0 & 0 \\ 1/4 & 0 & 1/4 & 0 & 1/4 & 1/4 \\ 1/6 & 1/6 & 1/6 & 1/6 & 1/6 & 1/6 \end{bmatrix}.$$

It is clear that E_1 and E_2 form a closed set (that is, no transition is

---

[5] For analogous formulas in the case of one reflecting and one absorbing barrier cf. M. Kac, Random Walk and the Theory of Brownian Motion, *American Mathematical Monthly*, vol. 54 (1947), pp. 369–391. The definition of the reflecting barrier is there modified so that the particle may reach $x = 0$; whenever this occurs, the next step takes it to $x = 1$. The explicit formulas are then more complicated. Kac also found formulas for $p_{jk}^{(n)}$ in the Ehrenfest model (example VII of Chapter 15).

possible to any of the remaining four states; cf. Chapter 15, section 4). Similarly E_3 and E_4 form another closed set. Finally, E_5 and E_6 are transient states. After finitely many steps the system passes into one of the two closed sets and remains there.

The matrix P has the form of a partitioned matrix

$$(4.2) \qquad P = \begin{bmatrix} A & 0 & 0 \\ 0 & B & 0 \\ U & V & T \end{bmatrix}$$

where each letter stands for a two by two matrix and each zero for a matrix with four zeros. For example, A has the rows $(1/3, 2/3)$ and $(2/3, 1/3)$; this is the matrix of transition probabilities corresponding to the chain formed by the two states E_1 and E_2. This matrix can be studied by itself, and the powers A^n can be obtained from example $(2.a)$ with $p = \alpha = 2/3$. When the powers $P^2, P^3, \ldots$ are calculated, it will be found that the first two rows are in no way affected by the remaining four rows. More precisely, P^n has the form

$$(4.3) \qquad P^n = \begin{bmatrix} A^n & 0 & 0 \\ 0 & B^n & 0 \\ U_n & V_n & T^n \end{bmatrix}$$

where A^n, B^n, T^n are the nth powers of A, B, and T, respectively, and can be calculated [6] by the method of section 1 [cf. example $(2.a)$ where all calculations are performed]. Instead of six equations with six unknowns we are confronted only with systems of two equations with two unknowns each.

It should be noted that the matrices U_n and V_n in (4.3) are not powers of U and V and cannot be obtained in the same simple way as A^n, B^n, and T^n. However, in the calculation of $P^2, P^3, \ldots$ the third and fourth columns never affect the remaining four columns. In other words, if in P^n the rows and columns corresponding to E_3 and E_4 are deleted, we get the matrix

$$(4.4) \qquad \begin{pmatrix} A^n & 0 \\ U_n & T^n \end{pmatrix}$$

---

[6] In T the rows do not add to unity so that T is not a stochastic matrix. However, the method of section 1 applies without change, except that $s = 1$ is no longer a root (so that $T^n \to 0$).

which is the nth power of the corresponding submatrix in P, that is, of

(4.5) $$\begin{pmatrix} A & 0 \\ U & T \end{pmatrix} = \begin{bmatrix} 1/3 & 2/3 & 0 & 0 \\ 2/3 & 1/3 & 0 & 0 \\ 1/4 & 0 & 1/4 & 1/4 \\ 1/6 & 1/6 & 1/6 & 1/6 \end{bmatrix}.$$

Therefore (4.4) can be calculated by the method of section 1, which in the present case simplifies considerably. The matrix V_n can be obtained in a similar way.

Usually the explicit forms of U_n and V_n are of interest only inasmuch as they are connected with *absorption probabilities*. If the system starts from, say, E_5, what is the *probability λ that it will eventually pass into the closed set formed by E_1 and E_2* (and not into the other closed set)? What is the *probability λ_n that this will occur exactly at the nth step?* Clearly $p_{51}^{(n)} + p_{52}^{(n)}$ is the probability that the considered event occurs at the nth step or before, that is,

$$p_{51}^{(n)} + p_{52}^{(n)} = \lambda_1 + \lambda_2 + \cdots + \lambda_n.$$

Letting $n \to \infty$, we get λ. A preferable way to calculate λ_n is as follows. The $(n-1)$st step must take the system to a state other than E_1 and E_2, that is, to either E_5 or E_6 (since from E_3 or E_4 no transition to E_1 and E_2 is possible). The nth step then takes the system to E_1 or E_2. Hence

(4.6)
$$\lambda_n = p_{55}^{(n-1)}(p_{51} + p_{52}) + p_{56}^{(n-1)}(p_{61} + p_{62})$$
$$= \frac{1}{4} p_{55}^{(n-1)} + \frac{1}{3} p_{56}^{(n-1)}.$$

It will be noted that λ_n is completely determined by the elements of T^{n-1}, and this matrix is easily calculated. In the present case

$$p_{55}^{(n)} = p_{56}^{(n)} = \frac{1}{4}\left(\frac{5}{12}\right)^{n-1} \quad \text{and hence} \quad \lambda_n = \frac{7}{48}\left(\frac{5}{12}\right)^{n-2}.$$

(b) *Brother-sister Mating.* As a second example we give a complete treatment of example XI of Chapter 15. A glance at the matrix shows that the states E_1 and E_5 form a closed set each (a fact which is clear from the biological meaning). If the system starts from any other state E_j, it will eventually pass either into E_1 or into E_5 and then remain there. The breeder desires to know the corresponding probabilities and the expected duration of the process.

Deleting the first and fifth column and row, we get the reduced matrix

(4.7) $$T = \begin{bmatrix} 1/2 & 1/4 & 0 & 0 \\ 1/4 & 1/4 & 1/4 & 1/8 \\ 0 & 1/4 & 1/2 & 0 \\ 0 & 1 & 0 & 0 \end{bmatrix}.$$

The powers T^n will now be calculated by the method of section 1. They represent the transition probabilities among transient states.

The equations (1.7a) reduce to

(4.8) $$x_1 = \frac{s(2x_1 + x_2)}{4}, \quad x_2 = \frac{s(2x_1 + 2x_2 + 2x_3 + x_4)}{8},$$

$$x_3 = \frac{s(x_2 + 2x_3)}{4}, \quad x_4 = sx_2.$$

It has a solution only if the determinant vanishes, and this condition leads to a fourth-degree equation in s. To simplify writing we put

(4.9) $$\theta_1 = 5^{1/2} - 1, \quad \theta_2 = 5^{1/2} + 1.$$

Then the four roots s_r are

(4.10) $$s_1 = 2, \quad s_2 = 4, \quad s_3 = \theta_1, \quad s_4 = -\theta_2,$$

and the corresponding solutions $(x_1^{(r)}, \cdots, x_j^{(r)})$ of (4.8) are

(4.11) $(1, 0, -1, 0), \quad (1, -1, 1, -4), \quad (1, \theta_1, 1, \theta_1^2), \quad (1, -\theta_2, 1, \theta_2^2).$

The system of linear equations for $y_k^{(r)}$ is obtained by specialization from (1.7b), and the four sets of solutions are in proper order

(4.12) $(1, 0, -1, 0), \quad (1, -1, 1, -1/2),$
$(1, \theta_1, 1, \theta_1^2/8), \quad (1, -\theta_2, 1, \theta_2^2/8).$

From (1.11) we find the four constants $c_1 = 1/2$, $c_2 = 1/5$, $c_3 = \theta_2^2/40$, $c_4 = \theta_1^2/40$. From (1.8) we get the $\rho_{jk}^{(r)}$; and finally (1.3) gives us $p_{jk}^{(n)}$ for all transient states, that is, for $j, k = 2, 3, 4, 6$. For fixed j, k the sequence $p_{jk}^{(n)}$ is the sum of four geometric series with ratios $s_1, \cdots, s_4$.

An absorption in E_1 exactly at the nth step is possible only if the $(n-1)$st step takes the system into either E_2 or E_3, and the nth step into E_1. The probability for this is $p_{j2}^{(n-1)}/4 + p_{j3}^{(n)}/16$. Similarly, the probability of absorption at E_5 is $p_{j3}^{(n-1)}/16 + p_{j4}^{(n-1)}/4$. Summing over all n we get the probabilities that the system will eventually pass into and stay in E_1 and E_5, respectively. The actual calculation of these probabilities requires only the summation of four geometric series.

5. Application to Recurrence Times

In problem 6 of Chapter 12 it is shown how the mean μ and the variance σ^2 of the recurrence time of a recurrent event $\mathcal{E}$ can be calculated in terms of the probabilities u_n that $\mathcal{E}$ occurs at the nth trial. If $\mathcal{E}$ is not periodic, then

$$(5.1) \qquad u_n \to \frac{1}{\mu} \quad \text{and} \quad \sum_{n=0}^{\infty} \left(u_n - \frac{1}{\mu} \right) = \frac{\sigma^2 - \mu + \mu^2}{2\mu^2},$$

provided that σ^2 is finite.

If we identify $\mathcal{E}$ with a recurrent state E_j, then $u_n = p_{jj}^{(n)}$ (and $u_0 = 1$). In a finite Markov chain all recurrence times have finite variance (cf. problem 8 of Chapter 15), so that (5.1) applies. Suppose that E_j is not periodic and that formula (1.3) applies. Then $s_1 = 1$ and $|s_r| > 1$ for $r = 2, 3, \ldots$, so that $p_{jj}^{(n)} \to \rho_{jj}^{(1)} = 1/\mu_j$. To the term $u_n - 1/\mu$ of (5.1) there corresponds

$$(5.2) \qquad p_{jj}^{(n)} - \frac{1}{\mu_j} = \sum_{r=2}^{a} \rho_{jj}^{(r)} s_r^{-n}.$$

This formula is valid for $n \geq 1$; summing the geometric series with ratio s_r^{-1}, we find

$$(5.3) \qquad \sum_{n=1}^{\infty} \left(p_{jj}^{(n)} - \frac{1}{\mu_j} \right) = \sum_{r=2}^{a} \frac{\rho_{jj}^{(r)}}{s_r - 1}.$$

Introducing this into (5.1), we find that if E_j is a *non-periodic recurrent state, then its mean recurrence time is given by* $\mu_j = 1/\rho_{jj}^{(1)}$, *and the variance of its recurrence time is*

$$(5.4) \qquad \sigma_j^2 = \mu_j - \mu_j^2 + 2\mu_j^2 \sum_{r=2}^{a} \frac{\rho_{jj}^{(r)}}{s_r - 1},$$

provided, of course, that formula (1.3) is applicable and $s_1 = 1$. The case of periodic recurrent events and the occurrence of double roots require only obvious modifications.

CHAPTER 17

THE SIMPLEST TIME-DEPENDENT STOCHASTIC PROCESSES [1]

1. General Orientation

Random walks and Markov chains are examples of stochastic processes [2] where changes can occur only at fixed times, say, $t = 1, 2, 3, \ldots$. On the other hand, in Chapter 6, sections 5 and 6, we were concerned with phenomena such as telephone calls, radioactive disintegrations, and chromosome breakages, where changes may occur at any time. It is clear that a complete description of these processes leads beyond the domain of discrete probabilities. To fix ideas, consider the incoming calls at a telephone exchange (or, rather, an idealized mathematical model of the actual process). Every instant t corresponds to a trial, and the result of an experiment may be described in terms of a function $X(t)$ giving the number of calls up to time t. If the first call occurs at time t_1, the second at t_2, etc., then the function $X(t)$ equals 0 for $0 < t < t_1$, 1 for $t_1 < t < t_2$, 2 for $t_2 < t < t_3$, etc. Conversely, every non-decreasing function $X(t)$, assuming only the values $0, 1, 2, \ldots$, represents a possible development at our telephone exchange. In other words, a complete description of our conceptual experiment calls for a sample space whose points are functions $X(t)$ (and not sequences as in the case of discrete trials). Further, we may consider compound events such as "seven calls within a minute on a certain day"; this is obviously the aggregate of those $X(t)$ which satisfy the condition that for some point t of a specified interval we have $X(t + h) - X(t) \geq 7$, where h represents the span of one minute.

We cannot deal here with such complicated sample spaces and must therefore defer the study of the more delicate aspects of the theory. Fortunately, certain interesting questions can be answered even with the simple means now at our disposal.

If we limit the consideration to the number of calls $X(t)$ within an arbitrary but fixed period of duration t, then $X(t)$ is a random variable of the familiar type, assuming the values $0, 1, 2, \ldots$. Let $P_n(t)$ be

---

[1] This chapter is almost independent of Chapters 10–16.
[2] Cf. the footnote on p. 337.

the probability that $X(t) = n$. It is true that the distribution $\{P_n(t)\}$ depends on the duration t, that is, on a continuous parameter. However, most of the probability distributions already introduced depend on a parameter, and we are not in an essentially new situation.

In Chapter 6, section 5, we used a limiting process to show that under certain conditions $X(t)$ has a Poisson distribution, that is,

$$(1.1) \qquad P_n(t) = e^{-\lambda t} \frac{(\lambda t)^n}{n!}.$$

A second derivation was given in Chapter 11, section 4, in connection with the Pascal distribution. We now derive this result by a new method which is more flexible and can be applied to more complicated processes. We start by translating the physical or intuitive description of a process into properties required of the probabilities $P_n(t)$. In this way we get a set of plausible and simple postulates on the distribution $\{P_n(t)\}$, from which analytic expressions for $P_n(t)$ can be derived.

The artificial limitation to discrete probabilities has unavoidable drawbacks. Consider, for example, the zero term in (1.1). We interpret

$$(1.2) \qquad P_0(t) = e^{-\lambda t}$$

as the probability that no call occurs within an observation period of duration t. This formulation suggests that $P_0(t)$ also might be interpreted as the probability that the waiting time (starting at an arbitrary moment) up to the first call exceeds t. It can be shown that this interpretation is correct, but it will be noticed that it involves probabilities in a continuum. The operational meaning of our first formulation is as follows: make a series of "identical observations" with a fixed observational period t. Each trial results in either "no call" (success) or "one or more calls" (failure). Then we have Bernoulli trials with the probability of success $e^{-\lambda t}$. With the second interpretation we are to wait until a call arrives. Every positive number is a possible waiting time, so that the sample space corresponding to each trial is the half-line $t > 0$. Formula (1.2) then represents a continuous probability distribution and as such will be treated in volume 2 (opening a new approach to the Poisson distribution).

2. The Poisson Process

The simplest stochastic processes are of the type considered in Chapter 6, section 5. A system is subject to instantaneous changes of state which can occur at any time; these changes are due to the occurrence of random events such as splitting of physical particles, arriving of telephone calls, or breakage of a chromosome under harmful irradiation. All changes are of the same kind, and we are concerned only with their total number. Each change is marked by a point on the time axis, so that we are studying certain random distributions of points on a line.

The physical processes which we have in mind are characterized by the two properties that they are stationary and that future changes are independent of past changes. By this we mean that the forces and influences which determine the process remain absolutely unchanged, so that the probability of any particular event is the same for all time intervals of length t, independent of where this interval is situated and of the past history of the system.[3]

We now translate this description into mathematical language. The process is to be described in terms of probabilities [4] $P_n(t)$ that exactly n changes occur during a time interval of length t. In particular, $P_0(t)$ is the probability of no change, and $1 - P_0(t)$ the probability of one or more changes. We shall assume that [5] as $t \to 0$

(2.1) $$\frac{1 - P_0(t)}{t} \to \lambda$$

where λ is a positive constant [the derivative of $-P_0(t)$ at $t = 0$]. Then for a small interval of length h the probability of one or more changes is $1 - P_0(h) = \lambda h + o(h)$, where the term $o(h)$ denotes a quantity which is of smaller order of magnitude than h. We now formulate our

---

[3] In a telephone exchange incoming calls are more frequent during the busiest hour of the day than, say, between midnight and 1 A.M.; the process is therefore not homogeneous in time. However, for obvious reasons telephone engineers are concerned mainly with the "busy hour" of the day and for that period the process can be considered homogeneous. Experience shows also that during the busy hour the incoming traffic follows the Poisson distribution with surprising accuracy. Similar considerations apply to automobile accidents, which are more frequent on Sundays, etc.

[4] Our notation implies that these probabilities are independent of where the interval of length t is taken. For a non-homogeneous process we should have to introduce the probability $P_n(t_1, t_2)$ that n changes occur in the interval $t_1 < t < t_2$.

[5] Instead of assuming (2.1) we may start from the following reasoning. The event "no change in the time interval $(0, t + s)$" requires that no change occurs within the two intervals $(0, t)$ and $(t, t + s)$. Because of the assumed independence this leads to the equation $P_0(s + t) = P_0(s)P_0(t)$ whose only positive bounded solution is $e^{-\lambda t}$. It follows that $P_0(t) = e^{-\lambda t}$, and this implies (2.1). However, we prefer to start from (2.1), since this procedure leads in a more natural way to various generalizations.

Postulates for the Poisson Process. Whatever the number of changes during $(0, t)$, the (conditional) probability that during $(t, t + h)$ a change occurs is $\lambda h + o(h)$, and the probability that more than one change occurs is of smaller order of magnitude than h.

These conditions easily lead to a system of differential equations for $P_n(t)$. Consider two contiguous intervals $(0, t)$ and $(t, t + h)$, where h is small. If $n \geq 1$, then exactly n changes can occur in the interval $(0, t + h)$ in three mutually exclusive ways: (1) no change during $(t, t + h)$ and n changes during $(0, t)$; (2) one change during $(t, t + h)$ and $n - 1$ changes during $(0, t)$; (3) $x \geq 2$ changes during $(t, t + h)$ and $n - x$ changes during $(0, t)$. According to our hypotheses, the probability of the first contingency is $P_n(t)$ times the probability of no change during $(t, t + h)$ and this last is $1 - \lambda h - o(h)$. Similarly, the second contingency has probability $P_{n-1}(t)\lambda h + o(h)$, while the last has a probability of smaller order of magnitude than h. This means that

(2.2) $$P_n(t + h) = P_n(t)(1 - \lambda h) + P_{n-1}(t)\lambda h + o(h)$$

or

(2.3) $$\frac{P_n(t + h) - P_n(t)}{h} = -\lambda P_n(t) + \lambda P_{n-1}(t) + \frac{o(h)}{h}.$$

As $h \to 0$, the last term tends to zero; hence the limit of the left side exists and

(2.4) $$P_n'(t) = -\lambda P_n(t) + \lambda P_{n-1}(t) \qquad (n \geq 1).$$

For $n = 0$ the second and third contingencies mentioned above do not arise, and therefore (2.4) is to be replaced by the simpler equation

(2.5) $$P_0(t + h) = P_0(t)(1 - \lambda h) + o(h),$$

which leads to

(2.6) $$P_0'(t) = -\lambda P_0(t).$$

From (2.6) and $P_0(0) = 1$ we get $P_0(t) = e^{-\lambda t}$. Substituting this $P_0(t)$ into (2.4) with $n = 1$, we get an ordinary differential equation for $P_1(t)$. Since $P_1(0) = 0$, we find easily that $P_1(t) = \lambda t e^{-\lambda t}$, in agreement with the Poisson distribution (1.1). Proceeding in the same way, we find successively all terms of (1.1).

The salient feature of this new derivation of the Poisson distribution is that it starts directly from plausible physical assumptions. The Poisson distribution no longer appears as an approximation to the binomial distribution or as a limiting distribution, but stands in its own right (or, one might say, as the expression of a physical law). The main advantage of the new derivation is that it lends itself to many generalizations.[6]

3. The Pure Birth Process

In the Poisson process the probability of a change during $(t, t + h)$ is independent of the number of changes during $(0, t)$. The simplest generalization consists of dropping this assumption. We then assume that, if n changes occur during $(0, t)$, the probability of a new change during $(t, t + h)$ equals $\lambda_n h$ plus terms of smaller order of magnitude than h; instead of a single constant λ characterizing the process, we have a sequence $\lambda_0, \lambda_1, \lambda_2, \ldots$.

It is convenient to introduce a more flexible terminology. Instead of saying that n changes occur during $(0, t)$, we shall say that the system is in state E_n. A new change then becomes a transition $E_n \to E_{n+1}$. In a pure birth process transitions from E_n are possible only to E_{n+1}. Such a process is characterized by the following

Postulates. If at time t the system is in state E_n ($n = 0, 1, 2, \ldots$), then the probability that during $(t, t + h)$ a transition to E_{n+1} occurs equals $\lambda_n h + o(h)$; the probability of several changes is of smaller order of magnitude than h.

The salient feature of this assumption is that the time which the system spends in any particular state plays no role: there are sudden changes of state but no aging as long as the system remains within a single state.

Again let $P_n(t)$ be the probability that at time t the system is in state E_n. The functions $P_n(t)$ satisfy a system of differential equations which can be derived by the argument of the preceding section, with the only change that (2.2) is replaced by

(3.1) $P_n(t + h) = P_n(t)(1 - \lambda_n h) + P_{n-1}(t)\lambda_{n-1} h + o(h).$

In this way we get the *basic system of differential equations*

(3.2) $\quad P_n'(t) = -\lambda_n P_n(t) + \lambda_{n-1} P_{n-1}(t) \quad\quad (n \geq 1),$

$\quad\quad\quad P_0'(t) = -\lambda_0 P_0(t).$

We can calculate $P_0(t)$ first and then, by recursion, all $P_n(t)$. If the state of the system represents the number of changes during $(0, t)$,

---

[6] The processes of this chapter and their relation to diffusion processes are treated in a lecture (of January 1946) by W. Feller, On the Theory of Stochastic Process with Particular Reference to Applications, *Proceedings of the Berkeley Symposium on Mathematical Statistics and Probability*, 1949, pp. 403–432. There the reader will also find further references.

then the initial state is E_0 so that $P_0(0) = 1$ and hence $P_0(t) = e^{-\lambda_0 t}$. However, it is not necessary that the system start from state E_0 [see example (3.b)]. If at time zero the system is in E_i, then we have

(3.3) $\qquad P_i(0) = 1, \qquad P_n(0) = 0, \qquad\qquad\text{for } n \neq i.$

These *initial conditions* uniquely determine the solution $\{P_n(t)\}$ of (3.2). [In particular, $P_0(t) = P_1(t) = \cdots = P_{i-1}(t) = 0$.] Explicit formulas for $P_n(t)$ have been derived independently by many authors, but are of no interest to us. It is easily verified that for arbitrarily prescribed λ_n the system $\{P_n(t)\}$ has all required properties, except that under certain conditions $\Sigma P_n(t) < 1$. This phenomenon will be discussed in section 4.

Examples. (a) *Radioactive Transmutations.* A radioactive atom, say uranium, may by emission of particles or γ-rays change to an atom of a different kind. Each kind represents a possible state of the system, and as the process continues, we get a succession of transitions $E_0 \to E_1 \to E_2 \to \cdots \to E_m$. According to accepted physical theories, the probability of a transition $E_n \to E_{n+1}$ remains unchanged as long as the atom is in state E_n, and this hypothesis is expressed by our starting supposition. The differential equations (3.2) therefore describe the process (a fact well known to physicists). If E_m is the terminal state from which no further transitions are possible, then $\lambda_m = 0$ and the system (3.2) terminates with $n = m$. [For $n > m$ one gets automatically $P_n(t) = 0$.]

(b) *The Yule Process.* Consider a population of members which can (by splitting or otherwise) give birth to new members but cannot die. Assume that during any short time interval of length h each member has probability $\lambda h + o(h)$ to create a new one; the constant λ determines the rate of increase of the population. If there is no interaction among the members and at time t the population size is n, then the probability of an increase during $(t, t + h)$ is $n\lambda h + o(h)$. The probability $P_n(t)$ that the population numbers exactly n elements therefore satisfies (3.2) with $\lambda_n = n\lambda$, that is,

(3.4) $\qquad P_n'(t) = -n\lambda P_n(t) + (n-1)\lambda P_{n-1}(t) \qquad\qquad (n \geq 1).$

If i is the population size at time $t = 0$, then the initial conditions (3.3) apply. The solution is easily found to be

(3.5) $\qquad P_n(t) = \binom{n-1}{n-i} e^{-i\lambda t}(1 - e^{-\lambda t})^{n-i}$

for $n \geq i$, while, of course, $P_n(t) = 0$ for $n < i$.

This type of process was first studied by Yule [7] in connection with the mathematical theory of evolution. The population consists of the species within a genus, and the creation of a new element is due to mutations. The assumption that each species has the same probability of throwing out a new species neglects the difference in species sizes. Since we have also neglected the possibility of a species dying out, (3.5) can be expected to give only a crude approximation. Furry [8] used the same model to describe a process connected with cosmic rays, but again the approximation is rather crude. The differential equations (3.4) apply strictly to a population of particles which can split into exact replicas of themselves, provided, of course, that there is no interaction among particles.

*4. Divergent Birth Processes

The solution $\{P_n(t)\}$ of the infinite system of differential equations (3.2) subject to initial conditions (3.3) can be calculated inductively, starting from $P_i(t) = e^{-\lambda_i t}$. The distribution $\{P_n(t)\}$ is therefore uniquely determined. From the familiar formulas for solving linear differential equations it follows also that $P_n(t) \geq 0$. The only question left open is whether $\{P_n(t)\}$ is an honest probability distribution, that is, whether or not

(4.1) $$\Sigma P_n(t) = 1$$

for all t. We shall see that this is not always so: if the coefficients λ_n increase sufficiently fast, then it may happen that

(4.2) $$\Sigma P_n(t) < 1.$$

At first sight this possibility appears surprising and, perhaps, disturbing, but it finds a ready explanation. The left side in (4.2) may be interpreted as the probability that during time t only a *finite number* of changes takes place. Accordingly, the difference between the two sides in (4.2) accounts for the possibility of infinitely many changes, or a sort of explosion. For a better understanding of this phenomenon

---

* Starred sections treat special topics and may be omitted at first reading.

[7] G. Udny Yule, A Mathematical Theory of Evolution, Based on the Conclusions of Dr. J. C. Willis, F.R.S., *Philosophical Transactions of the Royal Society, London*, Series B, vol. 213 (1924), pp. 21–87. Yule does not introduce the differential equations (3.4) but derives $P_n(t)$ by a limiting process similar to the one which we used in Chapter 6, section 5, for the Poisson process.

[8] Furry, On Fluctuation Phenomena in the Passage of High-energy Electrons through Lead, *Physical Reviews*, vol. 52 (1937), p. 569.

let us compare our probabilistic model of growth with the familiar deterministic approach.

The quantity λ_n in (3.2) could be called the average rate of growth at a time when the population size is n. For example, in the special case (3.4) we have $\lambda_n = n\lambda$, so that the average rate of growth is proportional to the actual population size. If growth is not subject to chance fluctuations and has a rate of increase proportional to the instantaneous population size, then the population size $x(t)$ varies in accordance with the deterministic differential equation

$$(4.3) \qquad \frac{dx(t)}{dt} = \lambda x(t).$$

It follows that at time t the population size is

$$(4.4) \qquad x(t) = ie^{\lambda t},$$

where $i = x(0)$ is the initial population size. The connection between (3.4) and (4.3) is not purely formal. It is readily seen that (4.4) actually gives the mean of the distribution (3.5), so that (4.3) describes the expected population size, while (3.4) takes account of chance fluctuations.

Let us now consider a deterministic growth process where the rate of growth increases faster than the population size. If the rate of growth is proportional to $x^2(t)$, we get the differential equation

$$(4.5) \qquad \frac{dx(t)}{dt} = \lambda x^2(t)$$

whose solution is

$$(4.6) \qquad x(t) = \frac{i}{1 - \lambda i t}.$$

Note that $x(t)$ increases over all bounds as $t \to 1/\lambda i$. In other words, the assumption that the rate of growth increases as the square of the population size implies an infinite growth within a finite time interval. Similarly, if in (3.4) the λ_n increase too fast, there is a finite probability that infinitely many changes take place in a finite time interval. A precise answer as to the conditions when such a divergent growth occurs is given by the

Theorem. *In order that* (4.1) *may hold for all t it is necessary and sufficient that the series*

$$(4.7) \qquad \sum \frac{1}{\lambda_n}$$

diverge.

Proof. Letting

(4.8) $$S_k(t) = P_0(t) + \cdots + P_k(t),$$

we get from (3.2)

(4.9) $$S_k'(t) = -\lambda_k P_k(t)$$

and hence for $k \geq i$

(4.10) $$1 - S_k(t) = \lambda_k \int_0^t P_k(\tau)\, d\tau.$$

Since all terms in (4.8) are non-negative, the sequence $S_k(t)$ (for fixed t) can only increase with k, and therefore the right side in (4.10) decreases monotonically with k. Call its limit $\mu(t)$. Then for $k \geq i$

(4.11) $$\lambda_k \int_0^t P_k(\tau)\, d\tau \geq \mu(t)$$

and hence

(4.12) $$\int_0^t S_n(\tau)\, d\tau \geq \mu(t) \left(\frac{1}{\lambda_i} + \frac{1}{\lambda_{i+1}} + \cdots + \frac{1}{\lambda_n} \right).$$

Because of (4.10) we have $S_n(t) \leq 1$, so that the left side in (4.12) is at most t. If the series (4.7) diverges, the second factor on the right in (4.12) tends to infinity, and the inequality can hold only if $\mu(t) = 0$ for all t. In this case the right side in (4.10) tends to 0 as $k \to \infty$, and therefore $S_n(t) \to 1$, so that (4.1) holds. Conversely, the left side of (4.12) is less than $\lambda_0^{-1} + \lambda_1^{-1} + \ldots + \lambda_n^{-1}$. If (4.7) converges, this expression is bounded and hence it is impossible that $S_n(t) \to 1$ for all t.

5. The Birth and Death Process

The pure birth process of section 3 provides a satisfactory description of radioactive transmutations, but it can obviously not serve as a realistic model for changes in the size of populations whose members can die (or drop out in any way). This suggests generalizing the model by permitting transitions from the state E_n not only to the next higher state E_{n+1}, but also to the next lower state E_{n-1}. (Still more general processes will be defined in section 9.) Accordingly we now start from the following

Postulates. *The system changes only through transitions from states to their next neighbors (from E_n to E_{n+1} or E_{n-1} if $n \geq 1$, but from E_0 to E_1 only). If at any time t the system is in state E_n, then the probability that during $(t, t+h)$ the transition $E_n \to E_{n+1}$ occurs equals $\lambda_n h + o(h)$,* and the probability of $E_n \to E_{n-1}$ (if $n \geq 1$) equals $\mu_n h + o(h)$. The probability that during $(t, t+h)$ more than one change occurs is of smaller order of magnitude than h.

It is easy to adapt the method of section 2 to derive differential equations for the probabilities $P_n(t)$ of finding the system at time t in state E_n. To calculate $P_n(t+h)$ we note that at time $t+h$ the system can be in state E_n only if one of the following conditions is satisfied: (1) at time t the system is in E_n and during $(t, t+h)$ no change occurs; (2) at time t the system is in E_{n-1} and a transition to E_n occurs; (3) at time t the system is in E_{n+1} and a transition to E_n occurs; (4) during $(t, t+h)$ two or more transitions occur. By assumption, the probability of the last situation tends to zero faster than h. The first three contingencies are mutually exclusive, so that their probabilities add. We get therefore

$$(5.1) \quad P_n(t+h) = P_n(t)\{1 - \lambda_n h - \mu_n h\} \\ + \lambda_{n-1} h P_{n-1}(t) + \mu_{n+1} h P_{n+1}(t) + o(h).$$

Transposing the term $P_n(t)$ and dividing the equation by h, we get on the left the difference ratio of $P_n(t)$. Letting $h \to 0$, we get

$$(5.2) \quad P_n'(t) = -(\lambda_n + \mu_n)P_n(t) + \lambda_{n-1}P_{n-1}(t) + \mu_{n+1}P_{n+1}(t).$$

This equation holds for $n \geq 1$. For $n = 0$ in the same way

$$(5.3) \quad P_0'(t) = -\lambda_0 P_0(t) + \mu_1 P_1(t).$$

If at time zero the system is in state E_i, then again the initial condition

$$(5.4) \quad P_i(0) = 1, \quad P_n(0) = 0 \quad \text{for } n \neq i$$

holds.

In (5.2)–(5.4) we have the fundamental equation of the birth and death process. In fact, *the coefficients λ_n and μ_n can be arbitrarily prescribed; the differential equations (5.2) and (5.3), together with the initial conditions (5.4), then uniquely determine the corresponding system of probabilities $P_n(t)$.* This assertion is by no means obvious. In the case of a pure birth process we had also an infinite system of differential equations; however, the system (3.2) had the form of recurrence relations where $P_n(t)$ can be calculated from $P_{n-1}(t)$, and $P_0(t)$ is determined by the first equation. The new system (5.2) is not of this form, and all $P_n(t)$ must be calculated simultaneously. A complete proof of the existence and uniqueness of a system of solutions $\{P_n(t)\}$ is lengthy and will be omitted.[9] It turns out that we have always

---

[9] This assertion is a special case of the more general theorem of section 9.

$P_n(t) \geq 0$ and

(5.5) $$\Sigma P_n(t) \leq 1.$$

In cases of practical interest the equality sign holds.

If $\lambda_0 = 0$, then the transition $E_0 \to E_1$ is impossible. In the terminology of Markov chains E_0 is an *absorbing state* from which no exit is possible; once the system is in E_0 it stays there. From (5.3) it follows that in this case $P_0'(t) \geq 0$, so that $P_0(t)$ increases monotonically. The limit $P_0(\infty)$ is the probability of ultimate absorption.

More generally, it can be shown that *the limits*

(5.6) $$\lim_{t \to \infty} P_n(t) = p_n$$

exist and are independent of the initial conditions (5.4); they satisfy the system of linear equations obtained from (5.2)–(5.3) on putting [10] $P_n'(t) = 0$.

This statement can be proved either from the explicit formulas [11] for the $P_n(t)$ or from general ergodic theories. Intuitively the theorem becomes almost obvious by a comparison of our process with a simple Markov chain with transition probabilities

(5.7) $$p_{n,n+1} = \frac{\lambda_n}{\lambda_n + \mu_n}, \quad p_{n,n-1} = \frac{\mu_n}{\lambda_n + \mu_n}.$$

In this chain the only direct transitions are $E_n \to E_{n+1}$ and $E_n \to E_{n-1}$, and they have the same conditional probabilities as in our process; the difference between the chain and our process lies in the fact that, with the latter, changes can occur at arbitrary times, so that the number of transitions during time t is a random variable. However, for large t this number is certain to be large, and hence it is plausible that for $t \to \infty$ the probabilities $P_n(t)$ behave as the corresponding probabilities of the simple chain.

The principal field of applications of the birth and death process is to problems of waiting times, trunking problems, etc. Such applications will be discussed in sections 6 and 7.

---

[10] This is the so-called "steady-state condition." It must be understood that despite the suggestive name no steady state ever is reached except when E_0 is an absorbing state. In general the chance fluctuations continue unabated, and the existence of the limits (5.6) only indicates that in the long run the influence of the initial state disappears. The steady state is of so-called statistical equilibrium.

[11] W. Feller, On the Integrodifferential Equations of Completely Discontinuous Markov Processes, *Transactions of the American Mathematical Society*, vol. 48 (1940), pp. 488–515.

Example. *Linear Growth.* Suppose that a population consists of elements which can split or die. During any short time interval of length h the probability for any living element to split into two is $\lambda h + o(h)$, while the corresponding probability of dying is $\mu h + o(h)$. Here λ and μ are two constants characteristic of the population. If there is no interaction among the elements, then we are led to a birth and death process with $\lambda_n = n\lambda$, $\mu_n = n\mu$. The basic differential equations take on the form

(5.8)
$$P_0'(t) = \mu P_1(t),$$
$$P_n'(t) = -(\lambda + \mu)nP_n(t) + \lambda(n-1)P_{n-1}(t) + \mu(n+1)P_{n+1}(t)$$

$(n = 1, 2, \ldots)$.

Explicit solutions can be found [12] (cf. problems 7–9), but we shall not discuss this aspect. It can be shown that the limits (5.6) exist. They obviously satisfy (5.8) with $P_n'(t) = 0$. From the first equation we find $p_1 = 0$, and then we see by induction from the second equation that $p_n = 0$ for all $n \geq 1$. If $p_0 = 1$, we may say that the probability of ultimate extinction is 1. If $p_0 < 1$, then the relations $p_1 = p_2 \ldots = 0$ imply that with probability $1 - p_0$ the population increases over all bounds; ultimately the population must either die out or increase indefinitely. To find the probability p_0 of extinction we compare the process to the related Markov chain. In our case the transition probabilities (5.7) are independent of n, and we have therefore an ordinary random walk in which the steps to the right and left have probabilities $p = \lambda/(\lambda + \mu)$ and $q = \mu/(\lambda + \mu)$, respectively. The state E_0 (or $x = 0$) is an absorbing barrier. We know from the classical ruin problem (Chapter 14, section 2) that the probability of extinction is 1 if $p \leq q$ and $(q/p)^r$ if $q < p$ and r is the initial state. We conclude that *in our process the probability $p_0 = \lim P_0(t)$ of ultimate extinction is 1 if $\lambda \leq \mu$, and $(\mu/\lambda)^r$ if $\lambda > \mu$.* (This is easily verified from the explicit solution; cf. problem 8.)

As in many similar cases, the explicit solution of (5.8) is rather complicated, and it is desirable to calculate the mean and the variance

---

[12] A systematic way consists in deriving a partial differential equation for the generating function $\Sigma P_n(t)s^n$. A more general process [where the coefficients λ and μ in (5.8) are permitted to depend on time] is discussed in detail in David G. Kendall, The Generalized "Birth and Death" Process, *Annals of Mathematical Statistics*, vol. 19 (1948), pp. 1–15. Cf. also a recent paper by the same author, Stochastic Processes and Population Growth (to appear in 1950 in the *Journal of the Royal Statistical Society*), where the theory is generalized so as to take account of the age distribution in biological populations.

of the distribution $\{P_n(t)\}$. Write for the mean

$$(5.9) \qquad M(t) = \sum_{n=1}^{\infty} n P_n(t).$$

We shall omit a formal proof that $M(t)$ is finite and that the following formal operations are justified (again both points follow readily from the solution given in problem 8). Multiplying the second equation in (5.8) by n and adding over $n = 1, 2, \ldots$, we find that the terms containing n^2 cancel, and we get

$$(5.10) \qquad M'(t) = \lambda \Sigma (n-1) P_{n-1}(t) - \mu \Sigma (n+1) P_{n+1}(t)$$
$$= (\lambda - \mu) M(t).$$

This is a differential equation for $M(t)$. At time $t = 0$ the population size is i, and hence $M(0) = i$. Therefore

$$(5.11) \qquad M(t) = i e^{(\lambda - \mu)t}.$$

We see that the mean tends to 0 or infinity, according as $\lambda < \mu$ or $\lambda > \mu$. The variance of $\{P_n(t)\}$ can be calculated in a similar way (cf. problem 10).

6. Exponential Holding Times

The principal field of applications of the pure birth and death process is connected with trunking in telephone engineering and various types of waiting lines for telephones, counters, or machines. This type of problem can be treated with various degrees of mathematical sophistication. The method of the birth and death process offers the easiest approach, but this model is based on a mathematical simplification known as the *assumption of exponential holding times*. We begin with a discussion of this basic assumption.

For concreteness of language let us consider a telephone conversation, and let us assume that its length is necessarily an integral number of seconds. We treat the length of the conversation as a random variable X and assume its probability distribution $p_n = Pr\{X = n\}$ known. The telephone line then represents a physical system with two possible states, "busy" (E_0) and "free" (E_1). If at an arbitrary moment t the line is busy, then the probability of a change in state during the next second depends on how long the conversation has been going on. In other words, the past has an influence on the future, and our process is therefore not a Markov process (cf. Chapter 15, section 10). This circumstance is the source of most difficulties in more complicated problems. However, there exists a simple exceptional case.

Imagine that the decision as to whether or not the conversation is to be continued is made each second at random by means of a skew coin. In other words, a sequence of Bernoulli trials with probability p of success is performed at a rate of one per second and continued until the first success. The conversation ends when this first success occurs. In this case the total length of the conversation, the "holding time," has the geometric distribution $p_n = q^{n-1}p$. If at any time t the line is busy, the probability that it will remain busy for more than one second is q, and the probability of the transition $E_0 \to E_1$ at the next step is p. These probabilities are now independent of how long the line was busy.

This situation has been discussed at length in Chapter 11, section 4. We found there that in passing to the limit we get a process with a continuous time parameter, and the geometric distribution approaches an exponential function (whence the name "exponential holding time.") In the limit we have then the following situation. The probability that a conversation starting at time 0 extends beyond time t is $e^{-\mu t}$; if at any time t the line is busy, the probability of a change in state during $(t, t + h)$ is μh plus terms which tend to zero faster than h.

The method of the birth and death process is applicable only if the transition probabilities in question do not depend on the past; for trunking and waiting line problems this means that all holding times must be exponential. From a practical point of view this assumption may at first sight appear rather artificial, but experience shows that it reasonably describes actual phenomena. In particular, many measurements have shown that telephone conversations within a city [13] follow the exponential law to a surprising degree of accuracy.

These remarks apply to holding times (e.g., length of telephone conversations, duration of machine repairs, etc.). We must also characterize the so-called incoming traffic (arriving calls, machine breakdowns, etc.). We shall assume that during any time interval of length h the probability of an incoming call is λh plus negligible terms, and that the probability of more than one call is in the limit negligible. According to the results of section 2, this means that the number of incoming calls has a Poisson distribution with mean λt. We shall describe this situation by saying that *the incoming traffic is of the Poisson type with intensity* λ.

---

[13] For conversations between cities, companies usually charge by intervals of 3 minutes, and the holding times are therefore likely to be multiples of 3 minutes. This is a systematic deviation from the exponential law, and our theory does not apply.

7. Waiting Line and Servicing Problems

(a) *The Simplest Trunking Problem.*[14] Suppose that infinitely many trunks or channels are available, and that the probability of a conversation ending during the interval $(t, t + h)$ is μh plus terms which are negligible as $h \to 0$ (exponential holding time). The incoming calls constitute a traffic of the Poisson type with parameter λ. The system is in state E_n if n lines are busy.

It is, of course, assumed that the durations of the conversations are mutually independent. If n lines are busy, the probability that one of them will be freed within time h is then $n\mu h + o(h)$. The probability that within this time two or more conversations terminate is obviously of the order of magnitude h^2 and therefore negligible. The probability of a new call arriving is $\lambda h + o(h)$. The probability of a combination of several calls, or of a call arriving and a conversation ending, is again of order of magnitude h^2. Thus, in the notation of section 5

(7.1) $$\lambda_n = \lambda, \qquad \mu_n = n\mu.$$

The basic differential equations (5.2)–(5.3) take the form

$$P_0'(t) = -\lambda P_0(t) + \mu P_1(t)$$

(7.2) $\quad P_n'(t) = -(\lambda + n\mu)P_n(t)$

$$+ \lambda P_{n-1}(t) + (n + 1)\mu P_{n+1}(t) \quad (n \geq 1).$$

Explicit solutions can be obtained by deriving a partial differential equation for the generating function (cf. problem 11). We shall only determine the limits (5.6). They satisfy the equations

(7.3)
$$\lambda p_0 = \mu p_1$$
$$(\lambda + n\mu)p_n = \lambda p_{n-1} + (n + 1)\mu p_{n+1}.$$

One finds by iteration that $p_n = p_0(\lambda/\mu)^n/n!$, and hence

(7.4) $$p_n = e^{-\lambda/\mu} \frac{(\lambda/\mu)^n}{n!}.$$

[14] C. Palm, Intensitätsschwankungen im Fernsprechverkehr, *Ericsson Technics* (Stockholm), no. 44 (1943), pp. 1–189, in particular p. 57. Palm studies approximations in the case where λ and μ are periodic functions of t. Problems of this type were studied by Erlang (1878–1929), whose work was a predecessor of the general theory of stochastic processes. See the recent book by E. Brockmeyer, H. L. Halström, and Arne Jensen, *The Life and Works of A. K. Erlang, Transactions of the Danish Academy Technical Sciences*, No. 2, Copenhagen, 1948. Independently valuable pioneer work has been done by T. C. Fry; his book quoted on p. 107, did much for the development of engineering applications of probability.

Thus, *the limiting distribution is a Poisson distribution with parameter* λ/μ. *It is independent of the initial state.*

It is easy to find the mean $M(t) = \Sigma n P_n(t)$. Multiplying the nth equation of (7.2) by n and adding, we get [taking into account that the $P_n(t)$ add to unity]

(7.5) $$M'(t) = \lambda - \mu M(t).$$

If the initial state is E_i, then $M(0) = i$, and

(7.6) $$M(t) = \frac{\lambda}{\mu}(1 - e^{-\mu t}) + i e^{-\mu t}.$$

As $t \to \infty$, we see that $M(t)$ approaches the mean of the Poisson distribution found above. Incidentally, the reader may verify that in the special case $i = 0$ the $P_n(t)$ are given exactly by a Poisson distribution with mean $M(t)$.

(b) *Waiting Lines for a Finite Number of Channels.*[15] We now treat the last example in a more realistic way. The assumptions are the same, except that *the number a of trunklines or channels is finite. If all channels are busy, each new call joins a waiting line and waits until a channel is freed.* This means that all trunklines have a *common* waiting line.

The word "trunk" may be replaced by *counter* at a postoffice and "conversation" by *service*. We are actually treating the general waiting line problem for the case where a person has to wait only if all a channels are busy.

We say that *the system is in state E_n if n is the total number of persons who are either being served or are in the waiting line.* A waiting line exists only when the system is in a state E_n with $n > a$, and then there are $n - a$ people in the waiting line.

As long as at least one channel is free, we are in exactly the same situation as in the preceding example. However, if the system is in a state E_n with $n > a$, then only a conversations are going on, and we have therefore $\mu_n = a\mu$, for $n \geq a$. The basic system of differential equations is therefore given by (7.2) for $n \leq a$, but for $n \geq a$ by

(7.7) $$P_n'(t) = -(\lambda + a\mu)P_n(t) + \lambda P_{n-1}(t) + a\mu P_{n+1}(t).$$

We again investigate the limits (5.6) (which can be shown to exist). They must satisfy the equations (7.3) if $n \leq a$ and

(7.8) $$(\lambda + a\mu)p_n = \lambda p_{n-1} + a\mu p_{n+1}$$

[15] A. Kolmogoroff, Sur le problème d'attente, *Recueil Mathématique* [*Sbornik*], Vol. 38, 1931, pp. 101–106.

if $n \geq a$. By recursion we find again that for $n \leq a$

(7.9) $$p_n = p_0 \frac{(\lambda/\mu)^n}{n!},$$

while for $n \geq a$

(7.10) $$p_n = \frac{(\lambda/\mu)^n}{a! a^{n-a}} p_0.$$

The series $\sum (p_n/p_0)$ converges only if

(7.11) $$\frac{\lambda}{\mu} < a.$$

Hence, if (7.11) does not hold, a limiting distribution $\{p_k\}$ cannot exist. *In this case $p_n = 0$ for all n, which means that gradually the waiting line grows over all bounds.* On the other hand, if (7.11) holds, then we can determine p_0 so that the sum of the expressions (7.9) and (7.10) equals unity. From the explicit expressions for $P_n(t)$ (which we have not derived, however), it can be shown that the p_n thus obtained really represent the *limiting distribution* of the $P_n(t)$. Table 1 gives a numerical illustration for $a = 3$, $\lambda/\mu = 2$.

TABLE 1

Limiting Probabilities in the Case of $a = 3$ Channels and $\lambda/\mu = 2$

n	0	1	2	3	4	5	6	7
Lines busy	0	1	2	3	3	3	3	3
People waiting	0	0	0	0	1	2	3	4
p_n	0.1111	0.2222	0.2222	0.1481	0.0988	0.0658	0.0439	0.0293

(c) *Servicing of Machines.*[16] The results derived in this and the next example are being successfully applied in Swedish industry. For orientation we begin with the simplest case and generalize it in the next example. The problem is as follows.

We consider automatic machines which normally require no human care. However, at any time a machine may break down and call for service. The time required for servicing the machine is again taken

[16] Examples (c) and (d), including the numerical illustrations, are taken from an article by C. Palm, The Distribution of Repairmen in Servicing Automatic Machines (in Swedish), *Industritidningen Norden*, vol. 75 (1947), pp. 75–80, 90–94, 119–123. Palm gives tables and graphs for the most economical number of repairmen.

as a random variable with an exponential distribution. In other words, the machine is characterized by two constants λ and μ with the following properties. If at time t the machine is in working state, the probability that it will call for service before time $t + h$ is λh plus terms which are negligible in the limit $h \to 0$. Conversely, if at time t the machine is being serviced, the probability that the servicing time terminates before $t + h$ and the machine reverts to the working state is $\mu h + o(h)$. For an efficient machine λ should be relatively small and μ relatively large. The ratio λ/μ is called the *servicing factor*.

We suppose that *m machines with the same parameters λ and μ are serviced by a single repairman*. If a machine breaks down, it will be serviced immediately unless the repairman is servicing another machine, in which case a waiting line is formed. We say that *the system is in state E_n if n machines are not working*. For $1 \leq n \leq m$ this means that one machine is being serviced and $n - 1$ are in the waiting line; in the state E_0 all machines work and the repairman is idle. The m machines are assumed to work independently.

A transition $E_n \to E_{n+1}$ is caused by a breakdown of one among the $m - n$ working machines, while a transition $E_n \to E_{n-1}$ occurs if the machine being serviced reverts to the working state. Hence we have a birth and death process with coefficients

$$\lambda_0 = m\lambda, \qquad \mu_0 = 0,$$
(7.12)
$$\lambda_n = (m - n)\lambda, \qquad \mu_n = \mu$$

$(0 < n \leq m)$, and the basic differential equations (5.2) and (5.3) become $(1 \leq n \leq m - 1)$:

$$P_0'(t) = -m\lambda P_0(t) + \mu P_1(t),$$

(7.13)
$$P_n'(t) = -\{(m - n)\lambda + \mu\}P_n(t) + (m - n + 1)\lambda P_{n-1}(t) + \mu P_{n+1}(t),$$

$$P_m'(t) = -\mu P_m(t) + \lambda P_{m-1}(t).$$

This is a finite system of differential equations and can be solved by ordinary methods. The limits (5.6) exist and satisfy the equations

$$m\lambda p_0 = \mu p_1,$$
(7.14) $\quad \{(m - n)\lambda + \mu\}p_n = (m - n + 1)\lambda p_{n-1} + \mu p_{n+1},$
$$\mu p_m = \lambda p_{m-1}.$$

It follows easily that the recursion formula

(7.15) $$(m - n)\lambda p_n = \mu p_{n+1}$$

holds. From it we get

(7.16) $$p_n = (m)_n \left(\frac{\lambda}{\mu}\right)^n p_0,$$

where $(m)_n = m(m - 1) \cdots (m - n + 1)$. The value of p_0 follows from the condition $\Sigma p_n = 1$. Table 2 gives the values of p_n for the case of $m = 6$ machines and a servicing factor $\lambda/\mu = 0.1$. The table also exhibits a simple way of calculating p_n.

The probability p_0 may be interpreted as the probability of the repairman being idle (in the example of Table 2 he is likely to be idle almost half the time). The *expected number of machines in the waiting line* is

(7.17) $$w = \sum_{k=2}^{m} (k - 1) p_k.$$

This quantity can be calculated by adding the relations (7.15) for $n = 0, 1, \cdots, m$. Using the fact that the p_n add to unity, we get

(7.17) $$m\lambda - \lambda w - \lambda(1 - p_0) = \mu(1 - p_0)$$

or

(7.18) $$w = m - \frac{\lambda + \mu}{\lambda}(1 - p_0).$$

In the example of Table 2 we have $w = 6(0.0549)$. Thus 0.0549 is the average contribution of a machine to the waiting line.

TABLE 2

PROBABILITIES p_n FOR THE CASE $\lambda/\mu = 0.1$, $m = 6$

n	Machines in Waiting Line	p_n/p_0	p_n
0	0	1	0.4845
1	0	0.6	.2907
2	1	0.3	.1454
3	2	0.12	.0582
4	3	0.036	.0175
5	4	0.0072	.0035
6	5	0.00072	.0003

$$\frac{1}{p_0} = \sum \frac{p_n}{p_0} = 2.06392$$

(d) *Continuation: Several Repairmen.* We shall not change the basic assumptions of the preceding problem, except that the m *machines are now serviced by r repairmen* $(r < m)$. Thus for $n \leq r$ the state E_n means that $r - n$ repairmen are idle, n machines are being serviced, and no machine is in the waiting line for repairs. For $n > r$ the state E_n means that r machines are being serviced and $n - r$ machines are in the waiting line. We can use the set-up of the preceding example except that (7.12) is obviously to be replaced by

$$\lambda_0 = m\lambda, \qquad \mu_0 = 0,$$
(7.19) $\qquad \lambda_n = (m-n)\lambda, \qquad \mu_n = n\mu \qquad (1 \leq n \leq r),$
$$\lambda_n = (m-n)\lambda, \qquad \mu_n = r\mu \qquad (r \leq n \leq m).$$

We shall not write down the basic system of differential equations, but only the equations for the limiting probabilities p_n. They are

$$m\lambda p_0 = \mu p_1,$$
(7.20) $\quad \{(m-n)\lambda + n\mu\}p_n = (m-n+1)\lambda p_{n-1} + (n+1)\mu p_{n+1}$
$$(1 \leq n < r),$$
$$\{(m-n)\lambda + r\mu\}p_n = (m-n+1)\lambda p_{n-1} + r\mu p_{n+1}$$
$$(r \leq n \leq m).$$

From the first equation we get the ratio of p_1/p_0. From the second equation we get by induction for $n < r$

(7.21) $\qquad\qquad (n+1)\mu p_{n+1} = (m-n)\lambda p_n;$

finally, for $n \geq r$ we get from the last equation in (7.20)

(7.22) $\qquad\qquad r\mu p_{n+1} = (m-n)\lambda p_n.$

These equations permit calculating successively the ratios p_n/p_0. Finally, p_0 follows from the condition $\Sigma p_k = 1$. The values in Table 3 are obtained in this way.

A comparison of Tables 2 and 3 reveals surprising facts. Note that both tables refer to the same machines $(\lambda/\mu = 0.1)$, but in the second case we have $m = 20$ machines and $r = 3$ repairmen. The number of machines per repairman has increased from 6 to $6\frac{2}{3}$, but at the same time, the machines are serviced more efficiently. Let us define a *coefficient of loss for machines by*

(7.23) $$\frac{w}{m} = \frac{\text{average number of machines in waiting line}}{\text{number of machines}}$$

and a coefficient of loss for repairmen by

(7.24) $$\frac{\rho}{r} = \frac{\text{average number of repairmen idle}}{\text{number of repairmen}}.$$

For practical purposes we may identify the probabilities $P_n(t)$ with their limits p_n. In Table 3 we have then $w = p_4 + 2p_5 + 3p_6 + \cdots + 17p_{20}$ and $\rho = 3p_0 + 2p_1 + p_2$. Table 4 proves conclusively that for our particular machines for which $(\lambda/\mu = 0.1)$ *three repairmen per 20 machines are ever so much more economical than one repairman per 6 machines.* Palm's tables referred to in footnote 16 enable us to find the most economical ratio of repairmen per machine.

TABLE 3

Probabilities p_n for the Case $\lambda/\mu = 0.1$, $m = 20$, $r = 3$

n	Machines Serviced	Machines Waiting	Repairmen Idle	p_n
0	0	0	3	0.13625
1	1	0	2	.27250
2	2	0	1	.25888
3	3	0	0	.15533
4	3	1	0	.08802
5	3	2	0	.04694
6	3	3	0	.02347
7	3	4	0	.01095
8	3	5	0	.00475
9	3	6	0	.00190
10	3	7	0	.00070
11	3	8	0	.00023
12	3	9	0	.00007

TABLE 4

Comparison of Efficiencies of Two Systems Discussed in Examples (c) and (d)

	I	II
Number of machines	6	20
Number of repairmen	1	3
Machines per repairman	6	6⅔
Coefficient of loss for repairmen	0.4845	0.4042
Coefficient of loss for machines	0.0549	0.01694

(e) *A Power-supply Problem*.[17] One electric circuit supplies a welders who use the current only intermittently. If at time t a welder uses current, the probability that he ceases using it at time $t + h$ is $\mu h + o(h)$; if at time t he requires no current, the probability that he calls for current before $t + h$ is $\lambda h + o(h)$. The welders work independently of each other.

We say that the system is in state E_n if n welders are using current. Thus we have only finitely many states $E_0, \cdots, E_a$.

If the system is in state E_n, then $a - n$ welders are not using current and the probability for a new call for current within time h is $(a - n)\lambda h + o(h)$; on the other hand, the probability that one of the n welders ceases using current is $n\mu h + o(h)$. Hence we have a birth and death process with

(7.25) $\qquad \lambda_n = (a - n)\lambda, \quad \mu_n = n\mu, \qquad\qquad 0 \leq n \leq a.$

The basic differential equations become

$$P_0'(t) = -a\lambda P_0(t) + \mu P_1(t),$$
(7.26) $\quad P_n'(t) = -\{n\mu + (a - n)\lambda\}P_n(t) + (n + 1)\mu P_{n+1}(t)$
$$+ (a - n + 1)\lambda P_{n-1}(t),$$
$$P_a'(t) = -a\mu P_a(t) + \lambda P_{a-1}(t)$$

(with $1 \leq n \leq a - 1$).

It is easily verified that *the limiting probabilities are given by the binomial distribution*

(7.27) $\qquad p_n = \binom{a}{n}\left(\frac{\lambda}{\lambda + \mu}\right)^n \left(\frac{\mu}{\lambda + \mu}\right)^{a-n},$

a result which could have been anticipated on intuitive grounds.

8. The Backward (Retrospective) Equations

In the preceding sections we were studying the probabilities $P_n(t)$ of finding the system at time t in state E_n. This notation is convenient but misleading, inasmuch as it omits mentioning the initial state E_i of the system at time zero. For theoretical purposes it is therefore more natural to introduce the notation $P_{in}(t)$; *this is the probability that the system is at time t in state E_n, given that at time zero it was in E_i.* The $P_{in}(t)$ will be called *transition probabilities.*

---

[17] This example was suggested by the problem treated (inadequately) by H. A. Adler and K. W. Miller, A New Approach to Probability Problems in Electrical Engineering, *Transactions of the American Institute of Electrical Engineers*, vol. 65 (1946), pp. 630–632.

17.8]   BACKWARD EQUATIONS

It must be emphasized that we have been studying these transition probabilities all along and that nothing is changed but notation. If the initial state is known to be E_i, then $\{P_{in}(t)\}$ is the absolute probability distribution at time t. If at time zero we have only a probability distribution $\{q_i\}$ for the initial state, then the probability of E_n at time t is

$$(8.1) \qquad Q_n(t) = \sum_i q_i P_{in}(t).$$

In the case of the pure birth process and of the birth and death process, we found that *for an arbitrary fixed i the transition probabilities $P_{in}(t)$ satisfy the basic differential equations* (3.2) *and* (5.3). The subscript i appears only in the initial conditions, which should now be written

$$(8.2) \qquad P_{in}(0) = \begin{matrix} 1 & \text{if } n = i \\ 0 & \text{otherwise.} \end{matrix}$$

These basic differential equations were derived by prolonging the time interval $(0, t)$ to $(0, t + h)$ and considering the possible changes during the short time $(t, t + h)$. We could as well have prolonged the interval $(0, t)$ in the direction of the past and considered the changes during $(-h, 0)$. In this way we get a new system of differential equations in which n (instead of i) remains fixed.

Consider first the case of a pure birth process and let us neglect events whose probability tends to zero faster than h. If the system passed from E_i ($i > 0$) at time $-h$ to E_n at time t, then at time 0 it must be either at E_i or at E_{i+1}. By the method of sections 2 and 3 we conclude that

$$(8.3) \qquad P_{in}(t + h) = P_{in}(t)(1 - \lambda_i h) + P_{i+1,n}(t)\lambda_i h + o(h).$$

Hence for $i > 0$ the new basic system now takes the form

$$(8.4) \qquad P_{in}'(t) = -\lambda_i P_{in}(t) + \lambda_i P_{i+1,n}(t),$$

and

$$(8.5) \qquad P_{0n}'(t) = -\lambda_0 P_{0n}(t).$$

These equations are called the *backward equations,* and, for distinction, (3.2) are called the *forward equations.* The initial conditions are (8.2).

In the case of the birth and death process, if the system is at time $-h$ in E_i, then at time t it must be in E_{i+1}, E_i, or E_{i-1}, and the same argument leads to the *backward equations*

(8.6) $\quad P_{in}'(t) = -(\lambda_i + \mu_i)P_{i,n}(t) + \lambda_i P_{i+1,n}(t) + \mu_i P_{i-1,n}(t).$

These equations correspond to (5.3).

It should be clear that the forward and backward equations are not independent of each other: the solution of the backward equations with the initial conditions (8.2) automatically satisfies the forward equations. These connections are mentioned here only as a preparation for the general theory of the next section.

Example. *The Poisson Process.* In section 2 we have interpreted the Poisson expression (1.1) as the probability that exactly n calls arrive during any time interval of length t. Let us now measure time from an arbitrary moment, and let us say that the system is in state E_n if exactly n calls arrive up to time t. Then a transition from E_i at t_1 to E_n at t_2 means that $n - i$ calls arrived during (t_1, t_2). This is possible only if $n \geq i$, and hence we have for the transition probabilities of the Poisson process

$$P_{in}(t) = e^{-\lambda t} \frac{(\lambda t)^{n-i}}{(n-i)!} \qquad \text{if } n \geq i$$

(8.7) $\qquad\qquad P_{in}(0) = 0 \qquad\qquad\qquad\qquad \text{if } n < i.$

The forward and backward equations are, respectively,

(8.8) $\qquad\qquad P_{in}'(t) = -\lambda P_{in}(t) + \lambda P_{i,n-1}(t)$

and

(8.9) $\qquad\qquad P_{in}'(t) = -\lambda P_{in}(t) + \lambda P_{i+1,n}(t),$

and it is easily verified that (8.7) is a solution of either system and satisfies the initial condition (8.2).

9. Generalization; the Kolmogorov Equations

Up to now we considered exclusively processes in which direct transitions from a state E_n were possible only to the neighboring states E_{n+1} and E_{n-1}. Moreover, the processes were time-homogeneous, that is to say, the transition probabilities $P_{in}(t)$ were the same for all time intervals of length t. We now consider more general processes in which both assumptions are dropped.

As in the theory of ordinary Markov chains (Chapter 15), we shall permit direct transitions from any state E_i to any state E_n. The transition probabilities are permitted to vary in time. This necessitates stating both end points of any time interval instead of noting only its length. Accordingly, we shall write $P_{in}(\tau, t)$ *for the conditional probability of finding the system at time t in state E_n, given that at a previous instant τ the state was E_i*. The symbol $P_{in}(\tau, t)$ is meaningless unless $\tau < t$. If the process is homogeneous in time, then $P_{in}(\tau, t)$ depends only on the difference $t - \tau$, and we can write $P_{in}(t)$ instead of $P_{in}(\tau, \tau + t)$ (which is then independent of τ).

The principal property of our processes is the Markov property discussed in Chapter 15, section 10. It states that, given the state of the system at any time, future changes are independent of the past. More precisely, consider three moments $\tau < s < t$ and suppose that at time τ the system is in state E_i and at time s in state E_ν. For a general process the (conditional) probability of finding the system at time t in state E_n depends on both i and ν; in other words, not only the "present state" E_ν, but also the past state E_i, has an influence on the state at time t. However, for a Markov process this is not so. For it the considered probability equals $P_{\nu n}(s, t)$, the probability of a transition from E_ν at time s to E_n at time t; the knowledge that at time $\tau < s$ the system was in state E_i permits no inference as to the future. This assumption leads directly to an important conclusion. The passage from E_j at time τ to E_n at time t must occur via some state E_ν at time s, and for a Markov process the probability that the passage goes via a particular state E_ν is $P_{i\nu}(\tau, s)P_{\nu n}(s, t)$. It follows that we must have

$$(9.1) \qquad P_{in}(\tau, t) = \sum_\nu P_{i\nu}(\tau, s)P_{\nu n}(s, t)$$

identically for all $\tau < s < t$. This is the Chapman-Kolmogorov equation. It is the counterpart to the equation (10.3) of Chapter 15, which is valid when the time parameter assumes integral values only.

It was shown in Chapter 15, section 10, that the Chapman-Kolmogorov equation does not hold for all stochastic processes. For our purposes we could take (9.1) *as defining the class of processes with which we are concerned*.[18] In fact, we shall add only regularity restrictions and derive our basic differential equations from (9.1). There is a probabilistic background leading up to the Chapman-Kolmogorov equation, but we need not refer to it: once (9.1) is given we can easily

---

[18] The question of whether the Kolmogorov equation characterizes Markov processes poses difficult problems requiring the study of the actual sample functions $X(t)$. It should be borne in mind that we are using a short-cut to obtain differential equations for certain probabilities and are not analyzing the process in all its aspects.

derive differential equations which determine the probabilities $P_{in}(t)$, and can proceed in a purely analytical way.

In the case of time-homogeneous processes, (9.1) assumes the simpler form

$$(9.2) \qquad P_{in}(t+s) = \sum_{\nu} P_{i\nu}(t) P_{\nu n}(s).$$

For the Poisson process this equation reduces to the convolution property of the Poisson distribution stated in Chapter 11.

We now introduce our fundamental regularity conditions which in an obvious way generalize the starting assumptions of the birth and death process.

Assumption 1. To every state E_n there corresponds a continuous function $c_n(t) \geq 0$ such that as $h \to 0$

$$(9.3) \qquad \frac{1 - P_{nn}(t, t+h)}{h} \to c_n(t)$$

uniformly in t.

The probabilistic interpretation of (9.3) is obvious: if at time t the system is in state E_n, then the probability that during $(t, t+h)$ a change occurs is $c_n(t)h + o(h)$. Analytically, (9.3) requires that $P_{nn}(t,s) \to 1$ as $s \to t$, and that $P_{nn}(t,x)$ has at $x=t$ a derivative with respect to x. The function $c_n(t)$ plays the role of $\lambda_n + \mu_n$ in the birth and death process. In the case of a time-homogeneous process, c_n is a constant.

Assumption 2. To every pair of states E_j, E_k with $j \neq k$ there correspond transition probabilities $p_{jk}(t)$ (depending on time) such that as $h \to 0$

$$(9.4) \qquad \frac{P_{jk}(t, t+h)}{h} \to c_j(t) p_{jk}(t) \qquad (j \neq k)$$

uniformly in t. The $p_{jk}(t)$ are continuous in t, and for every fixed t, j

$$(9.5) \qquad \sum_k p_{jk}(t) = 1, \qquad\qquad p_{jj}(t) = 0.$$

Here $p_{jk}(t)$ can be interpreted as the conditional probability that, *if* a change from E_j occurs during $(t, t+h)$, this change takes the system from E_j to E_k. In the birth and death process

$$(9.6) \qquad p_{j,j+1}(t) = \frac{\lambda_j}{\lambda_j + \mu_j}, \quad p_{j,j-1}(t) = \frac{\mu_j}{\lambda_j + \mu_j},$$

while $p_{jk}(t) = 0$ for all other combinations of j and k. For every fixed t the $p_{jk}(t)$ can be interpreted as transition probabilities of a Markov chain.

The two assumptions suffice to derive a system of backward equations for the $P_{jk}(\tau, t)$, but for the forward equations we require in addition

Assumption 3. For fixed k the passage to the limit in (9.4) is uniform with respect to j.

The necessity of this assumption is of considerable interest for the theory of infinite systems of differential equations and will be discussed in the next section.

We now derive a system of differential equations for the $P_{ik}(\tau, t)$ as functions of t and n (forward equations). From (9.1) we have

$$(9.7) \qquad P_{ik}(\tau, t+h) = \sum_j P_{ij}(\tau, t) P_{jk}(t, t+h).$$

If we express the term $P_{kk}(t, t+h)$ on the right in accordance with (9.3), we get

$$(9.8) \qquad \frac{P_{ik}(\tau, t+h) - P_{ik}(\tau, t)}{h} = -c_k(t) P_{ik}(\tau, t) + \frac{1}{h} \sum_{j \neq k} P_{ij}(\tau, t) P_{jk}(t, t+h) + \cdots$$

where the neglected terms tend to 0 with h, and the sum extends over all j except $j = k$. We can now apply (9.4) to the terms of the sums. Since (by assumption 3) the passage to the limit is uniform in j, the right side has a limit. Hence also the left side has a limit, which means that $P_{ik}(\tau, t)$ has a partial derivative with respect to t, and

$$(9.9) \qquad \frac{\partial P_{ik}(\tau, t)}{\partial t} = -c_i(t) P_{ik}(\tau, t) + \sum_j P_{ij}(\tau, t) c_j(t) p_{jk}(t).$$

This is the basic system of forward differential equations. Note that, in it, i and τ are fixed so that we have (despite the formal appearance of a partial derivative) a system of *ordinary* differential equations for the infinite system of functions $P_{ik}(\tau, t)$, $k = 0, 1, 2, \ldots$. The parameters i and τ appear only in the initial condition

$$(9.10) \qquad P_{ik}(\tau, \tau) = \begin{cases} 1 & \text{if } k = i \\ 0 & \text{otherwise.} \end{cases}$$

In like manner we can derive a system of backward equations, starting from

(9.11) $$P_{ik}(\tau - h, t) = \sum_{\nu} P_{i\nu}(\tau - h, \tau) P_{\nu k}(\tau, t)$$

and applying our assumptions to the $P_{i\nu}(\tau - h, \tau)$. We get

(9.12) $$\frac{P_{ik}(\tau - h, t) - P_{ik}(\tau, t)}{h} = -c_i(\tau) P_{ik}(\tau, t)$$
$$+ \frac{1}{h} \sum_{\nu \neq i} P_{i\nu}(\tau - h, \tau) P_{\nu k}(\tau, t) + \frac{o(h)}{h}.$$

Here $P_{i\nu}(\tau - h, \tau)/h \to c_i(\tau) p_{i\nu}(\tau)$, and the passage to the limit is always uniform, since (without using assumption 3) by (9.4) and (9.3)

(9.13) $$\frac{1}{h} \sum_{\nu \neq i} P_{i\nu}(t, t + h) = \frac{1 - P_{ii}(t, t + h)}{h}$$
$$\to c_i(t) = \sum_{\nu} c_i(t) p_{i\nu}(t).$$

[This means that (9.4) may be summed over k.] It follows then from (9.12) that

(9.14) $$\frac{\partial P_{ik}(\tau, t)}{\partial \tau} = c_i(\tau) P_{ik}(\tau, t) - c_i(\tau) \sum_{\nu} p_{i\nu}(\tau) P_{\nu k}(\tau, t).$$

This is the *basic system of backward differential equations.* In it k and t are fixed, and we have the initial condition

(9.15) $$P_{ik}(t, t) = \begin{matrix} 1 & \text{if } i = k \\ 0 & \text{otherwise.} \end{matrix}$$

The two systems of differential equations were first derived by A. Kolmogorov in a celebrated paper [19] in which he laid the foundations of the theory of Markov processes (of more general types than here considered). It can be shown [20] that each of the two systems uniquely

---

[19] Über die analytischen Methoden in der Wahrscheinlichkeitsrechnung, *Mathematische Annalen*, vol. 104 (1931), pp. 415–458.

[20] Cf. W. Feller, On the Integrodifferential Equations of Purely Discontinuous Markoff Processes, *Transactions of the American Mathematical Society*, vol. 48 (1940), pp. 488–515. There also necessary and sufficient conditions for (9.16) are given. Unfortunately, the paper treats a more general class of processes, so that our simple differential equations are replaced by much more complicated integrodifferential equations. Previously Kolmogorov gave a partial existence proof, but under very restrictive conditions.

determines a system of transition probabilities $P_{jk}(\tau, t)$ satisfying all our conditions, including the Chapman-Kolmogorov equation (9.1). We know from the case of the pure birth process (section 4) that the $P_{jk}(\tau, t)$ are not always a proper probability distribution, but that sometimes

(9.16) $$\sum_k P_{ik}(\tau, t) < 1,$$

where the difference between the two sides accounts for the possibility of infinitely many transitions in a finite time interval. From the point of view of applications the possibility (9.16) can be safely disregarded, but it is of interest both for the theory of stochastic processes and for the theory of infinite systems of differential equations.

Example. *The Compound Poisson Process.* Consider the case where all $c_i(t)$ equal the same constant

(9.17) $$c_i(t) = \lambda$$

and where the p_{jk} are independent of t. In this case they define an ordinary Markov chain, and we denote (as in Chapter 15) its higher transition probabilities by $p_{jk}^{(n)}$.

From (9.17) it follows that the probability of a transition occurring during $(t, t + h)$ is independent of the state of the system at time t and equals $\lambda h + o(h)$. Accordingly the number of transitions within the time interval (τ, t) has a Poisson distribution with parameter $\lambda(t - \tau)$. The conditional transition probabilities under the assumption that there are n transitions are given by $p_{jk}^{(n)}$. Hence we have

(9.18) $$P_{jk}(\tau, t) = e^{-\lambda(t-\tau)} \sum_{n=0}^{\infty} \frac{\{\lambda(t-\tau)\}^n}{n!} p_{jk}^{(n)}$$

[where $p_{jk}^{(0)}$ equals 1 or 0 according to whether $k = j$ or $k \neq j$.] It is easily verified that (9.18) is in fact a solution of our two systems of differential equations.

*10. Degenerate Processes

The theorems concerning Kolmogorov's two systems of differential equations round off the theory and are satisfactory except for two somewhat mystifying points. First, the possibility of the inequality (9.16) is disturbing, since in this case the transition probabilities do not form a proper probability distribution. Second, the derivation of the forward equations required an assumption which was not

---

* Starred sections treat special topics and may be omitted at first reading.

necessary for the backward equations. Doob [21] discovered that the two facts are related to each other and connected with the existence of a certain type of degenerate process. This discovery is of theoretical interest and also reveals new facts concerning infinite systems of ordinary differential equations. The situation is particularly simple and intuitive in the case of a pure birth process, and a detailed study of this case will contribute to an understanding of the general theory.

We are concerned with the pure birth process of sections 3 and 4. The states are $E_0, E_1, \ldots$, and direct transitions are possible only from E_n to E_{n+1}. The process is homogeneous in time, so that the transition probabilities $P_{ik}(t)$ depend only on the duration t of the time interval, not on its position. For fixed i the $P_{ik}(t)$ satisfy the *forward equations*

$$(10.1) \qquad P_{ik}'(t) = -\lambda_k P_{ik}(t) + \lambda_{k-1} P_{i,k-1}(t)$$

(where we put $\lambda_{-1} = 0$); for fixed k we have the *backward equations*

$$(10.2) \qquad P_{ik}'(t) = -\lambda_i P_{ik}(t) + \lambda_i P_{i+1,k}(t).$$

We know from section 4 that the condition

$$(10.3) \qquad \sum \frac{1}{\lambda_k} < \infty$$

is necessary and sufficient for the degenerate case to occur, that is, that

$$(10.4) \qquad \Sigma P_{ik}(t) < 1$$

(at least for large t). The difference of the two sides in (10.4) may be interpreted as the probability that infinitely many transitions occur in a finite time. In this case the state of the system has, so to speak, moved out to infinity. From a "practical" point of view it stays at infinity forever after, and this concludes the story. However, Doob remarked that the pure mathematician may introduce the assumption that, whenever the state has moved out to infinity, it instantaneously returns to E_0 (or some other state). The process then continues and has all the essential properties of our processes except that we have a new type of path. If the system is at time 0 at E_0 and at time t at E_6, it may have undergone six transitions or infinitely many, moving one or more times out to infinity and starting each time afresh at E_0. The transition probabilities of this new process have a curious property. They obviously satisfy assumptions 1 and 2 of the preceding section,

[21] J. L. Doob, Markoff Chains—Denumerable Case, *Transactions of the American Mathematical Society*, vol. 58 (1945), 455–473.

and therefore they satisfy the *backward* equations.[22] They can*not* satisfy the forward equations, since the solution of this system is unique.[23] This explains why a special assumption is required for the derivation of the forward equations: this assumption eliminates the possibility of a state E_k being reached via infinity. If this assumption is dropped, we cannot derive (10.1), but our method leads to (10.1) with the equality sign replaced by the sign $\geq$. The same statement holds in the general case. The *backward equations are always satisfied; however, in the forward equations the equality sign holds only when $P_{ik}(t)$ is interpreted as probability of a transition from E_i to E_k in finitely many steps. If also transitions "via infinity" are possible, then the equality sign in the forward equations must be replaced by the sign $\geq$.*

It must be understood that this discussion concerns only a very exceptional (and in some respects artificial) case. In general,[24] the common solution $\{P_{ik}(\tau, t)\}$ of the forward and backward equation satisfies, for every fixed t, the natural condition $\Sigma P_{ik}(t) = 1$. In this case the solution is unique, and no wandering out to infinity can occur.

For the theory of infinite systems of differential equations our discussion supplies interesting examples where the solution of a system is not unique. In the simple case of (10.2) we get the following

Theorem. *If* (10.3) *holds, then the infinite system of differential equations*

(10.5) $$y_i' = -\lambda_i y_i + \lambda_i y_{i+1}$$

has a non-zero solution with $y_i(0) = 0$ for all i and for which

(10.6) $$\Sigma y_i(t) \leq 1, \qquad y_i(t) \geq 0$$

for all t.

We conclude with a purely analytical proof of this theorem and indications of how the theorems of the preceding section can be proved in the special case of the pure birth process.

The solutions $P_{ik}(t)$ of (10.1) (where i is fixed) satisfy the initial condition

(10.7) $$P_{ii}(0) = 1, \quad P_{ik}(0) = 0 \quad \text{for} \quad k \neq i.$$

---

[22] Note that this implies that the backward equations have several systems of solutions satisfying the same initial conditions.

[23] Equations (10.1) for $k = 0, 1, \ldots, i-1$ form a system of i linear equations with initial values $P_{ik}(0) = 0$. Hence $P_{ik}(t) = 0$ for $k < i$ by familiar theorems on finite systems. For $k = i$ we get $P_{ii}'(t) = -\lambda_i P_{ii}(t)$, and all $P_{ik}(t)$ with $k > i$ can be calculated recursively. By contrast, (10.2) is *not* a recursive system since the ith equation involves the next higher term $P_{i+1,k}(t)$.

[24] Cf. the *Transaction* paper cited in section 9, where the necessary and sufficient conditions for the occurrence of (10.4) are given.

It has been noted before that the equations can be solved recursively. It is shown in elementary textbooks (and it is easily verified) that the solution is

(10.8)
$$P_{ik}(t) = 0 \quad (k < i)$$
$$P_{ii}(t) = e^{-\lambda_i t}$$
$$P_{ik}(t) = \lambda_{k-1} \int_0^t e^{-\lambda_k(t-\tau)} P_{i,k-1}(\tau) \, d\tau \qquad (k > i).$$

We prove first that these $P_{ik}(t)$ satisfy the backward equations (10.2). This assertion is equivalent to

(10.9)
$$P_{ik}(t) = \lambda_i \int_0^t e^{-\lambda_i(t-\tau)} P_{i+1,k}(\tau) \, d\tau \qquad (k > i).$$

This relation is readily verified for $k = i + 1$, and the general proof then proceeds by induction. Suppose that (10.9) is correct for $k < i + r$, and let $k = i + r$. Then the term $P_{i,k-1}(\tau)$ occurring under the integral in (10.8) can be expressed by means of (10.9) (where k, t, τ are to be replaced by $k-1$, τ, x respectively). If in the resulting double integral the order of integrations is reversed, we get, putting $\tau - x = y$,

(10.10)
$$P_{ik}(t) = \lambda_i \int_0^t e^{-\lambda_i x} \, dx \int_0^{t-x} \lambda_{k-1} e^{-\lambda_k(t-x-y)} P_{i+1,k-1}(y) \, dy.$$

The inner integral equals $P_{i+1,k}(t-x)$, and thus (10.10) reduces to (10.9) with τ replaced by $t - x$. This proves (10.9).

Now put

(10.11)
$$L_i(t) = 1 - \sum_{k=i}^{\infty} P_{ik}(t).$$

From section 4 we know that (10.3) implies $L_i(t) > 0$, at least for large t. Hence $L_i(t)$ does not vanish identically, and $L_i(0) = 0$. We now show that $L_i(0)$ satisfies the equations (10.5). For that purpose we introduce (10.9) into (10.11) and remember that $P_{ii}(t) = e^{-\lambda_i t}$. We see then that

(10.12)
$$L_i(t) = \lambda_i \int_0^t e^{-\lambda_i(t-\tau)} L_{i+1}(\tau) \, d\tau,$$

and by differentiation we find that the $L_i(t)$ satisfy (10.5).

11. Problems for Solution

1. In the pure birth process defined by (3.2) let $\lambda_n > 0$ for all n. Prove that for every fixed $n \geq 1$ the function $P_n(t)$ first increases, then decreases to 0. If t_n is the place of the maximum, then $t_1 < t_2 < t_3 < \ldots$
   Hint: Use induction; differentiate (3.2).

2. *Continuation.* Show that $t_n \to \infty$.
   Hint: If $t_n \to \tau$, then for fixed $t > \tau$ the sequence $\lambda_n P_n(t)$ increases. Use (4.10).

3. *The Yule process.* Derive the mean and the variance of the distribution defined by (3.4). [Use only the differential equations, not the explicit form (3.5).]

4. *Pure death process.* Find the differential equations of a process of the Yule type with transitions only from E_n to E_{n-1}. Find the distribution $P_n(t)$, its mean, and its variance, assuming that the initial state is i.

5. *The Polya process.*[25] This is a non-stationary pure birth process with λ_n depending on time:

(11.1) $$\lambda_n(t) = \frac{1 + an}{1 + at}.$$

Show that the solution with initial condition $P_0(0) = 1$ is

(11.2) $$P_0(t) = (1 + at)^{-1/a}$$
$$P_n(t) = t^n (1 + at)^{-n-1/a} \frac{(1 + a)(1 + 2a) \cdots \{1 + (n-1)a\}}{n!}.$$

Show from the differential equations that the mean and variance are t and $t(1 + at)$, respectively.

6. *Continuation.* The Polya process can be obtained by a passage to the limit from the Polya urn scheme of Chapter 5, example (2.b). If the state of the system is defined as the number of red balls, then the transition probability $E_k \to E_{k+1}$ at the $n + 1$st drawing is

(11.3) $$p_{k,n} = \frac{r + kc}{r + b + nc} = \frac{p + k\gamma}{1 + n\gamma}$$

where $p = r/(r+b)$, $\gamma = c/(r+b)$.

As in the passage from Bernoulli trials to the Poisson distribution, let drawings be made at the rate of one in time h and let $h \to 0$, $n \to \infty$ so that $np \to t$, $n\gamma \to at$. Show that in the limit (11.3) leads to (11.1). Show also that the Polya distribution (2.5) of Chapter 5 passes into (11.2).

7. *Linear growth.* If in the process defined by (5.8) $\lambda = \mu$, and $P_1(0) = 1$, then

(11.4) $$P_0(t) = \frac{\lambda t}{1 + \lambda t}, \quad P_n(t) = \frac{(\lambda t)^{n-1}}{(1 + \lambda t)^{n+1}}.$$

The probability of ultimate extinction is 1.

8. *Continuation.* Assuming a trial solution to (5.8) of the form $P_n(t) = A(t)B^n(t)$, prove that the solution with $P_1(0) = 1$ is

(11.5) $$P_0(t) = \mu B(t), \quad P_n(t) = \{1 - \lambda B(t)\}\{1 - \mu B(t)\}\{\lambda B(t)\}^{n-1}$$

with

(11.6) $$B(t) = \frac{1 - e^{(\lambda - \mu)t}}{\mu - \lambda e^{(\lambda - \mu)t}}.$$

9. *Continuation.* The generating function $P(s, t) = \Sigma P_n(t) s^n$ satisfies the partial differential equation

(11.7) $$\frac{\partial P}{\partial t} = \{\mu - (\lambda + \mu)s + \lambda s^2\} \frac{\partial P}{\partial s}.$$

10. *Continuation.* Let $M_2(t) = \Sigma n^2 P_n(t)$ and $M(t) = \Sigma n P_n(t)$ (as in section 5). Show that

(11.8) $$M_2'(t) = 2(\lambda - \mu) M_2(t) + (\lambda + \mu) M(t).$$

[25] O. Lundberg, *On Random Processes and Their Applications to Sickness and Accident Statistics*, Uppsala, 1940. In this book many properties of the Polya process are discussed mainly in relation to compound Poisson processes.

Deduce that when $\lambda > \mu$ the *variance* of $\{P_n\}$ is given by

(11.9) $\qquad e^{2(\lambda-\mu)t}\{1 - e^{(\mu-\lambda)t}\}(\lambda + \mu)/(\lambda - \mu).$

11. For the process defined by (7.2) the generating function $P(s, t) = \Sigma P_n(t)s^n$ satisfies the partial differential equation

(11.10) $\qquad \dfrac{\partial P}{\partial t} = (1 - s)\left\{-\lambda P + \mu \dfrac{\partial P}{\partial s}\right\}.$

Its solution is

$$P = e^{-\lambda(1-s)(1-e^{-\mu t})/\mu}\{1 - (1 - s)e^{-\mu t}\}^i.$$

For $i = 0$ this is a Poisson distribution with parameter $\lambda(1 - e^{-\mu t})/\mu$. As $t \to \infty$, the distribution $\{P_n(t)\}$ tends to a Poisson distribution with parameter λ/μ.

12. For the process defined by (7.26) the generating function $P(s, t) = \Sigma P_n(t)s^n$ satisfies the partial differential equation

$$(\mu + \lambda s)\dfrac{\partial P}{\partial s} = a\lambda P,$$

with the solution $P = \{(\mu + \lambda s)/(\lambda + \mu)\}^a$.

13. Show that the transition probabilities of the pure birth process and of the birth and death process satisfy the Chapman-Kolmogorov equations.

14. Consider a stationary process with finitely many states, that is, suppose that the system of differential equations (9.9) is finite and that the coefficients c_j and p_{jk} are constants. Prove that the solutions are linear combinations of exponential terms $e^{\lambda(t-\tau)}$ where the real part of λ is negative unless $\lambda = 0$.

ANSWERS TO PROBLEMS

CHAPTER 1

1. (a) 3/5; (b) 3/5; (c) 3/10.
2. The space contains the two points HH and TT with probability 1/4; the two points HTT and THH with probability 1/8; and generally two points with probability 2^{-n} when $n \geq 2$. These probabilities add to 1, so that there is no necessity to consider the possibility of an unending sequence of tosses. The required probabilities are 15/16 and 2/3, respectively.
3. $Pr\{AB\} = 1/6$, $Pr\{A \cup B\} = 23/36$, $Pr\{A\bar{B}\} = 1/3$.

6. $x = 0$ in the events (a), (b), and (g).
   $x = 1$ in the events (e) and (f).
   $x = 2$ in the event (d).
   $x = 4$ in the event (c).

9. (a) A; (b) AB; (c) $B \cup AC$.
10. Correct are (c), (d), (e), (f), (h), (i), (k), (l). The statement (a) is meaningless unless $C \subset B$. It is in general false even in this case, but is correct in the special case $C \subset B$, $AC = 0$. The statement (b) is correct if $C \supset AB$. The statement (g) should read $(A \cup B) - A = \bar{A} B$. Finally (k) is the correct version of (j).
11. (a) $A \bar{B} \bar{C}$; (b) $A B \bar{C}$; (c) ABC; (d) $A \cup B \cup C$; (e) $AB \cup AC \cup BC$;
(f) $A \bar{B} \bar{C} \cup \bar{A} B \bar{C} \cup \bar{A} \bar{B} C$;
(g) $A B \bar{C} \cup A \bar{B} C \cup \bar{A} B C = (AB \cup AC \cup BC) - ABC$; (h) $\bar{A} \bar{B} \bar{C}$, (i) $\overline{ABC}$.
12. $A \cup B \cup C = A \cup (B - AB) \cup \{C - C(A \cup B)\} = A \cup B\bar{A} \cup C \bar{A} \bar{B}$.

CHAPTER 2

1. (a) 26^3; (b) $26^2 + 26^3 = 18{,}252$; (c) $26^2 + 26^3 + 26^4$. In a city with 20,000 inhabitants either some people have the same set of initials or at least 1748 people have more than three initials.
2. $64 \cdot 14 = 896$. For a chess board with n^2 fields the formula is $n^2(2n - 2)$.
3. $2(2^{10} - 1) = 2046$.

4. $\binom{n}{2} + n = \dfrac{n(n+1)}{2}$.

5. (a) $\dfrac{1}{n}$; (b) $\dfrac{1}{n(n-1)}$.

6. $p_1 = 0.01$, $p_2 = 0.27$, $p_3 = 0.72$.
7. $p_1 = 0.001$, $p_2 = 0.063$, $p_3 = 0.432$, $p_4 = 0.504$.
8. $p_r = (10)_r 10^{-r}$. For example, $p_3 = 0.72$, $p_{10} = 0.00036288$. Stirling's approximation (7.7) gives $p_{10} = 0.0003598\ldots$

9. (a) $(9/10)^k$; (b) $(9/10)^k$; (c) $(8/10)^k$; (d) $2(9/10)^k - (8/10)^k$; (e) AB and $A \cup B$.

10. (a) $\binom{k}{3}\dfrac{9^{k-3}}{10^k}$; (b) $\left(\dfrac{9}{10}\right)^k \sum_{j=0}^{3} \binom{k}{j}\left(\dfrac{1}{9}\right)^j$.

11. (a) $1/1\cdot 3\cdot 5 \cdots (2n-1) = 2^n n!/(2n)!$; (b) $(n!)/1\cdot 3 \cdots (2n-1) = 2^n/\binom{2n}{n}$.

12. On the assumption of randomness the probability that all of 12 tickets come either on Tuesdays or Thursdays is $(2/7)^{12} = 0.0000003\ldots$. There are only $\binom{7}{2} = 21$ pairs of days, so that the probability remains extremely small even for any two days. Hence it is reasonable to assume that the police have a system.

13. Assuming randomness, the probability of the event is $(6/7)^{12} = 1/6$ appr. No safe conclusion is possible.

14. (a) $\dfrac{(100-r)_n}{(100)_n}$; (b) $\left(1 - \dfrac{r}{100}\right)^n$.

For $n = r = 3$ the probabilities are (a) $0.911812\ldots$ and (b) $0.912673\ldots$. For $n = r = 10$ they are (a) $0.330476\ldots$ and (b) $0.348678\ldots$.

15. Cf. problem 14 with $n = r = 10$.

16. $25!(5!)^{-5}5^{-25} = 0.00209\ldots$.

17. $\dfrac{2(n-2)_r(n-r-1)!}{n!} = \dfrac{2(n-r-1)}{n(n-1)}$.

18. (a) $1/216$; (b) $83/3888$.

19. The probabilities are $1 - (5/6)^4 = 0.517747\ldots$ and $1 - (35/36)^{24} = 0.491404\ldots$.

20. $(5/6)^x \leq 1/3$ or $x \geq 7$.

21. $12!/12^{12} = 0.000054$.

22. $\binom{12}{2}(2^6 - 2)12^{-6} = 0.00137\ldots$.

23. $\dfrac{30!}{2^6 6^6}\binom{12}{6} 12^{-30} \approx 0.00035\ldots$.

24. $\binom{n}{2r} 2^{2r} \div \binom{2n}{2r}$.

25. (a) $n\binom{n-1}{2r-2} 2^{2r-2} \div \binom{2n}{2r}$; (b) $\binom{n}{2}\binom{n-2}{2r-4} 2^{2r-4} \div \binom{2n}{2r}$.

26. $p = \binom{2N}{N}^2 \div \binom{4N}{2N} \approx \{2/(N\pi)\}^{1/2}$.

27. $\binom{26}{k}\binom{26}{13-k} \div \binom{52}{13}$.

28. $\dfrac{\binom{13}{k}\binom{39}{13-k}}{\binom{52}{13}}$.

ANSWERS TO PROBLEMS   399

29. $p = \dfrac{\binom{4}{k}\binom{48}{13-k}\binom{39}{13}\binom{26}{13}}{\binom{52}{13}\binom{39}{13}\binom{26}{13}} = \dfrac{\binom{4}{k}\binom{48}{13-k}}{\binom{52}{13}}.$

30. Cf. problem 29. The probability is

$$\binom{13}{m}\binom{39}{13-m}\binom{13-m}{n}\binom{26+m}{13-n} \div \binom{52}{13}\binom{39}{13}.$$

31. $\binom{4}{k}\binom{48}{26-k} \div \binom{52}{26}.$

32. $p_{a,b,c,d}$

$$= \dfrac{\binom{13}{a}\binom{39}{13-a}\binom{13-a}{b}\binom{26+a}{13-b}\binom{13-a-b}{c}\binom{13+a+b}{13-c}}{\binom{52}{13}\binom{39}{13}\binom{26}{13}}.$$

33. (a) $24p(5, 4, 3, 1)$; (b) $4p(4, 4, 4, 1)$; (c) $12p(4, 4, 3, 2)$.

34. $\dfrac{\binom{13}{a}\binom{13}{b}\binom{13}{c}\binom{13}{d}}{\binom{52}{13}}$. [Cf. problem 33 for the probability that the hand contains a cards of some suit, b of another, etc.]

35. $p(4, 0, 0, 0) = 4\binom{48}{9} \div \binom{52}{13}$ $= 0.010\,564\,\ldots$

$p(3, 1, 0, 0) = 12 \cdot 4 \binom{48}{10}\binom{38}{12} \div \binom{52}{13}\binom{39}{13}$ $= .164\,802\,\ldots$

$p(2, 2, 0, 0) = 6 \cdot 6 \binom{48}{11}\binom{37}{11} \div \binom{52}{13}\binom{39}{13}$ $= .134\,838\,\ldots$

$p(2, 1, 1, 0) = 4 \cdot 3 \cdot 12 \binom{48}{11}\binom{37}{12}\binom{25}{12}$

$\div \binom{52}{13}\binom{39}{13}\binom{26}{13}$ $= .584\,298\,\ldots$

$p(1, 1, 1, 1) = 4! \binom{48}{12}\binom{36}{12}\binom{24}{12} \div \binom{52}{13}\binom{39}{13}\binom{26}{13} = .105\,498\,\ldots$

36. (a) $\binom{4}{k}\binom{4-k}{k}\binom{48}{r-k}\binom{48-r+k}{r-k} \div \binom{52}{r}\binom{52-r}{r}$; with $k \leq 2$;

(b) $\left\{\binom{4}{k}\binom{48}{r-k} \div \binom{52}{r}\right\}^2$, with $k \leq 4$.

37. Cf. problem 31.

400   ANSWERS TO PROBLEMS

38. Let $q = 1/\binom{52}{5}$. The probabilities are: (a) $4q = 1/649,740$; (b) $36q = 3/216,580$; (c) $13 \cdot 12 \cdot 4q = 1/4165$; (d) $13 \cdot 12 \cdot 4 \cdot 6q = 6/4165$; (e) $\binom{13}{5} 4q = 33/16,660$; (f) $9 \cdot 4^5 q = 768/216,580$; (g) $13 \binom{12}{2}\binom{4}{3} 4^2 q = 88/4165$; (h) $\binom{13}{2} 11 \cdot 6 \cdot 6 \cdot 4q = 198/4165$; (i) $13\binom{12}{3} \cdot 6 \cdot 4^3 q = 1760/4165$.

39.
$Pr\{(7)\}$ $= 10 \cdot 10^{-7}$ $= 0.000\,001$.
$Pr\{(6, 1)\}$ $= 10 \cdot 9 \binom{7}{6} \cdot 10^{-7}$ $= .000\,063$.
$Pr\{(5, 2)\}$ $= 10 \cdot 9 \binom{7}{5} \cdot 10^{-7}$ $= .000\,189$.
$Pr\{(5, 1, 1)\}$ $= 10 \binom{9}{2}\binom{7}{5} 2 \cdot 10^{-7}$ $= .001\,512$.
$Pr\{(4, 3)\}$ $= 10 \cdot 9 \binom{7}{4} \cdot 10^{-7}$ $= .000\,315$.
$Pr\{(4, 2, 1)\}$ $= 10 \cdot 9 \cdot 8 \binom{7}{4}\binom{3}{2} \cdot 10^{-7}$ $= .007\,560$.
$Pr\{(4, 1, 1, 1)\}$ $= 10 \binom{9}{3}\binom{7}{4} 3 \cdot 2 \cdot 10^{-7}$ $= .017\,640$.
$Pr\{(3, 3, 1)\}$ $= \binom{10}{2} 8 \binom{7}{3}\binom{4}{3} \cdot 10^{-7}$ $= .005\,040$.
$Pr\{(3, 2, 2)\}$ $= 10 \binom{9}{2}\binom{7}{3}\binom{4}{2} \cdot 10^{-7}$ $= .007\,560$.
$Pr\{(3, 2, 1, 1)\}$ $= 10 \cdot 9 \binom{8}{2}\binom{7}{3}\binom{4}{2} 2 \cdot 10^{-7}$ $= .105\,840$.
$Pr\{(3, 1, 1, 1, 1)\}$ $= 10 \binom{9}{4}\binom{7}{3} 4 \cdot 3 \cdot 2 \cdot 10^{-7}$ $= .105\,840$.
$Pr\{(2, 2, 2, 1)\}$ $= \binom{10}{3} 7 \binom{7}{2}\binom{5}{2}\binom{3}{2} \cdot 10^{-7}$ $= .052\,920$.
$Pr\{(2, 2, 1, 1, 1)\}$ $= \binom{10}{2}\binom{8}{3}\binom{7}{2}\binom{5}{2} 3 \cdot 2 \cdot 10^{-7}$ $= .317\,520$.
$Pr\{(2, 1, 1, 1, 1, 1)\}$ $= 10 \binom{9}{5}\binom{7}{2} 5 \cdot 4 \cdot 3 \cdot 2 \cdot 10^{-7}$ $= .317\,520$.
$Pr\{(1, 1, 1, 1, 1, 1, 1)\} = \binom{10}{7} 7 \cdot 6 \cdot 5 \cdot 4 \cdot 3 \cdot 2 \cdot 10^{-7}$ $= .060\,480$.

40. $p_1(r) = \binom{39}{r-13} \div \binom{52}{r}$.

$p_2(r) = \binom{26}{r-26} \div \binom{52}{r}$.

$p_3(r) = \binom{13}{r-39} \div \binom{52}{r}$.

r	$p_1(r)$	$p_2(r)$	$p_3(r)$
52	1	1	1
51	0.75	0.5	0.25
50	.558 82	.245 10	.058 82
49	.413 53	.117 65	.012 94
48	.303 82	.055 22	.002 64
47	.221 53	.025 31	.000 50
46	.160 26	.011 31	.000 08
45	.114 97	.004 92	.000 01
44	.081 76	.002 08	
43	.057 60	.000 85	
42	.040 19	.000 34	
41	.027 75	.000 13	
40	.018 95	.000 05	
39	.012 79	.000 02	
38	.008 53	.000 005	
37	.005 61	.000 002	

CHAPTER 3

1. $\binom{r_1 + n - 1}{r_1}\binom{r_2 + n - 1}{r_2}$.

2. $\binom{r_1 + 5}{5}(r_2 + 1)$.

3. $\dfrac{(r_1 + r_2 + r_3)!}{r_1! r_2! r_3!}$.

4. $\alpha_0 = n! n^{-n} \sim (2\pi n)^{1/2} e^{-n}$ and $\alpha_1 = n(n-1)\alpha_0/2$.

5. $\dfrac{r!}{k_1! k_2! \cdots k_n!} n^{-r}$.

10. Use lemma 2 of section 1.
11. For p_k, τ_k, ρ_k use, respectively, formulas (6.6), (9.14), (9.5) of Chapter 2.
12. Select the ν alphas and the single beta which must follow it. The remaining elements can be arranged in $(r_1 - \nu + r_2 - 1)!$ ways.
13. Follows from problems 12 for $\nu = 0$.
14. Consider $P_{2\nu}/P_{2\nu-2}$.
15. The hint and the argument used to prove the theorem show that

$$\pi_k = \binom{r_1 - 1}{k - 1}\left\{\binom{r_2 - 1}{k - 2} + 2\binom{r_2 - 1}{k - 1} + \binom{r_2 - 1}{k}\right\} \div \binom{r_1 + r_2}{r_1}.$$

Use formula (6.5) of Chapter 2 twice.

16. The alpha runs can be put in arbitrary order between, preceding, or following the betas; we have, therefore $\binom{r_2 + 1}{k}$ selections for the places of the alpha runs. The first factor gives the number of ways in which the k places can be assigned to runs of different lengths.

CHAPTER 4

1. $99/323$.
2. $0.21\ldots$.
3. $1/4$.
4. $7/2^6$.
5. $1/81$ and $31/6^6$.
6. If A_k is the event that (k, k) does not appear, then from (1.5)

$$1 - p_r = 6\left(\frac{35}{36}\right)^r - \binom{6}{2}\left(\frac{34}{36}\right)^r + \binom{6}{3}\left(\frac{33}{36}\right)^r - \binom{6}{4}\left(\frac{32}{36}\right)^r + 6\left(\frac{31}{36}\right)^r - \left(\frac{30}{36}\right)^r.$$

7. $u_r = \sum_{k=0}^{N} (-1)^k \binom{N}{k} \left(1 - \frac{k}{n}\right)^r$.

9. $p_r = \sum_{k=0}^{N} (-1)^k \binom{N}{k} \frac{(n-k)_r}{(n)_r}$.

10. The general term is $a_{1k_1}a_{2k_2}\cdots a_{Nk_N}$, where $(k_1, k_2, \cdots, k_N)$ is a permutation of $(1, 2, \cdots, N)$. For a diagonal element $k_\nu = \nu$.

12. Note that, by definition, $u_r = 0$ for $r < n$ and $u_n = n!s^n/(ns)_n$.

14. $u_r - u_{r-1} = \sum_{k=1}^{n} (-1)^{k-1} \binom{n-1}{k-1} \frac{(ns - ks)_{r-1}}{(ns-1)_{r-1}} \to$

$$\sum_{k=0}^{n-1} (-1)^k \binom{n-1}{k} \left(1 - \frac{k+1}{n}\right)^{r-1}.$$

15.

r	$Q_1(r)$	$Q_2(r)$	$Q_3(r)$
51	0	0	....
50	0	0.76471	0.23529
49	0.39765	.55059	.05177
48	.58430	.29964	.01056
47	.58836	.14592	.00198
46	.50634	.06684	.00034
45	.40102	.02935	.00005
44	.30214	.01243	.00001
43	.22021	.00509	
42	.15672	.00201	
41	.10946	.00077	
40	.07524	.00028	
39	.05097	.00010	

16. $\binom{N}{2}^{-r} \binom{N}{m} \sum_{k=2}^{m} (-1)^{m-k} \binom{m}{k} \binom{k}{2}^r$.

17. Use $\binom{52}{5} S_k = \binom{4}{k} \binom{52 - 13k}{5}$.

$P_{[0]} = 0.264$, $P_{[1]} = 0.588$, $P_{[2]} = 0.146$, $P_{[3]} = 0.002$, approximately.

18. Use $\binom{52}{13} S_k = \binom{4}{k}\binom{52-2k}{13-2k}$.

$$P_{[0]} = 0.780217,\ P_{[1]} = 0.204606,\ P_{[2]} = 0.014845,$$
$$P_{[3]} = 0.000330,\ P_{[4]} = 0.000002,\ \text{approximately.}$$

19. $m!N!u_m = \sum\limits_{k=0}^{N-m} (-1)^k (N-m-k)!/k!$.

20. Cf. the following formula with $r = 2$.

21. $(rN)!x = \binom{N}{2} r^2(rN-2)! - \binom{N}{3} r^3(rN-3)! + - \cdots$
$$+ (-1)^N r^N (rN-N)!.$$

23. For $n > r$ we have $P_1 = 1$. In (7.1) let $k = n - \nu - 1$.

26. $P_{[m]} = \dfrac{\binom{n}{m}}{\binom{n+r-1}{r}} \sum\limits_{k=0}^{n-m} (-1)^k \binom{n-m}{k} \binom{n-m+r-1-k}{r}$.

27. Use (9.11) and (9.1) of Chapter 2.

CHAPTER 5

1. $1 - \dfrac{(5)_3}{(6)_3} = \dfrac{1}{2}$.

2. $p = 1 - \dfrac{10 \cdot 5^9}{6^{10} - 5^{10}} = 0.61 \ldots$.

3. $0.41 \ldots$.

4. $2 \cdot \dfrac{\binom{23}{10}}{\binom{26}{13}} = \dfrac{11}{50}$.

5. $2 \cdot \dfrac{\binom{23}{12}}{\binom{26}{13}} = \dfrac{13}{50}$.

6. $\dfrac{125}{345};\ \dfrac{140}{345};\ \dfrac{80}{345}$.

7. $1 - p^2$.

8. $\dfrac{20}{21}$.

10. $\dfrac{p}{2-p}$.

11. $p = (1-p_1)(1-p_2) \cdots (1-p_n)$.

12. Use $1 - x < e^{-x}$ for $0 < x < 1$ or Taylor's series for $\log(1-x)$; cf. (6.9) of Chapter 2.

15. $\dfrac{b+c}{b+c+r}$.

16. If the statement is true for the nth drawing regardless of b, r, and c, then the probability of red at the $(n+1)$st trial is

$$\frac{b}{b+r} \cdot \frac{b+c}{b+r+c} + \frac{r}{b+r} \cdot \frac{b}{b+r+c} = \frac{b}{b+r}.$$

17. The preceding problem states that the assertion is true for $m = 1$ and all n. For induction, consider the two possibilities at the first trial.
20. From (5.2) $2v = 2p(1-p) \leq 1/2$.
22. (a) u^2; (b) $u^2 + uv + v^2/4$; (c) $u^2 + (25uv + 9v^2 + vw + 2uw)/16$.
27. $p_{11} = p_{32} = 2p_{21} = p$, $p_{12} = p_{33} = 2p_{23} = q$, $p_{13} = p_{31} = 0$, $p_{22} = 1/2$.

CHAPTER 6

1. 5/16.
2. The probability is 0.02804 ....
3. $(0.9)^x \leq 0.1$, $x \geq 22$.
4. $q^x \leq 1/2$ and $(4q)^x \leq 1/2$ with $p = \binom{48}{9} \div \binom{52}{13}$. Hence $x \geq 263$ and $x \geq 66$, respectively.
5. $1 - (0.8)^{10} - 2(0.8)^9 = 0.6242 \ldots$.
6. $\{1 - (0.8)^{10} - 2(0.8)^9\}/\{1 - (0.8)^{10}\} = 0.6993$.
7. $\binom{26}{2}\binom{26}{11} \div \binom{52}{13} = 0.003954\ldots$, and $\binom{13}{2}\dfrac{1}{2^{13}} = 0.00952\ldots$.
8. $\binom{12}{2}\{6^{-6} - 2 \cdot 12^{-6}\}$.

9. True values: $0.6651\ldots$, $0.40187\ldots$, and $0.2009\ldots$; Poisson approximations: $1 - e^{-1} = 0.6321\ldots$, $0.3679\ldots$, and $0.1839\ldots$.

10. $e^{-2} \sum\limits_{4}^{\infty} 2^k/k! = 0.143 \ldots$.

11. $e^{-1} \sum\limits_{3}^{\infty} 1/k! = 0.080 \ldots$.

12. $e^{-x/100} \leq 0.05$ or $x \geq 300$.
13. $e^{-1} = 0.3679\ldots$, $1 - 2 \cdot e^{-1} = 0.264\ldots$.
14. $e^{-x} \leq 0.01$, $x \geq 5$.
15. $1/p = 649{,}740$.
16. $\sum\limits_{k=0}^{n} \binom{n}{k}^2 2^{-2n} = \binom{2n}{n} 2^{-2n} \approx \left(\dfrac{1}{\pi n}\right)^{1/2}$ for large n [cf. (9.8) of Chapter 2].

17. $\sum\limits_{k=a}^{a+b-1} \binom{a+b-1}{k} p^k q^{a+b-1-k}$. This can be written in the alternative form

ANSWERS TO PROBLEMS

$p^a \sum_{k=0}^{b-1} \binom{a+k-1}{k} q^k$, where the kth term equals the probability that the ath success occurs directly after $k \leq b-1$ failures.

19. The successive terms decrease faster than a geometric sequence with ratio $(n-k)p/kq$ when $k > np$ and $kq/(n-k)p$ when $k < np$.
20. Use $1-x < e^{-x}$ for $0 < x < 1$.
21. Use the obvious symmetry between successes and failures.
30. $p = p_1 q_2 / (p_1 q_2 + p_2 q_1)$.

CHAPTER 7

1. Proceed as in section 1.
2. Use (1.7).
3. Write the integral in the form

$$e^{-x^2/2} \int_0^{h/2} (e^{xy} + e^{-xy}) e^{-y^2/2} \, dy.$$

4. 0.99.
5. 500.
6. 66,400.
7. Most certainly. The inequalities of Chapter 6 suffice to show that an excess of more than 8 standard deviations is exceedingly improbable.
8. $(2\pi n)^{-1} \{p_1 p_2 (1 - p_1 - p_2)\}^{-1/2}$.

CHAPTER 8

1. $\beta = 21$.
2. $x = pu + qv + rw$, where u, v, w are solutions of

$$u = p^{\alpha-1} + (qv + rw) \frac{1 - p^{\alpha-1}}{1 - p},$$

$$v = (pu + rw) \frac{1 - q^{\beta-1}}{1 - q},$$

$$w = pu + qv + rw = x.$$

3. $u = p^{\alpha-1} + (qv + rw) \frac{1 - p^{\alpha-1}}{1 - p}$

$v = (pu + rw) \frac{1 - q^{\beta-1}}{1 - q}$

$w = (pu + qv) \frac{1 - r^{\gamma-1}}{1 - r}$.

4. Note that $Pr\{A_n\} < (2p)^n$, but

$$Pr\{A_n\} > 1 - (1 - p^n)^{2^n/2n} > 1 - e^{-(2p)^n/2n}.$$

If $p = 1/2$, the last quantity is $\sim 1/2n$; if $p > 1$, then $Pr\{A_n\}$ does not even tend to zero.

CHAPTER 9

1. In the joint distribution of X, Y the rows are 32^{-1} times (1, 0, 0, 0, 0, 0), (0, 5, 4, 3, 2, 1), (0, 0, 6, 6, 3, 0), (0, 0, 0, 1, 0, 0); of X, Z: (1, 0, 0, 0, 0, 0), (0, 5, 6, 1, 0, 0), (0, 0, 4, 6, 1, 0), (0, 0, 0, 3, 2, 0), (0, 0, 0, 0, 2, 0), (0, 0, 0, 0, 0, 1); of Y, Z: (1, 0, 0, 0), (0, 5, 6, 1), (0, 4, 7, 0), (0, 3, 2, 0), (0, 2, 0, 0), (0, 1, 0, 0). Distribution of $X + Y$: (1, 0, 5, 4, 9, 8, 5) all divided by 32, and the values of $X + Y$ ranging from 0 to 6; of XY: (1, 5, 4, 3, 8, 1, 6, 0, 3, 1) all divided by 32, the values ranging from 0 to 9. $E(X) = 5/2$, $E(Y) = 3/2$, $E(Z) = 31/16$, $\mathrm{Var}(X) = 5/4$, $\mathrm{Var}(Y) = 3/8$, $\mathrm{Var}(Z) = 303/256$.

2. $\binom{n}{k} 364^{n-k} 365^{1-n}$.

3. (a) $365\{1 - 364^n \cdot 365^{-n} - n 364^{n-1} \cdot 365^{-n}\}$; (b) $n \geq 28$.

4. (a) $\mu = n$, $\sigma^2 = (n-1)n$; (b) $\mu = (n+1)/2$, $\sigma^2 = (n^2-1)/12$.

5. $-n/36$.

6. $\sigma^2 \approx \dfrac{nN^2}{(n+1)^2(n+2)}$.

7. $Pr\{X \leq r, Y \geq s\} = \left(\dfrac{r-s+1}{N}\right)^n$;

$Pr\{X = r, Y = s\} = N^{-n}\{(r-s+1)^n - 2(r-s)^n + (r-s-1)^n\}$.

8. $x = \dfrac{r^{n-2} - (r-1)^{n-2}}{r^n - (r-1)^n}$ if $j < r, k < r$.

$x = \dfrac{r^{n-2}}{r^n - (r-1)^n}$ if $j \leq r, k = r$, or $j = r, k \leq r$.

$x = 0$ if $j > r$ or $k > r$.

9. $p_k = p^k q + q^k p$; $E(X) = pq^{-1} + qp^{-1}$; $\mathrm{Var}(X) = pq^{-2} + qp^{-2} - 2$.

10. $q_k = p^2 q^{k-1} + q^2 p^{k-1}$; $Pr\{X = m, Y = n\} = p^{m+1}q^n + q^{m+1}p^n$ with $m, n, \geq 1$; $E(Y) = 2$; $\sigma^2 = 2(pq^{-1} + qp^{-1} - 1)$.

11. The distribution is given in Chapter 2, (5.8); the means and variances in example (5.e). We have $E(X) = \dfrac{rn_1}{n}$; $\mathrm{Var}(X) = \dfrac{rn_1(n-n_1)(n-r)}{n^2(n-1)}$; $\mathrm{Cov}(X, Y) = \dfrac{-rn_1 n_2(n-r)}{n^2(n-1)}$; $\rho(X, Y) = -\left(\dfrac{n_1 n_2}{(n-n_1)(n-n_2)}\right)^{1/2}$.

12. $E(X) = \dfrac{r_1(r_2+1)}{r_1+r_2}$; $\mathrm{Var}(X) = \dfrac{r_1 r_2(r_1-1)(r_2+1)}{(r_1+r_2-1)(r_1+r_2)^2}$.

13. $E(S_n) = \dfrac{nb}{b+r}$; $\mathrm{Var}(S_n) = \dfrac{nbr\{b+r+nc\}}{(b+r)^2(b+r+c)}$.

14. $E(Y_r) = \sum_{k=1}^{r} \dfrac{N}{r-k+1}$; $\mathrm{Var}(Y_r) = \sum_{k=1}^{r} \dfrac{N(N-r+k-1)}{(r-k+1)^2}$.

15. $E\left(\dfrac{K}{N+1}\right) = \Sigma k p_{k,n}/(n+1) = q^2 p' q' \sum_{n=1}^{\infty}\left(1 - \dfrac{1}{n+1}\right)(p + qq')^{n-1}$

$= \dfrac{qq'}{1-qp'} - \dfrac{q^2 p' q'}{(1-qp')^2}\log\dfrac{1}{qp'}.$

$E(K) = \dfrac{q'}{p'};\quad E(N) = \dfrac{(1-qp')}{qp'};\quad \operatorname{Cov}(K, N) = \dfrac{q'}{qp'^2};$

$\rho(K, N) = \left\{\dfrac{q'}{(1-qp')}\right\}^{\frac{1}{2}}.$

18. (a) $1 - q^k$; (b) $E(X) = N\left\{1 - q^k + \dfrac{1}{k}\right\}$; (c) $\dfrac{dE(X)}{dk} = 0.$

26. $E\left(\dfrac{r}{X}\right) = r\sum_{k=r}^{\infty} k^{-1}\binom{k-1}{r-1} p^r q^{k-r}$

$= \sum_{k=1}^{r-1}(-1)^{k-1}\dfrac{r}{r-k}\left(\dfrac{p}{q}\right)^k + \left(\dfrac{-p}{q}\right)^r r \log p.$

To derive the last formula from the first, put $f(q) = r\Sigma k^{-1}\binom{k-1}{r-1} q^k$. Using (9.1) of Chapter 2, one finds that $f'(q) = rq^{r-1}(1-q)^{-r}$. The assertion now follows by repeated integrations by part.

CHAPTER 11

1. $sP(s)$ and $P(s^2)$.
2. (a) $(1-s)^{-1}P(s)$; (b) $(1-s)^{-1}sP(s)$; (c) $\{1 - sP(s)\}/(1-s)$; (d) $p_0 s^{-1} + \{1 - s^{-1}P(s)\}/(1-s)$; (e) $\frac{1}{2}\{P(s^{1/2}) + P(-s^{1/2})\}.$
3. $U(s) = pqs^2/(1-ps)(1-qs)$. Mean $= 1/pq$, Var $= (1-3pq)/p^2q^2.$
4. $U(s) = \dfrac{1}{2(1-s)} + \dfrac{1}{2\{1-(q-p)s\}}$, $2u_n = 1 + (q-p)^n.$

5. Using the generating function for the geometric distribution of X_ν we have without computation

$P_r(s) = s^r\left(\dfrac{N-1}{N-s}\right)\left(\dfrac{N-2}{N-2s}\right)\cdots\left(\dfrac{N-r+1}{N-(r-1)s}\right).$

6. From (9.1) $P_r(s)\{N - (r-1)s\} = P_{r-1}(s)(N-r-1)s.$

7. $P_r(s) = \dfrac{s}{N-(N-1)s} \cdot \dfrac{2s}{N-(N-2)s} \cdot \cdots \cdot \dfrac{rs}{N-(N-r)s}.$

8. S_r is the sum of r independent variables with a common geometric distribution. Hence

$P_r(s) = \left(\dfrac{q}{1-ps}\right)^r,\quad p_{r,k} = q^r p^k \binom{r+k-1}{k}.$

408   ANSWERS TO PROBLEMS

9. $Pr\{R = r\} = \sum_{k=0}^{\nu-1} Pr\{S_{r-1} = k\} Pr\{X_r \geq \nu - k\}$

$= \sum_{k=0}^{\nu-1} q^{r-1} p^k \binom{r+k-2}{k} p^{\nu-k} = p^\nu q^{r-1} \binom{r+\nu-2}{\nu-1}.$

$$E(R) = 1 + \frac{q\nu}{p}, \quad \text{Var}(R) = \frac{\nu q}{p^2}.$$

14. $u_n = q^n + \sum_{k=3}^{n} \binom{k-1}{2} p^3 q^{k-3} u_{n-k}$ with $u_0 = 1$, $u_1 = q$, $u_2 = q^2$, $u_3 = p^3 + q^3$. Using the fact that this recurrence relation is of the convolution type,

$$U(s) = \frac{1}{1-qs} + \frac{(ps)^3}{(1-qs)^3} U(s).$$

15. $u_n = pw_{n-1} + qu_{n-1}$, $v_n = pu_{n-1} + qv_{n-1}$, $w_n = pv_{n-1} + qw_{n-1}$. Hence $U(s) - 1 = psW(s) + qsU(s)$; $V(s) = psU(s) + qs \cdot V(s)$; $W(s) = psV(s) + qsW(s)$.

20. From (6.2), $P_{n+1}''(1) = P''(1)P_n'^2(1) + P'(1)P_n''(1)$. Putting $\lambda^2 = \text{Var}(X_1)$, this equation becomes $\text{Var}(X_{n+1}) = \mu^{2n}\lambda^2 + \mu \text{Var}(X_n)$ and hence $\text{Var}(X_n) = \lambda^2(\mu^{2n-2} + \mu^{2n-3} + \cdots + \mu^{n-1})$.
If $\mu = 1$, then $\text{Var}(X_n) = n\sigma^2$; otherwise $\text{Var}(X_n) = \lambda^2 \mu^{n-1}(\mu^n - 1)/(\mu - 1)$.

CHAPTER 12

1. It suffices to show that for all roots $s \neq 1$ of $F(s) = 1$ we have $|s| \geq 1$, and that $|s| = 1$ is possible only in the periodic case.

2. $u_{2n} = \left\{\binom{2n}{n} 2^{-2n}\right\}^r \sim (\pi n)^{-r/2}$. Hence $\mathcal{E}$ is certain only for $r = 2$. For $r = 3$ the tangent rule for numerical integration gives

$$\sum_{n=1}^{\infty} u_{2n} \sim \pi^{-3/2} \int_{1/2}^{\infty} x^{-3/2} dx = \left(\frac{2}{\pi}\right)^{3/2} \approx \frac{1}{2}.$$

Hence by (3.7) the probability of $\mathcal{E}$ ever occurring is, approximately, $x = 1/3$. A more precise evaluation of the sum is 0.47 and leads to $x = 0.32$.

3. The zero preceding the first negative value of S_n may be the first, second, etc., zero. Hence the required generating function is

$$\frac{s}{2}\left\{1 + \frac{1}{2}F(s) + \frac{1}{2^2}F^2(s) + \cdots\right\} = \frac{s}{2 - F(s)} = \frac{F(s)}{s}.$$

6. Note that $1 - F(s) = (1-s)Q(s)$ and $\mu - Q(s) = (1-s)R(s)$, whence $Q(1) = \mu$, $2R(1) = \sigma^2 - \mu + \mu^2$. The power series for $Q^{-1}(s) = \Sigma(u_n - u_{n-1})s^n$ converges for $s = 1$.

CHAPTER 13

1. $N_n^* \approx (N_n - 714.3)/22.75$; $\Phi(\tfrac{2}{3}) - \Phi(-\tfrac{2}{3}) \approx \tfrac{1}{2}$.

4. If a_n is the probability that an A-run of length r occurs at the nth trial, then $A(s)$ is given by (1.5) with p replaced by α and q by $1 - \alpha$. Let $B(s)$ and $C(s)$ be the corresponding functions for B- and C-runs. The required generating functions

are $F(s) = 1 - U^{-1}(s)$, where in case (a) $U(s) = A(s)$; in (b) $U(s) = A(s) + B(s) - 1$; in (c) $U(s) = A(s) + B(s) + C(s) - 2$.

5. Use a straightforward combination of the method in example (2.b) and problem 4.

9. $u_n = Np$, $v_k(\infty) = Npq^k$.

CHAPTER 15

1. (a) The chain is irreducible and ergodic; $p_{jk}^{(n)} \to 1/3$ for all j, k. (Note that P is doubly stochastic.) (b) The chain has period 3, with G_1 containing E_1 and E_2; the state E_4 forms G_2, and E_3 forms G_3. We have $u_1 = u_2 = 1/2$, $u_3 = u_4 = 1$. (c) The states E_1 and E_3 form a closed set S_1, and E_4, E_5 another closed set S_2, while E_2 is transient. The matrices corresponding to the closed sets are two by two matrices with elements $1/2$. Hence $p_{jk}^{(n)} \to 1/2$ if E_j and E_k belong to the same S_r; $p_{j2}^{(n)} \to 0$; finally $p_{2k}^{(n)} \to 1/2$ if $k = 1, 3$, and $p_{2k}^{(n)} \to 0$ if $k = 2, 4, 5$.

2. $p_{jj}^{(n)} = (j/6)^n$, $p_{jk}^{(n)} = (k/6)^n - ((k-1)/6)^n$ if $k > j$, and $p_{jk}^{(n)} = 0$ if $k < j$.

3. $x_k = (3/4, 1/2, 1/4, 1/2)$, $y_k = (1/4, 1/2, 3/4, 1/2)$.

4. $p_{jj} = 2j(N-j)/N^2, p_{j,j+1} = (N-j)^2/N^2, p_{j,j-1} = j^2/N^2, u_k = \binom{N}{k}^2 \div \binom{2N}{N}$.

10. Note that the matrix is doubly stochastic and use example (6.d).

15. Let M be the maximum of x_j. Consider the states E_r for which $x_r = M$.

18. If $N \geq m - 2$, the variables $X^{(m)}$ and $X^{(n)}$ are independent, and hence the three rows of the matrix $p_{jk}^{(m,n)}$ are identical with the distribution of $X^{(n)}$, namely, $(1/4, 1/2, 1/4)$. For $n = m + 1$ the three rows are $(1/2, 1/2, 0)$, $(1/4, 1/2, 1/4)$, $(0, 1/2, 1/2)$.

CHAPTER 17

3. $E(X) = ie^{\lambda t}$; $\text{Var}(X) = ie^{\lambda t}(e^{\lambda t} - 1)$.
4. $P_n' = -\lambda n P_n + \lambda(n+1)P_{n+1}$.

$$P_n = \binom{i}{n} e^{-i\lambda t}(e^{\lambda t} - 1)^{i-n} \quad (n \leq i).$$

$$E(X) = ie^{-\lambda t}; \quad \text{Var}(X) = ie^{-\lambda t}(1 - e^{-\lambda t}).$$

14. The standard method of solution leads to a system of linear equations. Use the hint to problem 15 of Chapter 15.

INDEX

Absorption: birth and death process 373; diffusion 296, 304; Markov chains 332, 360, 373; random walk 279 (several dimensions 299); sequential sampling 281, 300.
Acceptance 281, 300, 314.
Accidents 117, 189, 234, 365 (distr. of damage 222).
ADLER, H. A., and K. W. MILLER 384.
Age distribution 276 (stable 278).
Ages of a couple 12, 14.
Aggregates, self-renewing 275.
ANDERSEN, E. SPARRE 252.
Approximation: binomial distr. by normal (individual terms 135, 146; central part 137, tails 144, 147), binomial distr. by Poisson 110, 125 (error estimate 115); birthday distr. 29, 72; Bonferoni's inequalities 75, 101; hypergeometric distr. (by binomial 47, 108, 125, by Poisson 114, by normal 146); multinomial by Poisson 127; $n!$ 43, 50; normal distr. (tails) 131, 145; by partial fraction expansions 229, 237 (numerical examples 230, 267, 276; error estimate 269); sequential sampling (= generalized random walk) 303. [Cf. *Limit Theorems.*]
Arc sine law 252, 257; counterpart 262.
Assignable causes 56.
Atomic bombs 223.
Average of distr. 172.
Averages, moving 339, 340, 346.

$b(k; n, p)$ 106.
BACHELIER, L. 293.
Backward equations 385, 390, 392.
Bacteria counts 122.
Banach's match-box problem 108, 176.
Barriers 279, 280 (in several dimensions 299, 345).

BARTKY, W. 281, 314.
Bayes's rule 85.
BERNOULLI, D. 199.
BERNOULLI, J. 104.
BERNOULLI *trials* 104; (billiards 235, 236); gambling systems 151; infinite sequences 149, 282; iterated logarithm 157, 163; multiple 125, 127, 168; number theoretical interpretation 161; return to equilibrium 244, 305; (even number of successes 235). [Cf. *Binomial distribution, Coin tossing, Random walk, Recurrent events, Ruin, Runs.*]
BERNSTEIN, S. 88, 140.
Beta function 127.
Betting, on runs 149, 268, 278; — systems 151, 282. [Cf. *Coin tossing, Duration of games, Games, Ruin.*]
Billiards 235, 236.
Bingo 47, 76.
Binomial coefficients 30, 40; identities with 47, 76; integrals for 292, 305.
Binomial distribution 106; as beta function 127; central term 109; combination with Poisson 128, 221, 344; convolutions 126, 216; expectation 173, 216 (absolute 189); generating fct. 216; — and hypergeometric 47, 108, 125; negative — 218; normal approximation (individual terms 135, 146; central part 137, 139; tails 144, 147); in number theory 161; in occupancy problems 55, 69; Poisson approximation 110, 125 (error estimate 115; comparison with normal 143); tails 126, 144; variance 178, 180, 216. [Cf. *Bernoulli trials, Random walk, Ruin.*]
Binomial formula 41.
Biological applications, cf. *Birth and death process, Breeding, Chromosomes, Genes, Larvae, Renewal, Survival.*

Birth process 367, 394, 396 (divergent 369, 391).
Birth and death process 371, 395.
Birthdays 29; expectations 174, 187; as occupancy problem 52; Poisson distr. 72, 112; special problems 45, 125.
BISHOP, D. J. 276.
Blood: counts 122; tests 189.
BOLTZMANN-MAXWELL *statistics* 53.
Bomb hits 120.
BONFERONI'S *inequalities* 75, 101.
BOOLE'S *inequality* 20.
BOREL, E. 157, 163; Borel-Cantelli lemmas 154, 159, 160, 209.
BOSE-EINSTEIN *statistics* 53, 59, 77.
BOTTEMA, O. 235.
Branching process, cf. *Chain reactions.*
Breeding 101; (as Markov chain problem 315, 345, 360).
Bridge 9; aces (among r cards 35, 46; joint distr. 166); algebra of events 15, 21; bingo 47, 76; composition of hands 31, 33, 39, 46, 62, 63, 65, 125, 187; conditional prob. 79, 100. [Cf. *Shuffling.*]
Brother-sister mating 101; (as Markov chain 315, 345, 360).
Brownian motion 279, 293; Ehrenfest model 312, 327, 345.
Busy hour 365.

CAMPBELL, N. R. 276.
CANTELLI, F. P. 157 (Borel-Cantelli lemmas 154, 159, 160, 209).
CANTOR, G. 17, 233.
Cards, cf. *Bridge, Matching, Poker, Shuffling.*
CATCHESIDE, D. G. 45, 76, 120, 222.
CENSUS, BUREAU OF 188.
Centenarians 113.
Central force in diffusion 313.
Central limit theorem: for arbitrary distr. 202, 209; binomial distr. 140; identical distr. 192; Markov processes 342; recurrent events 248; runs 266; (infinite moment analogue 253; frequency of decimals 162; permutations 205).
CHANDRASEKHAR, S. 345.
Chain reactions 223, 237.
Chains (polymer molecules) 190.

CHAPMAN-KOLMOGOROV *equations* for Markov chains 338, 341; for stochastic processes 387, 396.
Characteristic equation 302.
Characteristic values 350.
CHEBYSHEV, P. L. 183; — inequality 183; generalized 189.
Chess problems 44, 76.
Chi-square test: mentioned in connection with tabular material, but not defined.
Chromosomes 92, 96; breakages and interchanges 45, 76, 120, 128, 222.
CHUNG, K. L. 189, 252.
CLARKE, R. D. 120.
Classification, multiple 24.
Closed set (in Markov chains) 318.
COCHRAN, W. G. 57.
Coin tossing: arc sine law 252, 257 (counterpart 262); distr. of leads 250, 255; return to equilibrium 238, 245 (limit theorem 253, 258; random walk 288, 304); ties in multiple — 246, 261.
Collector's problems 52, 64, 76, 174; moments 181, 188.
Colorblindness 96, 98, 100, 126.
Combinatorial product space 91.
Competition problem 139.
Complementary event 13.
Composite Markov process (shuffling) 340.
Compound distributions 221, 237; binomial and geometric 223; binomial and Poisson 128, 221, 344.
Compound experiments 81.
Compound Poisson distr. and process 237, 391, 395.
Confidence level 142.
Contagion 56, 83, 128, 223.
Continuity theorem 232, 278.
Convergence (almost everywhere and in measure) 162, 207.
Convolutions 215, 236; (binomial distr. 126; Poisson 127).
Correlation coefficient 186.
Cosmic rays 369.
Counters (waiting lines) 378.
Coupons, collecting of, 52, 64, 76, 174; moments 181, 188.
Covariance 179.
CRAMÉR, H. 119.

Craps 16.
Cumulative chance effects 251.
Cumulative distribution 133.
Cycles 205.
Cyclical random walk 311, 352, 353.
Cylindrical sets 91.

DAHLBERG, G. 100.
Death process 394.
Decimals, distribution of 161; (e, 29, 124; π, 124). [Cf. *Random digits*.]
Defectives 45, 100, 112; blood tests 189; (Bartky's sampling scheme 315; Dodge's 168, 188).
Degenerate processes 369, 391.
DEMOIVRE, A. 133, 212, 236.
DEMOIVRE-LAPLACE *limit theorem* 137 (traditional form 139).
Density fluctuations (Ehrenfest model 327; particles in space 345).
Density function 133.
Derivatives, number of 52.
Descendants: in breeding 315, 360; in chain reactions 224; family relations 102; genetics 92, 101, 315; renewal 275.
Determinants 76.
Dice, ace runs 150, 163, 266; distr. of scores 167, 178 (generating fct. 236); equalization of ones, twos, . . . , 239, 247; normal distr. 146, 192; as occupancy problem 51; special problems 33, 45, 76, 100, 106, 124, 125, 187, 344; (dice illustrating compound experiments 83; pairwise independence 87).
Difference, nth, of zero 77.
Difference equations: method of particular solutions 283, 288, 302; passage to limit of differential eqns. 294, 304; several dimensions 299, 306; (for Ehrenfest model 327, 345; Polya's urn scheme 101, 395; reflecting barriers 326). [Cf. *Renewal equation*.]
Differential equations, ordinary; backward and forward 385, 389, 392; generating functions 395; Kolmogorov's — 386; recursive, without uniqueness 393; special (birth process 367, 390, 394; birth and death 372, 395; compound Poisson 391; Poisson 366, 386; Polya 395; power supply 384, 396; radioactive process 368; servicing 380; trunking 377, 396; waiting lines 378; Yule process 368, 394).
Diffusion 279, 293; absorption and first passage 296, 304; — coefficients 295; Ehrenfest model 313, 327, 345.
DIRAC-FERMI *statistics* 53.
Discrete sample spaces 16.
Disorder and chance fluctuations 189.
Dispersion 178.
Distinguishable objects 11, 51.
Distribution: function 133, 165; normal 129; probability distr. 165 (joint, marginal 166).
DODGE'S *inspection plan* 168, 188.
DOEBLIN, W. 342, 344.
Dominant gene 92.
Domino 44.
DOOB, J. L. 152, 344, 392.
DORFMAN, R. 189.
Double generating functions 255.
Double sampling 168, 188, 314.
Doubling system 285.
Drift: diffusion 295; random walk 279.
Duration of games: Bernoulli trials 280 (expectation 286; generating function 289, 305; explicit expressions 292, 304); Markov chains 335, 358; sequential sampling 303, 306. [Cf. *Extinction*.]

e (distr. of decimals) 29, 124.
EGGENBERGER, F. 83.
EHRENFEST, P. and T. 313; — model of diffusion 312, 345 (stationary distr. 327).
Eigenvalue 350.
EINSTEIN-BOSE *statistics* 53, 59, 77.
EINSTEIN-WIENER *diffusion* 293.
EISENHART, C., and F. S. SWED 56.
Elastic: barrier 280; — force in diffusion 313.
Elevator problem 30, 47, 52.
ELLIS, R. E. 292.
Equilibrium: coin tossing 238; Ehrenfest model 327; macroscopic 329; statistical 373.
ERDÖS, P. 163, 244, 252.
Ergodic (properties of aperiodic chains

324; periodic 329; stochastic processes 373, 396); — states 321; mean — theorem 346; non-stochastic matrices 343.
ERLANG, A. K. 377.
Error function 133.
Estimation, statistical 37.
Events: compatible 60; compound and simple 9, 13; independent 86 (pairwise and mutual 88); relations between — 13; simultaneous realization 15 (at least one 60; m among N 64, 74, 101); — in repeated trials 90, 91. [Cf. *Recurrent events.*]
Evolution 369.
Expectation 171; — and generating fcts. 213; infinite 214 (recurrence times 242); of reciprocals 188, 189, 190.
Experiments: conceptual 4, 9; compound 81; repeated 89; — and random variables 164.
Exponential: distribution 220; holding times 375.
Extinction: birth and death process 374, 395; chain reactions 224; — of genes 333.
Extra Sensory Perception (ESP) 45, 336.

F, for failure 104.
Factorials 30; Stirling's formula 41, 50.
Family: relations 102; names, survival 224; (problems on sex distr. 79, 81, 86, 100, 125, 222).
Favorable cases 20, 23.
FERMI-DIRAC *statistics* 53.
Fire accidents 189, 234; (damage distr. 222).
First passages: diffusion 296; Markov chains 324; random walk 280 (generating fct. 290, 305; explicit expression 292); recurrent events 243.
Fish catches 37.
FISHER, R. A. 6, 38, 107, 224, 315.
Fission 224.
Flaws in material 118.
FOKKER-PLANCK *equations* 295, 296, 304.
Forward equations 385, 389, 392.
FRÉCHET, M. 60, 75, 343, 347, 351.
Frequency function 133.
FRIEDMAN, B. 313.

FROBENIUS' *theory of matrices* 343.
FRY, T. C. 107, 377.
FURRY, W. H. 369.
FÜRTH, R. 339; (—'s formula for first passages 296, 304).

GALTON, F. 204, 224.
Gambling systems 151, 282. [Cf. *Coin tossing, Ruin.*]
Games, fair 196, 284, 287; generalized — 200; unfavorable 200, 210. [Cf. *Billiards, Duration of games, Ruin.*]
Gamma function 49.
GAUSSIAN *distribution* 133.
Generating functions 212; of compound distr. 223; continuity theorem 232, 278; of differences 236; use in solving differential equations 395, 396; double — 255; for first passages 290, 305; for Markov chains 347; moment — 236; for recurrent events 243; renewal 272; ruin 288, 305; runs 265, 268, 278; sequential sampling 302, 306; sums 215; tails 219.
Genes and genotypes 92; distributions 94, 101, 315, 333; inheritance 92, 204; mutations 224, 369; sex-linked 96.
Genetics, cf. *Chromosomes, Genes*.
Geometric distribution 174 (generating fct. 217; variance 181); composition with binomial 223; exponential limit 220; holding times 219; limit in Bose-Einstein statistics 59; special applications Dodge's plan 168, 188; family size 100, 224; mortality distr. 278).
GONČAROV, V. 206.
GREENWOOD, J. A. 45, 336.
Growth 277, 370, 374, 395.
Guessing 66, 182.
GUMBEL, E. J. 113.

HARDY, G. H. 95, 162; Hardy's law 95 (for pairs of genes 102).
HARRIS, T. E. 227.
HAUSDORFF, F. 157, 162.
HELLY'S *theorem* 233.
Higher sums 339.
Holding times, exponential 218, 375.
HOSTINSKY, B. 343.
Hypergeometric distribution 33, 46, 167;

INDEX

binomial approximation 47, 108; double 39, 187; normal appr. 146; Poisson appr. 114; variance and mean 183.

Images, method of, 304.
Independent events 86 (pairwise and mutually 88); experiments 90; random variables 169, 190; trials 88.
Indistinguishable objects 11, 51.
Initials 44.
Insect survivors 128, 222.
Intersection of events 13.
Inversions 205.
Iterated logarithm for Bernoulli trials 157; generalized 163; number theoretical interpretation 162.

KAC, M. 45, 252, 313, 358.
KAKUTANI, S. 346.
KELVIN, LORD 304.
KENDALL, D. G. 374.
KENDALL, M. G. and B. SMITH 26.
Key problem 187.
KHINTCHINE, A. 147, 157, 191.
KOLMOGOROV, A. 6, 161, 293, 343, 378, 390; Chapman-Kolmogorov eqns. 338, 341, 387, 396; — criterion 207 (converse 211); — differential eqns. 389, 390; — inequality 184.
KOOPMAN, B. O. 4.

LAGRANGE, J. L. 292.
LAPLACE, P. S. 62, 84, 133, 212, 236, 344; (law of succession 84); DeMoivre — limit theorem 137, 139.
Largest observation 175, 187.
Larvae 128, 222.
Latent root 350.
Law of the arc sine 252, 257; counterpart 262.
Law of the iterated logarithm 157, 162, 163.
Law of large numbers: Bernoulli trials (weak 141; strong 156; number theoretical interpretations 161); dependent variables 209; indep. random variables (identically distr. 191; infinite expectation 200; arbitrary variables 202, 209; strong 207, 210, 211); Markov processes 342; permutations 205.

LAWRENCE, T. E. 6.
LEA, D. E. 76, 120.
Leads, distribution of 250.
Lefthanders 125.
LÉVY, P. 252, 253.
Limit theorems: arc sine law 252, 262; average recurrence times 253; Bose-Einstein statistics 59; continuity theorem 232, 278; distributions with infinite moments 200, 252; geometric distr. 220; matching distr. 67, 77; occupancy 72; Pascal 221, 233; Polya 128, 395; sampling 76; uniform distr. 236. [Cf. *Approximation, Central limit theorem, Law of large numbers, Markov chains, Normal approximation, Steady state.*]
LINDEBERG, J. W. 192; — condition 202.
LITTLEWOOD, J. E. 162.
LJAPUNOV, A. 192; — condition 209.
Loss, coefficient of 383.
LOTKA, A. J. 100, 224.
Lunch-counter example 56, 58.
LUNDBERG, O. 395.

McCREA, W. H. 297, 300.
Machine servicing 376, 379.
MALÉCOT, G. 315.
MARGENAU, H. 54.
Marginal distribution 166.
MARKOV, A. 192, 307.
MARKOV *chains* 309; absorption 332, 358; associated with continuous time processes 373, 391; classification 320, 345; closed sets 318; decomposable 316, 320; ergodic properties 324, 330, 345; finite 324, 342; general 337; irreducible 318; limit theorems 342; periodic 316, 329, 351; probabilities (absolute 318, initial 309, inverse 341, stationary 328, 331, transition 309, 317, 338); recurrence times 320, 324; reversible 342; superposition of 340.
MARKOV *process* 337; with continuous time parameter 386; (waiting times 219).
MARKOV *property* 338, 387.
Match box problem 108, 176.
Matching of cards 62, 66; multiple 76; variance 181.

Mating 93, 101; (as Markov chain problem 315, 345, 360).
Matrix: notation 103, 317; partitioned 316, 323, 359; stochastic 309 (doubly 327, 336, 345; canonical decomposition 347, 350; application to non-stochastic matrices 343, 359).
Maximum likelihood 38.
MAXWELL-BOLTZMANN *statistics* 53.
Mean = *expectation* 171 (in terms of generating fct. 213); normal distr. 133; number of successes 138.
Measure in product spaces 91; convergence in measure 162, 207.
Median 36.
MENDEL, G. 92.
MÉRÉ'S *paradox* 45.
MISES, R. VON 6, 72, 152, 157, 266, 278.
Misprints 113, 126.
Mixed populations 82, 237.
Molecules, long-chain 190.
MOLINA, E. C. 113, 143.
Moment generating function 236.
Moments 177.
MONTMORT, P. R. 62.
MOOD, A. 146.
Morse alphabet 44.
Mortality, cf. *Renewal.*
Moving averages 339, 340, 346.
Multinomial coefficients 32.
Multinomial distribution 124, 167; (maximal term 126, 146).
Multiplets 24.
MURPHY, G. M. 54.
Mutations 224, 368.

$(n)_r$ 25.
NATIONAL BUREAU OF STANDARDS 28, 130.
Negative binomial distribution 218.
Neighbors, unlike 56.
NEYMAN, J. 123.
*Non-*MARKOVIAN *processes* 338.
Normal approximation to binomial distr. (individual terms 135, 146, central part 137, 139, tails 144, 147); combinatorial runs 146; hypergeometric distr. 146; permutations 205; Poisson distr. 143, 146, 193; success runs 205. [Cf. *Central limit theorem.*]

Normal density and *distribution* 129; tables 132; tails 131, 145.
Normal numbers 163.
Normalized variables 179.
Nuclear chain reaction 223.
Null state 321.
Number theoretical interpretations 161.

Occupancy problem 54, 69, 174; limit theorem 72; treatment by Markov chains 313, 354.
Optional stopping 140, 190, 197.
ORNSTEIN, L. S. 293.

$p(k; \lambda)$ 111.
π, *decimals of*, 124.
Pairs 23.
PALM, C. 377, 379, 383.
Parapsychology 45, 336.
Parking tickets 45.
Partial fraction expansions 227, 237; for Markov chains 348; numerical examples 230, 266, 276; recurrent events 261; renewal 276; success runs 266 (numerical estimate 269).
Particles: in chain reactions 223; random walk 279; splitting — 365, 369, 374; statistics of — 53.
Particular solutions, method of 283, 289, 302.
Partitioning of matrices 316, 323.
Partitions, combinatorial 30.
PASCAL *distribution* 174, 217, 237; in game of billiards 236; Poisson limits 221, 233; and Polya distr. 218; reciprocal 190; variance 181, 217; and waiting times 221.
PEARSON, K. 127, 204.
Pedestrians as non-Markovian process 339.
Periods: Markov chains 316, 321; recurrent events 241, 244; renewal theory 273.
Permutations 90, 205.
Petersburg paradox 7, 199, 255.
Petri plate 122.
Phase space 12.
POISSON, S. D. 110.
POISSON *distribution* 115; compound 237, 391 (combined with binomial

128, 221, 344); convolutions 127, 216; empirical examples 111, 119, 125, 225; generating fct. 216; holding times 220; limiting distr. for (binomial 110, 115; fluctuation 344; hypergeometric 114; matching 67; multinomial 127; occupancy 59, 72; Pascal 221, 233; Poisson trials 233; Polya 128; long runs 278; traffic problem 340; trunking 378, 396); mean 174, 216; multiple 127; normal approximation 143, 146, 193; spatial 118; time dependent 117, 364; variance 178 (216).
POISSON *process* 364, 386; compound 376, 391, 395.
POISSON *traffic* 376.
POISSON *trials* 189, 233.
Poker 9, 46; special problems 31, 76, 126.
POLLARD, H. 244.
POLYA, G. 83, 174, 297; — distributions 128 (mean 188; as Pascal distr. 218; limiting forms 128, 210, 395); — process, 395; — urn scheme 83, 101 (as non-Markovian process 338).
Polymer molecules 190.
Population theory, cf. *Birth and death process*, *Family*, *Genes*, *Renewal*.
Power-supply problems 108, 384, 396.
Probability 17 (in product spaces 91); absolute 79 (for Markov chains 318, 338); — of causes 85; compound — 81; conditional 78, (random variables 168; Markov chains 307; processes 338, 387); — distributions 164; initial 307; inverse 341.
Product measure and *space* 91.

Quality control 34, 56, 125; (Bartky's sampling 281, 314; Dodge's 168, 188).

RADEMACHER, H. 6.
Radiation effects 45, 76, 120, 222.
Radioactive disintegrations 119, 368.
Railroad problem 139.
Raisins, distribution of 113, 118, 126, 237.
Random chains 190.
Random choice 25.
Random digits: counts 26, 27, 55 (π and e 29, 124); k distinct among $n - 27$, 70; distr. (binomial and Poisson 54, 112; normal 141, 142); frequency of decimals 161; as occupancy problem 54; special problems 44, 125.
Random mating 93, 101.
Random variables 164; generalized (= improper) 242; integral valued 212; Markovian 337; normalized 179; time-dependent 363, 387.
Random walk 279, 304; absorbing barriers 279 (generating fct. 288, 304, explicit expression 292, 304, as Markov chain 310, 333); cyclical — 311, 353; diffusion 293, 304; first passages 280, 292; generalized (= sequential sampling) 281, 300, 306, 333; more dimensional 281, 297, 306, 345; reflecting barriers 280, 304, 305, 311, 326 (explicit expressions 355, more dimensions 345); renewal method 305.
Randomness: of sequences 157; tests of — 56, 68.
Recessive 92, 98, 102.
Recurrence times 241; limit theorems 248, 253; in Markov chains 320, 362; mean — 242; moments 262; in random walks 280, 297, 305.
Recurrent events (patterns) 238; classification 241; criteria 244.
Recurrent states 320.
Reduced number of successes 138.
Reflecting barriers 280, 304, 305, 311; explicit solution 355; in plane 345; stationary distr. 326.
Rejection 281, 300, 314.
Rencontre 62.
Renewal: coefficients 276; equation 272; method in random walks 305; of populations 275.
Repairs of machines 379, 382.
Repeated trials 88 (product spaces 91); infinite sequences 149, 282, 307; random variables representing — 169.
Replacements 275. [Cf. *Sampling*.]
Retrospective equations = backward equations.
Reversible Markov chains 342.
ROBBINS, H. 223.
ROMANOVSKY, V. 343.

Ruin problem 279, 282; generating fct. 288, 292, 304; in Markov chains 332; numerical illustration 287; sequential sampling 281, 300, 306.

Runs, combinatorial 56, 59; mean 188; normal approximation 146.

Runs in repeated trials 239, 264; in game of billiards 235, 236; generating fct. 265, 268; as Markov chain problem 310, 318; normal distr. 265; partial fraction method 266 (error estimate 269, special cases 229, 278); Poisson distr. of long runs 278; r successes before n failures, etc. 149, 163, 268, 278; special problems 187, 235.

RUTHERFORD-CHADWICK-ELLIS 119.

S, for success 104.
Sample average 193.
Sample point 10, 13.
Sample size, required 142, 146, 192.
Sample space 5, 10, 12; discrete 16; in terms of random variables 169; repeated trials 89 (infinite sequences 149).
Sampling with and without replacements 24, 91 (comparison 47, 108); distinct elements in samples 63, 76, 174, 181, 188, 235; of fish 37; inspection — 34, 125 (Bartky's scheme 281, 314; Dodge's 168, 188); largest observation 175, 187; required size 142, 146, 192; sequential 281, 300, 306, 313; stratified 188.

SCHROEDINGER, E. 223.

Seeds: Poisson distr. 118; survival 224.
Selection, genetic: 93, 99, 102, 224.
Selections, combinatorial 30.
Self-renewing aggregates 275.
Senator problem 32, 34.
Sequential sampling 281, 300, 306, 313.
Servicing factor 380; — problems 377, 379; (power supply 108, 384, 396).
Sets, closed (in Markov chains) 318; — cylindrical 91. [Cf. *Events.*]
Seven-way lamps 24.
Sex distribution in families 79, 81, 86, 100, 125, 222.
Sex-linked 96, 98, 102.

SHEWHART, W. A. 56.

Shoe problems 46, 75.
Shooting 80, 125.
Shuffling 335; composite — 340.

SMIRNOV, N. 147.

Stable distribution: age, 276, 278; genotype 94, 98, 102.
Stakes, effect of changing 285.
Standard deviation 178; (normal distr. 133; successes 138).
States in Markov chains 309; classification 320 (unessential — 344).
Stationary distribution for aperiodic chains 328; periodic 331. [Cf. *Stable distribution.*]
Steady state: age distr. 276, 278; birth and death process 373; genotype distr. 95; power supply 384; servicing problems 381, 382; waiting lines 379.

STEINHAUS, H. 6, 108.

Stirling's formula 41, 134; alternative — 50.
Stochastic matrix 309; doubly — 327, 336, 345; generalization 343, 359.
Stochastic process 337, 363; stationary 396. [Cf. *Differential equations.*]
Stratified sampling 188.

STUART, E. E. 45, 336.

Success 104; reduced number 138. [Cf. *Runs.*]
Succession, law of 84.
Summation formula 61.
Survival, birth and death process 374, 395; insect eggs, etc. 128; family names, genes 224.
Systems of gambling 151, 282.

Tauberian theorem 259, 272.
Telephone statistics 122, 234, 365, 376; — trunking 143, 377; (holding times 219).
Tests, of grouping 56; sequential 127; statistical 45.
Theta functions 304.

THODAY, J. M. 76, 120.
THORNDIKE, F. 122.

Ties: billiards 235; multiple coin games 246, 261; dice 247.
Time-homogeneous process 386.

TIPPETT, L. H. C. 26.

Traffic, incoming 376.

Traffic problem 339.
Transient states 320; ergodic properties 332; finite chains 358.
Transition probabilities 309; general process 337; higher — 317; stationary 338; stochastic processes 385, 386.
Trials, independent 89 (as Markov chain 310); repeated 88, 91 (— and random variables 169; infinite sequences 149, 282, 307).
Truncation method 195, 200, 202, 210, 212.
Trunking problems 143, 377, 396.

UHLENBECK, G. E. 293, 313.
Uniform distribution 236.
Union of events 13.
Urn schemes as Markov chains 308; (Ehrenfest 312, 327, 345; Laplace 83, 344; Polya 83, 101, 128). [Cf. *Sampling*.]
USPENSKY, J. V. 137, 140, 269.

Variance 177 (from generating fct. 214); normal distr. 133.
VEEN, S. C. VAN 235.

Waiting lines 377, 378.
Waiting times 218; (card drawing 35); exponential — 375.
WALD, A. 127, 146, 197, 281, 300.
WANG, MING CHEN 313.
Welders problem 108, 384.
Weldon's dice data 106.
WHIPPLE, F. J. W. 297, 300.
WHITWORTH, W. A. 23.
WIENER, N. 293.
WILKS, S. S. 57.
WOLFOWITZ, J. 146.
WRIGHT, S. 315.

X-rays, effect on cells 45, 76, 120, 222.

YOSIDA, K. 346.
YULE, G. U. 369; — process 368, 394.